CISM COURSES AND LECTURES

Series Editors:

The Rectors of CISM
Sandor Kaliszky - Budapest
Mahir Sayir - Zurich
Wilhelm Schneider - Wien

The Secretary General of CISM
Giovanni Bianchi - Milan

Executive Editor
Carlo Tasso - Udine

The series presents lecture notes, monographs, edited works and
proceedings in the field of Mechanics, Engineering, Computer Science
and Applied Mathematics.
Purpose of the series is to make known in the international scientific
and technical community results obtained in some of the activities
organized by CISM, the International Centre for Mechanical Sciences.

INTERNATIONAL CENTRE FOR MECHANICAL SCIENCES

COURSES AND LECTURES - No. 390

RECENT ADVANCES IN BOUNDARY LAYER THEORY

EDITED BY

ALFRED KLUWICK
TECHNICAL UNIVERSITY OF VIENNA

DEDICATED TO PROF. W. SCHNEIDER
ON THE OCCASION OF HIS 60TH BIRTHDAY

Springer-Verlag Wien GmbH

Le spese di stampa di questo volume sono in parte coperte da
contributi del Consiglio Nazionale delle Ricerche.

This volume contains 116 illustrations

ISBN 978-3-211-83136-6   ISBN 978-3-7091-2518-2 (eBook)
DOI 10.1007/978-3-7091-2518-2
© 1998 by Springer-Verlag Wien
Originally published by CISM, Udine in 1998.

SPIN 10682537

PREFACE

For many years boundary-layer theory was considered a classical discipline and advances in this area were almost synonymous with the development of more powerful numerical codes to solve the well known and well established equations. This situation changed drastically with the advent of the triple deck. This concept and other recent advances in boundary-layer theory showed how modern analytical and computational techniques can and should be used to deepen the understanding of high Reynolds number flows and to design effective calculation strategies. This is the unifying theme of the present volume which contains the lecture notes presented during a course held July 7-11, 1997 at the Centre for Mechanical Sciences in Udine. Topics covered by these lecture notes include laminar boundary-layers, turbulent boundary-layers, interacting boundary-layers and the stability of laminar boundary-layers.

Laminar boundary-layers serve as the ideal starting point to introduce the basic physical concepts and mathematical techniques of boundary-layer theory. These lead to a systematic derivation of the Prandtl equations which can be extended to higher order and clearly display the asymptotic nature of boundary-layer theory. This certainly contributes to a better insight into wall bounded high Reynolds number flows but in addition opens the possibility for a simplified yet rigorous treatment of a number of effects which are of direct relevance in engineering problems as for example variable properties effects.

The two-fold role of asymptotic analysis to elucidate the dominant physical processes and structure of wall bounded flows in the limit of large Reynolds number and to provide a self consistent description based on first principles is extremely useful also if the assumption of laminar flow is no longer justified. In fact, it has become increasingly clearer in the past years that asymptotic methods represent very powerful tools to elucidate important properties of turbulent shear flows which are independent of specific closure models. Results of this generality are of course of great significance in their own but can - even more important - be exploited also to simplify engineering calculations. Therefore, special emphasis is placed on the minimum information and the numerical algorithms that are required to structure a prediction scheme for such flows.

Classical boundary-layer theory rests on the assumption that the pressure inside the viscous wall layer is imposed by the external inviscid flow. This concept fails if the boundary-layer is subjected to rapid changes of the boundary conditions as for example near the rear end of a body of finite length. In the case of laminar flow, triple deck theory and the theory of marginal separation have shown that this breakdown can be avoided if the flow properties inside and outside the boundary-layer are allowed to interact. Related multideck problems which arise in studies of turbulent boundary layers are also outlined.

The stability of laminar boundary-layer flows and the transition from a laminar to a fully turbulent state has been but still is the subject of intense study. Without doubt, early classical theories such as the Orr-Sommerfeld equation approach have provided important first insight into basic instability mechanisms. Furthermore, in combination with other empirical methods they have generated useful tools for analyzing engineering problems involving boundary-layer transition. Nevertheless, it should not be overlooked that the underlying assumptions are ad hoc in nature and cannot be justified rigorously. Similar comments apply to more recently developed techniques based on linear equations which have proved quite successful in applications. Owing to their ad hoc nature it is not clear how they can be improved in a systematic manner to include, for example, nonlinear effects which are of importance in later stages of transition. This difficulty is avoided by an alternative approach using asymptotic methods to construct self consistent approximations of the full Navier-Stokes equations holding in the limit of large Reynolds number. The analysis is carried out first for the linear range of disturbances and then extended to the weakly nonlinear and nonlinear regimes. Finally, aspects of the receptivity problem are also addressed.

The authors would like to thank the members of CISM and in particular Prof. Bianchi, Prof. Kaliszky, Prof. Schneider and Prof. Tasso for their advice and help during the preparation of this volume.

Alfred Kluwick

CONTENTS

Page

CONTENTS

INTRODUCTION TO BOUNDARY-LAYER THEORY

K. Gersten

Ruhr University of Bochum, Bochum, Germany

ABSTRACT

The basic concepts of Prandtl's boundary–layer theory and the extension to a general asymptotic theory for flows at high Reynolds numbers are described, and their growing importance is discussed.

The concept and theory of boundary layers have been developed by Ludwig Prandtl and presented in a historic paper in 1904. At that time the theory of inviscid flows (potential theory) had reached already a high standard. It was possible by this theory for example, to calculate the pressure distribution past an airfoil reasonably well, as shown in Figure 1. However, this theory had some major shortcomings. It predicted no frictional resistance or drag of immersed bodies, and it could only predict lift on a body if a circulation about it were postulated, but the theory could say nothing about the origin of the circulation.

These deficits of the then existing theory were removed by Prandtl's *boundary–layer theory*. This theory rested on certain basic observations.

1. Although the inviscid flow is the limiting solution for vanishing viscosity ($\nu = 0$ or $Re = VL/\nu = \infty$), for any finite viscosity (however small it may be) the viscosity effects cannot be ignored.

2. For high Reynolds numbers $Re = VL/\nu$ the viscosity effects are restricted to a very thin layer close to the wall called *boundary layer*. Its thickness decreases with

increasing Reynolds number. The flow field has for $Re \to \infty$ a layer structure: it consists of the inviscid outer flow and the viscous boundary layer.

3. In the thin boundary layer the Navier–Stokes equations can be destinctly simplified which lead to so–called *boundary–layer equations*, which can be solved much easier. The solutions of the boundary–layer equations which are independent of the Reynolds number, lead to formulas for the skin friction of the form

$$c_f = \frac{1}{\sqrt{Re}} f\left(\frac{x}{L}\right) . \tag{1}$$

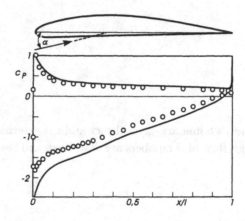

Figure 1. Pressure distribution on airfoil NACA 4412 ($\alpha = 8°$) at low speed.
  o   Experiment at $Re = 3 \cdot 10^6$
  ——   Theory, $Re = \infty$ (inviscid)

The main characteristic of the boundary–layer theory is, that the solution for the inviscid outer flow and the solution for the boundary layer are being determined separately and matched properly. The former leads to the pressure distribution and lift, whereas the latter yields the skin–friction distribution and the drag.

Prandtl's boundary–layer theory had a tremendous effect on the development of fluid mechanics. It not only provided a basic ingredient for the subsequent rapidly developing science of aerodynamics and its application, but it also became a great stimulus to related developments in other branches of engineering involving the flow of fluids, e.g. turbomachinery design, hydraulics, heat and mass transfer etc.

The application of the basic boundary–layer theory for airfoils in transonic flows resulted in some discrepancies shown in Figure 2. The location of the shock indicated by the deep slope of the pressure distribution is not properly described by the inviscid solution for $Re = \infty$. As Prandtl already acknowledged the boundary layer has a displacement effect on the inviscid outer flow such that the outer flow has to be calculated past a *fictitious* airfoil which is generated by thickening the given airfoil with the so–called displacement–thickness distribution. The full line in Fig.2 is the pressure distribution for this fictitious airfoil. Also in Fig.1 a complete agreement between theory and experiment could be achieved when the displacement effect is taken into account. The displacement effect is not included in the basic boundary–layer theory and is therefore called higher–order boundary–layer effects.

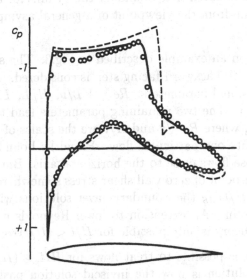

Figure 2. Pressure distribution on airfoil RAE 2822 ($\alpha = 2.8°$) at $Ma = 0.73$.
  o   Experiment at $Re = 6.5 \cdot 10^6$
  - - - Limiting solution, $Re = \infty$
  —— Theory for $Re = 6.5 \cdot 10^6$

In the mid fiftees a strong mathematical theory for the boundary–layer theory and its extension to high–order effects has been developed. Connected with the names Kaplun (1957), Kaplun, Lagerstrom (1957) and Van Dyke (1964). According to that the boundary–layer theory is an *asymptotic theory* of the full Navier–Stokes equations for high Reynolds numbers. It leads to a *singular* perturbation problem, which can be solved by the *method of matched asymptotic expansions*. The solution for finite but high Reynolds numbers is considered as a small perturbation of the limiting solution for the inviscid flow. Since the latter does not satisfy the no–slip condition the perturbation problem is called *singular*. The above mentioned layer structure of the flow field is typical for singular perturbation problems. This asymptotic theory put the boundary–layer theory on a strong mathematical basis, but showed also how it could be systematically extended to high–order theory for lower Reynolds numbers. Among several higher–order effects, see Schlichting, Gersten (1997, p.418), the displacement effect and the curvature effects are the most important ones.

Prandtl's original boundary–layer equation is identical with the first–order equation of this general asymptotic theory. This equation is solved for a given pressure distribution found from the inviscid theory. A fundamental feature of the boundary–layer equation is that its solution becomes singular in the point of vanishing wall shear stress (separation point) and hence cannot be continued beyond that point. Therefore, for long time it was believed that boundary theory fails for flows with separation. This is, however, not true from the viewpoint of a general asymptotic theory for high Reynolds numbers.

This will be shown on an example described in Fig.3. The steady incompressible laminar flow past a smooth backward facing step is considered. The flow is characterized by three nondimensional parameters: $Re = VL/\nu$, H/L, L/l. For simplification it is assumed $L/l =$const. The two remaining parameters lead to the diagram in the upper part of the figure, where $Re^{-1/2}$ and H/l are the scales of the axes. Each point in this diagram represents one particular flow. Prandtl's boundary–layer theory describes in particular those flows close to the horizontal axis. But at the point MS the solution will just reach a point of zero wall shear stress somewhere on the contour. For all parameters $H/l > (H/l)_{MS}$ the boundary–layer solutions will show a singularity and are therefore not useful. An extension to lower Reynolds numbers by the above mentioned asymptotic theory is only possible for $H/l < (H/l)_{MS}$.

By a new concept it is possible to treat flows for $H/l > (H/l)_{MS}$ by asymptotic theory. The starting solution is now the inviscid solution past a flat plate (origin $H = 0$) rather than past the smooth backward facing step. In the limiting process $Re \to \infty$ is also $H/l \to 0$, in other words, the geometry is properly coupled with the Reynolds number. This leads to the theory of *interacting boundary layers* or the *triple deck theory*. This theory was developed independently by Stewartson (1969), Neiland (1969) and Messiter (1970). The flow field has a three–layer structure near the wall. Furthermore, the effect of the inviscid outer flow on the boundary layer via the pressure distribution and the effect of the boundary layer on the inviscid outer flow via displacement have to be taken into account simultaneously (interacting boundary layers).

It turns out that the solution according to the triple–deck theory is not unique for the smooth backward facing–step flow. In the shaded area of Fig.3 at least two solutions can be found. In fact, there exist also two inviscid flow solutions for the backward facing step as shown in Fig.4. The solution in Fig.4b has a discontinuity and a "dead water" which is typical for separated flows, whereas the solution in Fig.4a can be used as limiting solution only for attached flows.

One can get an overall view on the structure of the solutions including multiple solution by considering the maximum of the pressure coefficient $c_{p\,max}$ as shown in the lower part of Fig.3. At the boundaries of the shaded area the solution jumps which

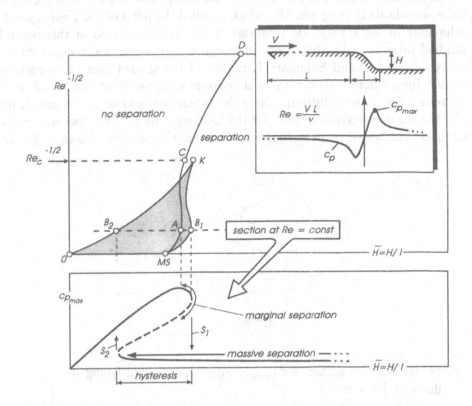

Figure 3. Diagram for flows past smooth backward facing steps, L/l =const.

      D–C–A–MS : Boundary between unseparated flows and flows with separation

      K–B₁–MS :   Line for jumps from local to massive separation

      K–B₂–O :   Line for jumps from massive separation to attached flows

      K :   Cusp

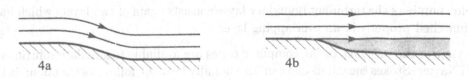

Figure 4. Two limiting inviscid solutions for smooth backward facing step flows.

leads to hysteresis effects. This special form of the solution surface for $c_{p\,max}$, as shown in Fig.5, refers to a so-called *cusp catastrophe*, one of the basic structures according to elementary catastrophe theory. The center of the cusp is the point K. As one can see from Fig.5 that the solutions are everywhere regular in the cusp area except the singularity at the point MS, which is called the point of *marginal separation*. The behaviour of the asymptotic solutions in the neighbourhood of this point has been studied intensively by the theory of marginal separation, see Stewartson et al. (1982). At the right–hand boundary K-B_1-MS of the shaded area the structure of the solution jumps from a flow with *local separation* (upper solution) to a flow with *massive separation* (lower solution), where the extent of the separated region is much larger for the massive separation than for the local separation. The flows with massive separation are considered as perturbations of the limiting solution shown in Fig.4b.

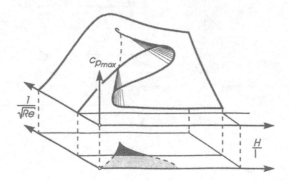

Figure 5. Folded solution surface for $c_{p\,max}$. Smooth backward facing step
         flows at $L/l = $ const.

All the features described so far in connection with Fig.3 are features of the Navier–Stokes equations, at high Reynolds numbers of course.

So far only laminar boundary layers have been discussed. The same concepts can also be applied to *turbulent boundary layers*. Mainly two new aspects appear: First, when the theory is based on the Reynolds–averaged Navier–Stokes equations, then additional equations have to be used to model the turbulence. Second, at high Reynolds numbers the turbulent boundary layer consists again of two layers which have to be matched properly in an overlapping layer.

In view of the fact that today computer codes are available to simulate solutions of the full Navier–Stokes equations one can occasionally hear the following question: Is the model of boundary–layer theory running out because first it is only an approximation, second it is too complicated to work with various layers, whereas the Navier-Stokes codes deal with one set of equations for the whole flow field and third it cannot properly handle separated flows.

To answer that question it is quite clear that boundary–layer theory or more general asymptotic theory is today more important than ever, which is justified by — among others — following reasons:

1. The layer structure is not a mathematical subtlety, but rather the description of the *physics* of the flow. When the layer structure is not taken into consideration important information about the physics of the flow get lost. Many flow phenomena can be explained only by the boundary–layer concept (e.g. secondary flows).

2. The asymptotic theory yields asymptotic formulas like Eq.(1) rather than just calculation points.

3. The flow equations become independent of Reynolds number. Only one calculation is necessary rather than calculations for each Reynolds number separately. Hence, the problem is reduced by one dimensionless parameter.

4. The equations of motion are simplified such that capacities and running times of computers are reduced drastically.

5. There is no problem for the asymptotic theory to treat flows with separation as long as the proper limiting inviscid solution can be identified, so that the singular perturbation problem can be carried out.

6. The dividing into two layers of turbulent boundary layers is a great advantage rather than disadvantage because universal laws, so–called *wall functions*, can be formulated a priori in the overlapping layer without using any turbulence model. They lead to conditions which asymptotically correct turbulence models have to satisfy.

REFERENCES

Gersten, K.; Herwig, H. (1992): Strömungsmechanik. Grundlagen der Impuls–Wärme– und Stoffübertragung aus asymptotischer Sicht. Vieweg–Verlag, Braunschweig, Wiesbaden.

Kaplun, S. (1957): Low Reynolds number flow past a circular cylinder. J. Math. Mech.,Vol. 6, 595–603

Kaplun, S.; Lagerstrom, P.A. (1957): Asymptotic expansions of Navier–Stokes solutions for small Reynolds numbers. J. Math. Mech., Vol. 6, 585–593.

Messiter, A.F. (1970): Boundary-layer flow near the trailing edge of a flat plate. SIAM J. Appl. Math., Vol. 18, 241–257.

Neiland, V.Y. (1969): Theory of laminar boundary layer separation in supersonic flow. Izv. Akad. Nauk SSSR Mekh. Zhid. i Gaza, Vol. 4, 53–57. Engl. Übersetzung: Fluid Dynamics, Vol. 4, 1969, 33-35.

Prandtl, L. (1904): Über Füssigkeitsbewegung bei sehr kleiner Reibung, In: Krazer, A. (Hrsg.): Verh. III Intern. Math. Kongr., Heidelberg, 484–491, Teubner, Leipzig. 1905–Repr. Nendeln 1967.

Schlichting, H.; Gersten, K. (1997): Grenzschicht–Theorie, 9th Edition, Springer–Verlag, Berlin, Heidelberg.

Stewartson, K. (1969): On the flow near the trailing edge of a flat plate. II. Mathematika, Vol. 16, 106–121.

Stewartson, K. (1974): Multistructured boundary layers on flat plates and related bodies. Advances in Applied Mechanics, Vol. 14, 145–239.

Stewartson, K.; Smith, F.T.; Kaups, K. (1982): Marginal separation. Studies in Appl. Math., Vol. 67, 45–61.

Van Dyke, M. (1975): Perturbation Methods in Fluid Mechanics. The Parabolic Press, Stanford, California.

Young, A.D. (1989): Boundary Layers. BSP Professional Books, Oxford.

LAMINAR BOUNDARY LAYERS
ASYMPTOTIC AND SCALING CONSIDERATIONS

H. Herwig
Technical University of Chemnitz, Chemnitz, Germany

ABSTRACT

Laminar boundary layer theory is introduced with special emphasis on its underlying physics. The aim of this article is to reveal the asymptotic character of wall bounded flows as well as details of the asymptotic theory that can describe these flows for increasing Reynolds numbers in a unique and systematic way.

Scale analysis is a powerful tool in developing an adequate mathematical / physical model to describe high Reynolds number flows and is intimately related to an asymptotic description of these flows. It will be demonstrated that all necessary transformations in boundary layer theory can be deduced by scale analysis arguments, even in situations with a complicated multilayered structure of the flow and temperature field.

The following content will be covered with this asymptotic and scaling approach:

1. Introduction
2. The physics of laminar boundary layers
3. Mathematical description
4. Viscous/inviscid interaction theory
5. Thermal boundary layers
6. Variable property effects

1. INTRODUCTION

Laminar boundary layer theory dates back to the year 1904 when Ludwig Prandtl [1] for the first time systematically described the physics of what later was called a singular perturbation problem. The most important method for solving such problems, the method of matched asymptotic expansions, was used by Prandtl in the limit of (infinitelly) high Reynolds numbers for the first time. Later this method was also applied in the low Reynolds number limit as well as for many other momentum, heat and mass transfer problems.

The formal mathematical foundation for this theory was developed about 40 years ago by several scientists like Kaplun, Lagerstrom and Cole, to mention a few of them. Comprehensive descriptions may be found in Van Dyke [2], Schneider [3], Schlichting and Gersten [4], Gersten and Herwig [5].

In the following chapters emphasis is made on the physical background and, to a certain extent, on a rigorous mathematical description. Scale analysis may serve as the proper tool to relate these two aspects.

2. THE PHYSICS OF LAMINAR BOUNDARY LAYERS

2.1 Re$\rightarrow \infty$, a singular limit

In Fig. 2.1 a detail of the laminar flow field around a thin airfoil is shown for three different Reynolds numbers $Re = U_\infty^* L^* / v^*$. For increasing Reynolds numbers, for example due to an increasing velocity of the oncoming flow, close to the wall a more and more pronouced layer appears in which velocity gradients are very high. This layer is caused by the non-slip condition at the wall which holds for all Reynolds numbers, i.e. also in the limit Re$\rightarrow\infty$. Its impact is restricted to a near wall region because downstream convection "overwhelms" a lateral spreading of its influence. So far this is only a qualitative statement. It will be quantified in chapter 2.3 below.

With increasing Reynolds number the velocity gradients in the boundary layer grow (to infinity as Re$\rightarrow\infty$), i.e. the thickness of the boundary layer decreases (to zero as Re$\rightarrow\infty$), but : the boundary layer never disappears! In this sense it is a singular limit, providing the exact solution for Re=∞.

It is important to keep in mind that the corresponding solution of the laminar boundary layer theory is exact in the limit Re = ∞. However, in reality only finite, though often large Reynolds numbers exist. Moreover, increasingly high velocity gradients will inevitably provoke turbulence, i.e. real boundary layers will be turbulent for Reynolds numbers above a certain (critical) value.

For a laminar theory that is exact for Re=∞ to be of any use for real world flows, it must be as an approximation for finite Reynolds numbers below the critical value. Fig. 2.2 illustrates how the singular solution may serve to approximate finite Reynolds number laminar flows.

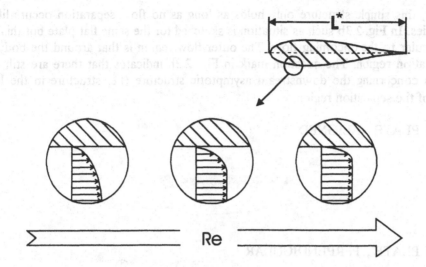

Fig. 2.1: Detail of the flow field close to the wall

2.2 The whole flow field

The near wall region has just been identified as the boundary layer. Physically the steep gradients in these layers arise through the effect of viscosity, i.e. of viscous forces. Their influence on the flow field is negligibly small everywhere, except within the boundary layers. Thus the flow field is composed of two regions:

(1) the viscous boundary layer and wake region (infinitesimally small for Re→∞)

(2) the inviscid outer flow region (infinitesimally close to the body for Re→∞)

This is illustrated in Fig. 2.3a for a flat plate aligned with the oncoming flow.

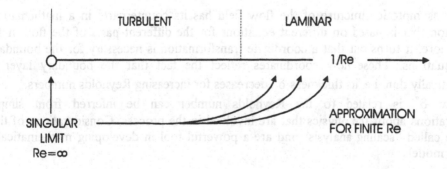

Fig. 2.2: Sketch of the approximation character of boundary layer theory

However, this simple structure only holds as long as no flow separation occurs like at bluff bodies. In Fig. 2.3b such as situation is sketched for the same flat plate but this time perpendicular to the oncoming flow. The outer flow region is that around the body and the separation region. The question mark in Fig. 2.3b indicates that there are still open questions concerning the downstream asymptotic structure (i.e. structure in the limite Re→∞) of the separation region.

(a) FLAT PLATE , ALIGNED

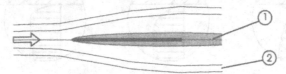

(b) FLAT PLATE , PERPENDICULAR

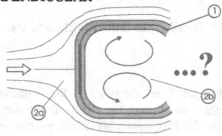

Fig. 2.3: (a) Flow regions without separation; ① boundary layers and wake,
② outer flow
(b) Flow regions with separation; ① boundary layers and free shear
layers, ②ⓐ outer flow ②ⓑ separation region

2.3 Scaling of the laminar boundary layer

The asymptotic structure of the flow field has its counterpart in a mathematical description that is based on different equations for the different parts of the flow field. Furthermore, it turns out that a coordinate transformation is necessary for the boundary layer equations. These new coordinates reflect the fact that the boundary layer is asymptotically thin, i.e. its thickness δ^* decreases for increasing Reynolds numbers.

How δ^* is related to the Reynolds number can be inferred from simple considerations about the physics that are involved in the process. Considerations of this kind are called "scaling analysis" and are a powerful tool in developing mathematical / physical models.

If we assume a flow field like the one sketched in Fig. 2.4 the δ^* – Re relation may be deduced from the question: which of the oncoming flow particles will be absorbed into the boundary layer and which will not? The answer is obvious: It will be absorbed if the oncoming particle is close enough to the wall so that during the time it is over the wall it can be reached by the lateral momentum transfer process of the viscous forces in the boundary layer, otherwise it will not.

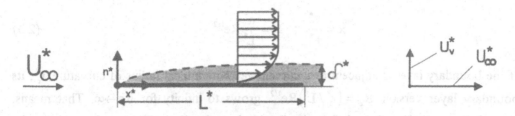

Fig. 2.4: Sketch of the boundary layer

A characteristic time for its presence above the plate is L^* / U_∞^* since it is convected downstream with a velocity U_∞^* over a distance L^*. A characteristic time for the lateral spreading of viscous effects is δ^* / U_v^* where U_v^* is a characteristic velocity for the momentum transport by viscous effects. Here, arguments of dimensional analysis tell us what U_v^* is: It must be related to the viscosity ν^* with dimensions $[\nu^*] = m^2 / s$ so that with the length δ^* over which the process occurs a characteristic „viscous velocity" is $U_v^* = \nu^* / \delta^*$.

Now we simply equate the two times (indicating this by the symbol ~ which means "of the same order of magnitude", not equal as far as numbers are concerned)

$$ \frac{L^*}{U_\infty^*} \sim \frac{\delta^*}{U_v^*} = \frac{\delta^{*2}}{\nu^*} \quad \rightarrow \quad \frac{\delta^*}{L^*} \sim \left[\frac{\nu^*}{U_\infty^* L^*} \right]^{\frac{1}{2}} \tag{2.1} $$

With the definition of the Reynolds number $Re = U_\infty^* L^* / \nu^*$ we thus have

$$ \frac{\delta^*}{L^*} \sim Re^{-\frac{1}{2}} \quad \text{for} \quad Re \to \infty \tag{2.2} $$

for all laminar boundary layers in the vicinity of solid walls that are provoked by an outer flow. Indeed, we must be this precise because in a different physical situation different exponents in the Reynolds number dependence of δ^* may occur.

If, for example, a narrow jet is blown along a wall, the flowfield of which can also be described by the boundary layer theory, the exponent is -3/4 instead of $-1/2$, see Gersten and Herwig [5, p. 154].

In order to have boundary layer coordinates that do not degenerate when Re is increased (Re→∞) the normal coordinate must be transformed according to (2.2). The nondimensional boundary layer coordinates thus are (c. f. Fig. 2.4):

$$ x = \frac{x^*}{L^*} \quad ; \quad N = \frac{n^*}{L^*} Re^{1/2} \tag{2.3} $$

If the boundary layer is adjacent to a curved wall with a local radius of curvature r_c^*, its boundary layer version $R_c = \left(r_c^* / L^* \right) Re^{1/2}$ grows to infinity for Re→∞. That means: from the perspective of the infinitesimally thin boundary layer the curvature cannot be felt, just like man from his perspective doesn't feel the curvature of the earth.

2.4 Hierarchical order ; matching

The basic idea of boundary layer theory is not only to find solutions in two different regions of the flow field by solving two different sets of equations but also to bring these two regions together *asymptotically*. This phrase describes a certain hierarchical order that holds in the limit Re→∞.

Since for Re→∞ the boundary layer is infinitesimally small the outer flow in a first step can be solved in a region that goes right to the wall. However, these equations cannot fulfill the no-slip condition at the wall. The physics at the wall in a second step are described by the boundary layer equations which are *matched* with the equations for the outer flow.

In a simple version this matching means:

$$ \lim_{y \to 0} U(x,y) = \lim_{N \to \infty} u(x,N) \quad \text{for} \ Re \to \infty \tag{2.4} $$

The left hand side of eq. (2.4) is the slip velocity of the outer flow at the wall, the right hand side the velocity at the outer edge of the boundary layer. "Outer edge" here does not mean a certain distance from the wall but: far enough outside, where $u(x,N)$ is independent of N.

Eq. (2.4) for large but finite Reynolds numbers does not describe a smooth transition between the outer flow and the boundary layer. Instead it gives an outer boundary condition for the boundary layer solution that can only be understood when it is seen asymptotically, i.e. in the limit Re→∞.

Fig. 2.5 may illustrate this matching process. It is shown in two ways:

(1) From the perspective of the outer flow (upper row in Fig. 2.5).
    With y as normal coordinate the outer flow remains unchanged for increasing Re, but
    the boundary layer thickness shrinks.

(2) From the perspective of the boundary layer (lower row in Fig 2.5).
    With N as normal coordinate the boundary layer remains unchanged for increasing
    Re, but the outer flow is "streched" in the normal direction.

In both cases it can be seen that the transition between the two regions is not smooth as
long as the Reynolds number is finite. Only in the limit of infinite Re number are no
descrepancies left. This is what is meant by "asymptotically correct" behaviour.

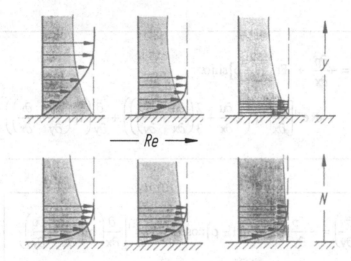

Fig. 2.5: Velocity profiles of the outer and the boundary layer flow for increasing
            Reynolds numbers
            outer flow: *shaded area*, boundary layer: *arrows*

            upper row: y-coordinate, outer flow unchanged
            lower row: N-coordinate, boundary layer unchanged

From Fig. 2.5 we may also infer that for finite Reynolds numbers there must be
improved procedures of matching. This is the scope of a higher order theory which will
be described later.

Note that often the matching process is described with the flat plate flow as an
example. For this special case, however, the outer flow is constant all over. Hence, a
smooth transition in a sketch like in Figure 2.5 occurs for finite Re numbers which
conceals the real situation for non-flat plate flows.

3. MATHEMATICAL DESCRIPTION

3.1 The Navier-Stokes equations

At the length scales at which the fluid can be looked upon as a continuum the Navier-Stokes equations are an adequate mathematical model of the fluid behaviour. Nondimensionalised according to table 3.1 and for the two-dimensional steady case they are given by:

$$\frac{\partial(\rho u)}{\partial x} + \frac{\partial(\rho v)}{\partial y} = 0 \tag{3.1}$$

$$\rho\left(u\frac{\partial u}{\partial x} + v\frac{\partial u}{\partial y}\right) = -\frac{\partial p}{\partial x} + Fr^{-2}[1-\rho]\sin\alpha$$
$$+ Re^{-1}\left[\frac{\partial}{\partial x}\left(\eta\left(2\frac{\partial u}{\partial x} - \frac{2}{3}\left(\frac{\partial u}{\partial x} + \frac{\partial v}{\partial y}\right)\right)\right) + \frac{\partial}{\partial y}\left(\eta\left(\frac{\partial u}{\partial y} + \frac{\partial v}{\partial x}\right)\right)\right] \tag{3.2}$$

$$\rho\left(u\frac{\partial v}{\partial x} + v\frac{\partial v}{\partial y}\right) = -\frac{\partial p}{\partial y} + Fr^{-2}[1-\rho]\cos\alpha + Re^{-1}\left[\frac{\partial}{\partial x}\left(\eta\left(\frac{\partial v}{\partial x} + \frac{\partial u}{\partial y}\right)\right)\right.$$
$$\left. + \frac{\partial}{\partial y}\left(\eta\left(2\frac{\partial v}{\partial y} - \frac{2}{3}\left(\frac{\partial u}{\partial x} + \frac{\partial v}{\partial y}\right)\right)\right)\right] \tag{3.3}$$

with the definition of the Reynolds and the Froude number, respectively,

$$Re = \frac{\rho_R^* U_R^* L_R^*}{\eta_R^*} \quad , \qquad Fr = \frac{U_R^*}{\sqrt{g^* L_R^*}} \tag{3.4}$$

Here, p_m^* is the modified pressure (deviation from the static pressure field), α is the angle of the x-coordinate against the horizontal.

For later use we also provide the two scalar equations for temperature T^* in terms of the nondimensional temperature $\Theta = (T^* - T_R^*)/T_R^*$ and for the concentration of a component i in terms of its partial density ρ_i^* which in nondimensional form is written as $\Phi = (\rho_i^* - \rho_{iR}^*)/\rho_{iR}^*$.

x	y	u	v	p	ρ	η
$\dfrac{x^*}{L_R^*}$	$\dfrac{y^*}{L_R^*}$	$\dfrac{u^*}{U_R^*}$	$\dfrac{v^*}{U_R^*}$	$\dfrac{p_m^*}{\rho_R^* U_R^{*2}}$	$\dfrac{\rho^*}{\rho_R^*}$	$\dfrac{\eta^*}{\eta_R^*}$

Table 3.1: Nondimensional quantities ; R=reference

$$
\rho c_p \left(u \frac{\partial \Theta}{\partial x} + v \frac{\partial \Theta}{\partial y} \right) = Re^{-1} Pr^{-1} \left[\frac{\partial}{\partial x}\left(\lambda \frac{\partial \Theta}{\partial y} \right) + \frac{\partial}{\partial y}\left(\lambda \frac{\partial \Theta}{\partial y} \right) \right]
$$

$$
+ \beta(1+\Theta) Re^{-1} \tilde{E}c\, Re \left[u \frac{\partial p}{\partial x} + v \frac{\partial p}{\partial y} \right] \tag{3.5}
$$

$$
+ \eta Re^{-1} \tilde{E}c \left[2\left(\frac{\partial u}{\partial x}\right)^2 + 2\left(\frac{\partial v}{\partial y}\right)^2 + \left(\frac{\partial v}{\partial x}+\frac{\partial u}{\partial y}\right)^2 - \frac{2}{3}\left(\frac{\partial u}{\partial x}+\frac{\partial v}{\partial y}\right)^2 \right]
$$

$$
\rho \left(u \frac{\partial \Phi}{\partial x} + v \frac{\partial \Phi}{\partial y} \right) = Re^{-1} Sc^{-1} \left[\frac{\partial}{\partial x}\left(D\rho \frac{\partial \Phi}{\partial x} \right) + \frac{\partial}{\partial y}\left(D\rho \frac{\partial \Phi}{\partial y} \right) \right] \tag{3.6}
$$

During the process of nondimensionalisation additional nondimensional parameters appear in these equations. They are the Prandtl, Eckert and Schmidt number, respectively:

$$\text{Pr} = \frac{\eta_R^* c_{PR}^*}{\lambda_R^*} = \frac{\nu_R^*}{a_R^*} \quad ; \quad \tilde{\text{Ec}} = \frac{U_R^{*2}}{c_{PR}^* T_R^*} \quad ; \quad \text{Sc} = \frac{\eta_R^*}{\rho_R^* D_R^*} = \frac{\nu_R^*}{D_R^*} \tag{3.7}$$

In all equations the physical properties are still assumed to be variable and obey property laws governing their dependence on pressure temperature and partial density. Constant properties are the special case $\rho = \eta = \lambda = c_p = D = 1$. For further details of the derivation see [5].

3.2 The Re → ∞ limit

The Navier-Stokes equations are a *complete* set of equations (under the assumptions under which they have been derived) and so they hold in the whole flow field, i. e. in the outer flow region as well as in the boundary layer. However, due to the special physical situation in each of the two regions special simplifications are possible in the equations, different for the two regions.

In the outer flow region viscous effects are negligible in the limit Re→∞. This corresponds to the fact that all terms related to η drop out of the equations (3.2) and (3.3) when Re^{-1} is set to zero (i.e. Re=∞). The inviscid equations left over are called the *Euler equations*. Compared to the Navier-Stokes equations they have lost their highest (second) derivative terms and thus can fulfil one boundary condition less than the Navier-Stokes equations. This deficiency will be "healed" by the boundary layer.

In the boundary layer the (x,y) coordinate system is not adequate since the normal coordinate must be transformed according to eq. (2.3). The need for this transformation was described in chapter 2.3: since the vertical extend of the boundary layer is ~ $\text{Re}^{-1/2}$ for Re→∞ only with the transformation of n the mathematical domain remains finite. A similar argument, however, holds for the vertical velocity v. A simple scale analysis applied to the continuity equation (4.1), with $\rho = 1$ for simplicity, reveals:

$$x^* \sim L^* \quad ; \quad n^* \sim L^* \text{Re}^{-1/2} \quad ; \quad u^* \sim U_R^* \quad \rightarrow \quad v^* \sim U_R^* \text{Re}^{-1/2} \tag{3.8}$$

i.e. v must also be transformed in order to be kept in the equation. We set

$$\bar{v} = \frac{v^*}{U_R^*} \text{Re}^{1/2} \tag{3.9}$$

This mathematical argument has its counterpart in the physics of the boundary layer: for the velocity vector with the components u^* and v^* not to stick out of the asymptotically thin boundary layer its vertical component v^* must be asymptotically small.

After the transformations (2.3) and (3.9) the boundary layer equations that hold for Re = ∞ emerge from the (transformed) Navier-Stokes equations by setting Re^{-1} to zero (i.e. Re=∞). They are called the *Prandtl boundary layer equations*, and read:

$$\frac{\partial(\rho u)}{\partial x} + \frac{\partial(\rho \bar{v})}{\partial N} = 0 \qquad (3.10)$$

$$\rho\left(u\frac{\partial u}{\partial x} + \bar{v}\frac{\partial u}{\partial N}\right) = -\frac{\partial p}{\partial x} + \frac{\partial}{\partial N}\left(\eta\frac{\partial u}{\partial N}\right) + \frac{1}{Fr^2}(1-\rho)\sin\alpha \qquad (3.11)$$

$$0 = -\frac{\partial p}{\partial N} + Re^{-1/2}\frac{1}{Fr^2}(1-\rho)\cos\alpha \qquad (3.12)$$

for the flow field and

$$\rho c_p\left(u\frac{\partial \vartheta}{\partial x} + \bar{v}\frac{\partial \vartheta}{\partial N}\right) = \frac{1}{Pr}\frac{\partial}{\partial N}\left(\lambda\frac{\partial \vartheta}{\partial N}\right) + Ec\left[\eta\left(\frac{\partial u}{\partial N}\right)^2 + \beta^*T^*\left(u\frac{\partial p}{\partial x} + \bar{v}\frac{\partial p}{\partial N}\right)\right] \qquad (3.13)$$

$$\rho\left(u\frac{\partial \Phi}{\partial x} + \bar{v}\frac{\partial \Phi}{\partial N}\right) = \frac{1}{Sc}\frac{\partial}{\partial N}\left(\rho D\frac{\partial \Phi}{\partial N}\right) \qquad (3.14)$$

for the two scalar quantities (temperature and concentration).

The associated boundary conditions for the flow field are:

$N=0$: $u = \bar{v} = 0$ (no slip, impermeable wall)   (3.15)

$N\rightarrow\infty$: $u = U(x)$ (matching the outer flow)   (3.16)

Those for the temperature and concentration will be given later.

3.3 Approximations for finite Re numbers

The solutions of the boundary layer equations matched to the inviscid outer flow Euler equations are exact solutions of the Navier-Stokes equations in the (singular) limit $Re^{-1}=0$. For finite Reynolds numbers they may serve as an approximation. For the boundary layer region we may write for the exact solutions u_{exact}, $\bar{v}_{exact}$:

$$u_{exact} = u_0(x, N) + \Delta u(x, N, Re) \tag{3.17}$$

$$\bar{v}_{exact} = \bar{v}_0(x, N) + \Delta\bar{v}(x, N, Re) \tag{3.18}$$

Here u_0, $\bar{v}_0$ are the solutions of the Prandtl boundary layer equations (3.10)-(3.12) which are exact for $Re^{-1}=0$. The deviations Δu, $\Delta\bar{v}$ are asymptotically small, i. e. they are zero for $Re^{-1}=0$ and non-zero but small for $Re^{-1}>0$, i. e. for finite Reynolds numbers.

An extended version of the boundary layer theory can systematically account for the terms Δu and $\Delta\bar{v}$ by (again asymptotically) approximating them. It is called *higher order boundary layer theory* (HOBL).

This is achieved when u_{exact} and $\bar{v}_{exact}$ are sought as series solutions. These series are asymptotic series of the general form:

$$u(x, y, Re) = u_0(x, N) + g_1(\varepsilon)\, u_1(x, N) + g_2(\varepsilon)\, u_2(x, N) + \dots \tag{3.19}$$

$$\bar{v}(x, y, Re) = \bar{v}_0(x, N) + \bar{g}_1(\varepsilon)\, \bar{v}_1(x, N) + \bar{g}_2(\varepsilon)\, \bar{v}_2(x, N) + \dots \tag{3.20}$$

with $\varepsilon = Re^{-1/2}$ and $g_{i+1}/g_i \to 0$ for $\varepsilon \to 0$ (i. e. for $Re \to \infty$).

Here ε is called a *perturbation parameter* (since it "perturbs" the limit solution u_0, $\bar{v}_0$) the functions g_i are called *gauge functions*.

To be precise with respect to the asymptotic character of the expansions two definitions of order-symbols will be introduced. They refer to the asymptotic character of two functions of a perturbation parameter ε. If these functions are $f(\varepsilon)$ and $\delta(\varepsilon)$ we say:

☐ $f(\varepsilon)$ is of the (asymptotic) order $\delta(\varepsilon)$, written as

$$f(\varepsilon) = O\left[\delta(\varepsilon)\right] \qquad \text{if} \qquad \lim_{\varepsilon \to 0} \frac{f(\varepsilon)}{\delta(\varepsilon)} = C < \infty \tag{3.21}$$

This means: $f(\varepsilon)$ goes to zero „equally fast" as $\delta(\varepsilon)$ does.

❑ $f(\varepsilon)$ is of smaller (asymptotic) order than $\delta(\varepsilon)$, written as

$$f(\varepsilon) = o\,[\delta(\varepsilon)] \qquad \text{if} \qquad \lim_{\varepsilon \to 0} \frac{f(\varepsilon)}{\delta(\varepsilon)} = 0 \qquad\qquad (3.22)$$

The symbols O(...) and o(...) are called *Landau symbols*.

With these definitions u_0, u_1, u_2 ... and $\bar{v}_0$, $\bar{v}_1$, $\bar{v}_2$ ... in (3.19), (3.20) are all of order one, written as O(1). That does not say anything about specific numbers, it only means that they do not depend on ε, i. e. they do not vanish for $\varepsilon \to 0$.

The gauge functions g_i, however, all are of order o(1), i.e. they vanish for $\varepsilon \to 0$ with the special feature $g_{i+1} = o(g_i)$.

Based on these expansions a rational theory can be formulated that approximates the exact solution for finite Reynolds numbers with increasing accuracy as more terms are taken into account. How these expansions are incorporated into the theory that subdivides the flow field into an outer flow and a boundary layer is shown next.

3.4 Matched asymptotic expansions; a 4-step procedure

It has been mentioned already that boundary layer theory is an example for a certain class of mathematical/physical models or theories called singular perturbation theories. Thus there is a certain scheme for applying this kind of theory. However, this is by no means a formal „recipe", easy to use with the success guaranteed. It is, however, a powerful tool when applied with a sound physical background knowledge of the problem.

With this warning in mind the "recipe" in Fig. 3.1 may be used as a guideline how to solve a problem by the theory of matched asymptotic expansions.

In the first step (S1) the basic equations of the whole problem should be provided in nondimensional form and in an adequate coordinate system. For boundary layer theory these are the Navier-Stokes equations in a body fitted coordinate system, for example a curvilinear orthogonal one see [5] [6]. An important step is to identify the perturbation parameter ε for the subsequent perturbation procedure. Often the nondimensional parameter with which it will be formed is obvious but the specific form of the perturbation parameter itself is not. For the boundary layer theory ε will be formed with the Reynolds number, however, that it will be $\varepsilon = Re^{-1/2}$ is not obvious from the beginning. Instead one could tentatively assume $\varepsilon = Re^{-n}$ and fix the exponent n in a later step.

In the second step (S2) the perturbation parameter is simply set equal to zero giving the so-called „naive approximation". It is this step that will show if the problem is a singular one or not. If the naive approximation has an error (with respect to the exact solution) which is of the same asymptotic order of magnitude in the whole flow field including the boundaries (in finite distance or at infinity) the problem is a regular one, otherwise its singular. In boundary layer theory the error of the naive approximation is asymptotically small, i. e. of order o(1), in the whole flow field, except in the regions close to the walls where the error is O(1), i.e. it does not vanish for Re→∞. Thus we have a singular perturbation problem. The asymptotic expansion $F_I = ...$ means that all dependent variables in this region I (outside the near wall region) are approximated by a series expansion with the naive approximation as leading order terms.

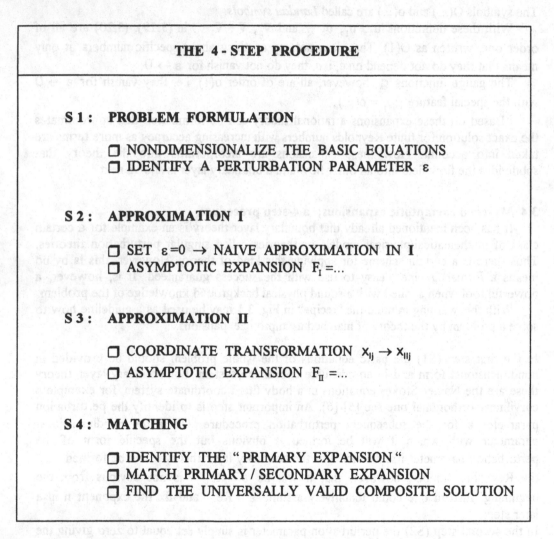

THE 4 - STEP PROCEDURE

S 1 :   PROBLEM FORMULATION

☐ NONDIMENSIONALIZE THE BASIC EQUATIONS
☐ IDENTIFY A PERTURBATION PARAMETER ε

S 2 :   APPROXIMATION I

☐ SET ε =0 → NAIVE APPROXIMATION IN I
☐ ASYMPTOTIC EXPANSION $F_I = ...$

S 3 :   APPROXIMATION II

☐ COORDINATE TRANSFORMATION $x_{Ij} \to x_{IIj}$
☐ ASYMPTOTIC EXPANSION $F_{II} = ...$

S 4 :   MATCHING

☐ IDENTIFY THE " PRIMARY EXPANSION "
☐ MATCH PRIMARY / SECONDARY EXPANSION
☐ FIND THE UNIVERSALLY VALID COMPOSITE SOLUTION

Fig. 3.1: The 4-step procedure of matched asymptotic expansions

In the third step (S3) the region which is not covered by I from the previous step is considered. Since this region, now called II, will not be of an O(1) extension in all directions a coordinate transformation is employed in order to make sure that the intrinsic coordinates in this region all are of the order O(1) after the transformation. In boundary layer theory the extension normal to the wall is asymptotically small. After the transformation (2.3) the boundary layer coordinate system (x,N) in region II is of order O(1). Like in region I all dependent variables are expanded in asymptotic series, now with the Prandtl boundary layer terms at leading order.

The fourth and last step (S4) in almost all cases is the most difficult one. It helps if one has a rough idea of how the physics in the problem behave in the imagined limit process with respect to the perturbation parameter under consideration.

At first one has to identify the so-called *primary expansion*. It is that expansion in either I or II for which the first term can be determined without information from the other region. In boundary layer theory it is the outer expansion (region I) since its first term (inviscid flow over the body alone, no boundary layer) is not influenced by the physics in region II (the boundary layer).

Next, both expansions have to be matched which in most cases is an alternating process of determining the inner and outer boundary conditions of subsequent orders in the inner and outer regions of the flow field. Here *inner* and *outer* characterize the two regions in an illustrative way. In boundary layer theory these terms are used from the perspective of the boundary layer region (inner region) for which region I is the outer flow region.

Matching still means that two regions must be brought together asymptotically (i.e. for $\varepsilon \rightarrow 0$), as was introduced in chapter 2.4 for the Prandtl boundary layer theory. However, now the flow field in both regions is approximated by asymptotic series which must be matched. Since the dependent variables in both regions not necessarily are expanded in a similar way with a one-to-one correspondence of successive terms the matching rule (in this form proposed by Van Dyke [2]) reads:

The secondary expansion up to $O\left(g_{\mathrm{II}n}(\varepsilon)\right)$ of the primary expansion

up to $O\left(g_{\mathrm{I}m}(\varepsilon)\right)$

is equal to   (3.23)

the primary expansion up to $O\left(g_{\mathrm{I}m}(\varepsilon)\right)$ of the secondary expansion

up to $O\left(g_{\mathrm{II}n}(\varepsilon)\right)$

Here m and n are the numbers of orders in I and in II, respectively. They should be equal or differ at least by one when (3.23) is applied to increasing orders. In (3.23) the phrase "secondary expansion of the primary expansion" for example means: rewrite the primary expansion in secondary variables and then expand it in series for $\varepsilon \rightarrow 0$.

With the help of (3.23) for the boundary layer theory, for example, boundary conditions are determined at N→∞ for the boundary layer and at n→0 for the outer flow. Physically that means: at some location off the wall for the boundary layer and at the wall for the outer flow. This, however, means that for finite Reynolds numbers there must be a certain *overlap region* in which both solutions are valid! This is an important feature in the theory of matched asymptotic expansions. Instead of using (3.23) one could equally well introduce a so-called *intermediate variable* which is of order one in this overlap region, rewrite both expansions in this variable and equate terms of equal asymptotic magnitude as a matching procedure.

Once the solutions in the two regions are found (in different coordinates, however) they can be combined by rewriting them in one common coordinate. In this coordinate both solutions can be added. After subtracting the common part (which after the addition is in the profile twice) a universal profile emerges. Since in the process of rewriting one coordinate the perturbation parameter appears explicitly, this common profile has the perturbation parameter as an explicit parameter. In the case of the boundary layer theory this is exactly the situation sketched in Fig. 2.1, which was the starting point for the boundary layer theory.

3.5 First and second order boundary layer equations

The 4-step procedure applied to the Navier-Stokes equations in the limit of Re→∞ is the systematic way to deduce the Prandtl boundary layer equations (3.10)-(3.12) as the leading order set of equations. The details of this procedure, as well as the application of the 4-step scheme in the other Re-limit, i.e. Re→0, are described in [5]. Here, only the first and second order boundary layer equations are shown. The first order equations with index II1 correspond to (3.10)-(3.12). The second order equations explicitly contain R which is the nondimensional local Radius of wall curvature. In laminar boundary layer theory curvature effects therefore are second order effects.

The equations are

1st order

$$\frac{\partial u_{II1}}{\partial x} + \frac{\partial \bar{v}_{II1}}{\partial N} = 0 \tag{3.24}$$

$$u_{II1} \frac{\partial u_{II1}}{\partial x} + \bar{v}_{II1} \frac{\partial u_{II1}}{\partial N} = -\frac{\partial p_{II1}}{\partial x} + \frac{\partial^2 u_{II1}}{\partial N^2} \tag{3.25}$$

$$\frac{\partial p_{III}}{\partial N} = 0$$

(3.26)

2nd order

$$\frac{\partial u_{II2}}{\partial x} + \frac{\partial \bar{v}_{II2}}{\partial N} = \frac{1}{R}\left(N\frac{\partial u_{III}}{\partial x} - \bar{v}_{III} \right)$$

(3.27)

$$u_{III}\frac{\partial u_{II2}}{\partial x} + u_{II2}\frac{\partial u_{III}}{\partial x} + \bar{v}_{III}\frac{\partial u_{II2}}{\partial N} + \bar{v}_{II2}\frac{\partial u_{III}}{\partial N} + \frac{\partial p_{II2}}{\partial x} - \frac{\partial^2 u_{II2}}{\partial N^2}$$
$$= \frac{1}{R}\left(N\frac{\partial^2 u_{III}}{\partial N^2} + \frac{\partial u_{III}}{\partial N} - \bar{v}_{III}\frac{\partial(Nu_{III})}{\partial N} \right)$$

(3.28)

$$\frac{\partial p_{II2}}{\partial N} = \frac{1}{R}u_{III}^2$$

(3.29)

In Fig. 3.2 the hierarchy of the matching process for the boundary layer theory can be seen. Outer boundary conditions (N→∞) for the inner expansion and wall boundary conditons (n→0) for the outer expansion are determined for successive orders in a hierarchical manner.

This scheme is an adequate description of the high Reynolds number physics if sudden changes in the boundary conditions only occur with respect to one coordinate (here with respect to n).

If, however, there is a sudden change also in a second direction like for the flow along a flat plate when it reaches the trailing edge and suddenly the no slip condition ceases to hold, there is a "break down" of the hierarchical order. Inviscid and viscous effects can no longer be separated into different orders for different parts of the flow field, but are closely related to each other by what is called viscous/inviscid interaction.

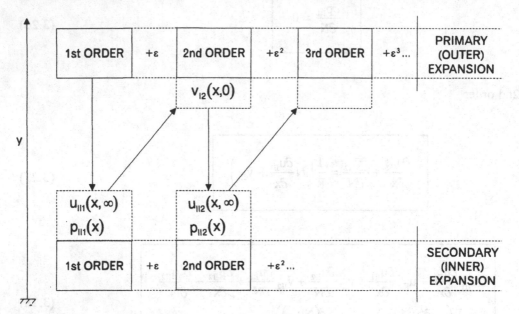

Fig. 3.2: Matching hierarchy

4.   VISCOUS / INVISCID INTERACTION THEORY

4.1 A typical example

High Reynolds number flow over the trailing edge of a (aligned) flat plate is a typical example for what will be called *asymptotic viscous/inviscid interaction theory*. It is typical in the sense that the asymptotic structure historically first found with this kind of flow turned out to be that of a general interaction process of inviscid and viscous parts of a flow field.

In Fig. 4.1 two effects are sketched as characteristic features of this flow:

(1) Since suddenly the no slip condition ceases to exist the wake profile is "filled up" with fluid in a region of length L_u^* and a lateral extend δ_u^* which are yet of unknown magnitude.

(2) A direct consequence of filling up the wake in the vicinity of the centerline is a lateral displacement Δ^* of the whole profile towards the centerline. This streamline displacement is also felt outside the boundary layer and causes pressure changes in the inviscid outer region.

With these two features in mind one can easily determine the asymptotic orders of the length L_u^* the thicknesses δ_u^* and Δ^* as well as pressure changes $p^* - p_\infty^*$ from a simple model of the flow.

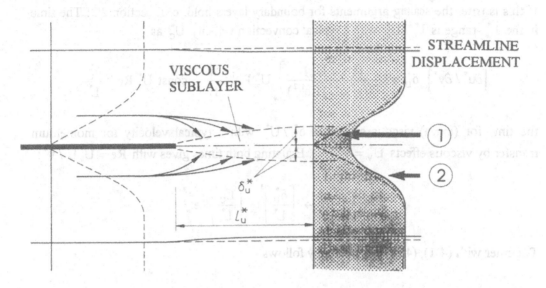

Fig. 4.1:   Flow field at the trailing edge of a flat plate; see for example Stewartson [7]

4.2 Scale analysis of viscous/inviscid interaction

If we expect an asymptotic structure of the viscous/inviscid interaction, δ_u^*, L_u^*, Δ^* and $p^* - p_\infty^*$ will scale with the Reynolds number in a yet unknown way. It can be cast into the nondimensional form:

$$\frac{\delta_u^*}{L^*} \, , \frac{\Delta^*}{L^*} = O(Re^{n_\Delta}) \tag{4.1}$$

$$\frac{L_U^*}{L^*} = O(Re^{n_L}) \tag{4.2}$$

$$\frac{p^* - p_\infty^*}{\rho^* U_\infty^{*2}} = O(Re^{n_p}) \tag{4.3}$$

with the unknown exponents n_Δ, n_L and n_p. Since the displacement Δ^* is an integration over a layer of thickness δ_u^*, they have the same exponent n_Δ.

Three equations will fix the numbers of n_Δ, n_L and n_p. They follow from three hypothesis' about the physics of the flow under consideration:

(H1) *The sublayer of dimensions L_u^* and δ_*^* acts like a boundary layer within a boundary layer.*

If this is true, the scaling arguments for boundary layers hold, c. f. section 2.3: The time in the L_u^*-range is L_U^* / U_C^* with a typical convection velocity U_C^* as

$$\left[\partial u^* / \partial y^*\right]_w \delta_U^* = \left[\frac{\partial(u^* / U_\infty^*)}{\partial((y^* / L^*) \mathrm{Re}^{1/2})}\right]_w U_\infty^* \mathrm{Re}^{1/2} \frac{\delta_U^*}{L^*} = \mathrm{const}. U_\infty^* \mathrm{Re}^{1/2} \frac{\delta_U^*}{L^*} ;$$

the time for (extra) viscous effects is δ_U^* / U_V^* with a typical velocity for momentum transfer by viscous effects $U_V^* = v^* / \delta_U^*$. Equating both times gives with $\mathrm{Re} = U_\infty^* L^* / v^*$:

$$\frac{L_U^*}{U_C^*} \sim \frac{\delta_U^*}{U_V^*} \rightarrow \left[\frac{\delta_U^*}{L^*}\right]^3 \sim \left[\frac{L_U^*}{L^*}\right] \mathrm{Re}^{-\frac{3}{2}}$$

Together with (4.1), (4.2) it immediately follows

$$3n_\Delta = n_L - \frac{3}{2} \tag{4.4}$$

(H2) *The outer flow is that over a contour $\Delta^*(x^*)$*

The pressure distribution then is given by the so-called Hilbert integral, i. e.:

$$\frac{p^* - p_\infty^*}{\rho^* U_\infty^{*2}} = -\frac{1}{\pi} \int_{-\infty}^{+\infty} \frac{d\Delta^* / dx^*}{x^* - \overline{x}^*} d\overline{x}^*$$

with $x^* / L_U^* = O(1)$ and (4.2) it immediately follows

$$n_p = n_\Delta - n_L \tag{4.5}$$

(H3) *In the sublayer viscous and pressure forces are of the same order of*
      magnitude

If this is true typical terms which can be equated with respect to their asymptotic order
are:

$$O\left[\frac{\partial((p^* - p_\infty^*)/\rho^* U_\infty^{*2})}{\partial(x^*/L^*)}\right] = O\left[Re^{-1}\frac{\partial^2(u^*/U_\infty^*)}{\partial(y^*/L^*)^2}\right]$$

From this with (see H1):

$$O\left[\frac{u^*}{U_\infty^*}\right] = O\left[\frac{U_c^*}{U_\infty^*}\right] = Re^{1/2}\frac{\delta_U^*}{L^*}$$

it immediately follows

$$n_p - n_L = -1 + \left(\frac{1}{2} + n_\Delta\right) - 2n_\Delta = -\frac{1}{2} - n_\Delta \tag{4.6}$$

From (4.4)-(4.6) the exponents are determined:

$$n_L = -\frac{3}{8}\ ;\quad n_\Delta = -\frac{5}{8}\ ;\quad n_p = -\frac{2}{8}$$

so that we have the asymptotic scaling

$$\frac{x^*}{L^*} \sim Re^{-3/8}\ ;\quad \frac{\delta_U^*}{L^*} \sim Re^{-5/8}\ ;\quad \frac{p^* - p_\infty^*}{\rho^* U_\infty^{*2}} \sim Re^{-1/4} \tag{4.7}$$

which turns out to be universally valid whenever viscous/inviscid interactions occur. The
structure of this interaction is characterized by three different layers:

(1) the oncoming boundary layer of thickness $\sim Re^{-1/2}$

(2) an asymptotically thin sublayer of thickness $\sim Re^{-5/8}$

(3) a part of the inviscid outer flow field which is responsible for the pressure
     adjustement over the displacement contour. Since in the outer flow characteristic

lengths in both directions are of the same magnitude and the streamwise extend according to (4.7) is $\sim$ Re$^{-3/8}$, the thickness of this layer is $\sim$ Re$^{-3/8}$.

This three layer structure is called *triple deck*.

4.3 The triple deck structure of viscous/inviscid interaction

With the physics described in the previous section a triple deck structure emerges in the limit Re→∞ when sudden changes in the streamwise boundary conditions occur.

Fig. 4.2 shows the asymptotic orders of magnitude of the three decks. The interaction mechanism is a simultaneously developing displacement in the lower deck and pressure distribution in the upper deck. Both effects are transferred through the main deck: the displacement is felt in the upper deck and the pressure distribution is felt in the lower deck. The main deck is passive in character. It is only displaced laterally by the sublayer and it transfers the pressure distribution from the upper deck without change to the lower deck.

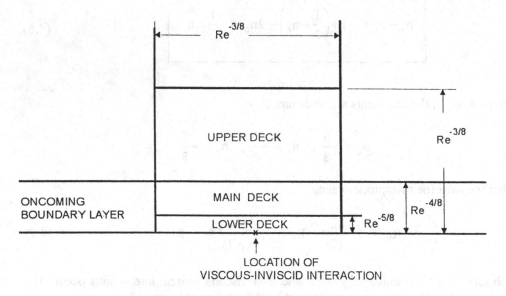

Fig. 4.2: Triple deck structure of viscous/inviscid interaction

Like in the conventional boundary layer theory the asymptotic scaling of the different regions (layers) is the basis for a coordinate transformation, expansion and subsequent deduction of the equations in the different layers.

Details of this procedure can be found in the vast literature dealing with viscous/inviscid interaction problems. A comprehensive early study of this kind is Stewartson [8].

5. THERMAL BOUNDARY LAYERS

5.1 The thermal energy equation

The thermal energy equation, equation (3.5), has been introduced in chapter 3.1. It is the complete energy equation for 2D flows, i.e. no boundary layer approximation has been made. Equation (3.5) emerges when the mechanical energy equation (momentum equation multiplied by the velocity vector) is substracted from the total equation of energy conservation (first law of thermodynamics). These two equations are not shown here, for details see for example [5].

Equation (3.5) is in nondimensional form. It is interesting to note that the first terms on the right hand side are multiplied by the combination of nondimensional numbers $(\text{RePr})^{-1}$. The product RePr is simply $\rho^* U_R^* L^* c_P^* / \lambda^*$. If we would not have boundary layers in mind this combination would by no means be related to the Reynolds number, but given its own name as it is. Indeed, this combination is known as Peclet number Pe. If, however, we rewrite it as $\text{Pe} = \text{RePr}$, which is nothing else but defining the Prandtl number

$$Pr = \frac{Pe}{Re} = \frac{\eta^* c_P^*}{\lambda^*} = \frac{\nu^*}{a^*} \tag{5.1}$$

in the boundary layer equations Pr will be left as the nondimensional parameter and Re will be absorbed into the coordinate transformation $y \to N$ according to equation (2.3).

Often, people are so familar with the Prandtl number (which has the nice feature of being a combination of fluid but not of flow quantities) that they forget its origin as a „secondary" parameter emerging when the Peclet number Pe is devided by the Reynolds number Re (which makes sense in the boundary layer theory as well as in the so-called slender channel theory, see [5]).

5.2 Thermal boundary layers

We expect thermal boundary layers to exist due to a similar physical argument as that for the momentum transfer in the flow boundary layers adjacent to a wall. If there is a different temperature at the wall to the far field there will be temperature gradients in the near wall region resulting in heat fluxes across the wall according to Fouriers law

$$\vec{q}^* = -\lambda^* \text{ grad } T^* \tag{5.2}$$

If the heat transferred across the wall is convected downstream before it can be conducted further away from the wall a thermal boundary layer of thickness δ_T^* appears. It turns out to be asymptotically thin in the limit Re→∞. Even though the ratio δ^* / δ_T^* is of order $O(1)$ with respect to the limit process Re→∞ the two thicknesses may differ considerably for different Prandtl numbers.

A simple scaling analysis can show how δ^* / δ_T^* depends on the Prandtl number.

5.3 Scaling of thermal boundary layers (forced convection)

The order of magnitude of δ_T^* can be found by the general procedure to determine boundary layer thicknesses: two times are equated, that for the presence of a particle over a streamwise distance L^* and that for the lateral spreading of the effect caused by the boundary condition at the wall.

A characteristic time for the particle presence in the L^*-range is L^* / U_C^*. Here U_C^* is the characteristic convection velocity which will be specified afterwards.

A characteristic time of the lateral spreading of temperature effects is δ_T^* / U_T^* with $U_T^* = a^* / \delta_T^*$ as a "velocity of the heat conduction" in total analogy to the "viscous velocity" $U_V^* = v^* / \delta^*$, in chapter 2.3.

Equating both times gives

$$\frac{L^*}{U_C^*} \sim \frac{\delta_T^*}{U_T^*} = \frac{\delta_T^{*2}}{a^*} \quad \rightarrow \quad \frac{\delta_T^*}{L^*} \sim \left[\frac{a^*}{U_C^* L^*} \right]^{\frac{1}{2}} \tag{5.3}$$

In equation (5.3) the convection velocity is not yet specified. If the ratio δ_T^* / δ^* is large, i.e. if $\delta_T^* >> \delta^*$ the characteristic convection velocity will be the outer flow velocity U_∞^*, if it is small it will be the velocity in the flow boundary layer at a wall distance δ_T^*. These two extreme cases correspond to the two limits of the Prandtl number $Pr = v^* / a^*$.

For $Pr \rightarrow 0$ the lateral spreading of temperature effects is much faster than that of viscous effects, thus $\delta_T^* >> \delta^*$ as illustraded in Fig. 5.1(a). For $Pr \rightarrow \infty$ it is vice versa and $\delta^* >> \delta_T^*$ holds, see Fig. 5.1(b).

Based on these considerations U_C^* in equation (5.3) can be specified:

(a) For $Pr \rightarrow 0$: $U_C^* = U_\infty^*$

Internal energy transferred across the wall is convected downstrean with U_∞^* as a characteristic velocity. From (5.3) immediately follows

$$\frac{\delta_T^*}{L^*} \sim Pe^{-\frac{1}{2}} \tag{5.4}$$

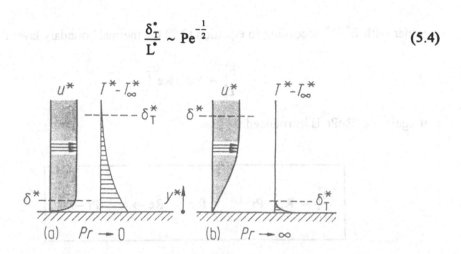

Fig. 5.1: Flow and temperature boundary layers for $Pr \rightarrow 0$ and $Pr \rightarrow \infty$

which again emphasizes that the Peclet number $Pe = \rho^* U_\infty^* L^* c_p^* / \lambda^* = U_\infty^* L^* / a^*$ appears in the heat transfer analysis (c. f. chapter 5.1). After replacing Pe by RePr which is approriate for boundary layers (5.4) becomes

$$\boxed{\frac{\delta_T^*}{L^*} \sim Re^{-\frac{1}{2}} Pr^{-\frac{1}{2}} \quad for \quad Re \rightarrow \infty \ , \ Pr \rightarrow 0} \tag{5.5}$$

(b) For $Pr \rightarrow \infty$: $U_c^* = \left.\frac{\partial u^*}{\partial y^*}\right|_w \delta_T^* \sim \frac{U_\infty^*}{\delta^*}\delta_T^*$

Internal energy is convected downstream by the near wall velocity boundary layer profile. This can be represented by the first non-zero term of a Taylor series expansion at the wall. For order of magnitude considerations $\left(\partial u^* / \partial y^*\right)_w$ can be replaced by the characteristic quantities U_∞^* / δ^*. With this U_c^*, from equation (5.3) follows:

$$\frac{\delta_T^*}{L^*} \sim Pe^{-\frac{1}{3}}\left(\frac{\delta^*}{L^*}\right)^{\frac{1}{3}} \tag{5.6}$$

Together with δ^* / L^* according to equation (2.2) the thermal boundary layer thickness is

$$\frac{\delta_T^*}{L^*} \sim Pe^{-\frac{1}{3}} Re^{-\frac{1}{6}}$$

or, if again Pe=RePr is introduced:

$$\frac{\delta_T^*}{L^*} \sim Re^{-\frac{1}{2}} Pr^{-\frac{1}{3}} \qquad \text{for} \qquad Re \to \infty \ , \ Pr \to 0 \tag{5.7}$$

For Prandtl numbers of order O(1) the Prandtl number dependence of δ_T^* / δ^* will be between $Pr^{-\frac{1}{2}}$ for $Pr \to 0$ and $Pr^{-\frac{1}{3}}$ for $Pr \to \infty$ though one cannot expect a certain exponent $-1/2 < n < -1/3$ in a power representation Pr^n.

The thermal boundary layer equation (first order in an asymptotic sense) has been introduced in chapter 3.2, see equation (3.13). It emerges from the energy equation (3.5) after the boundary layer transformation $y \to N$ in the limit $Re \to \infty$. For boundary layers it is more convenient to use

$$\vartheta = \frac{T^* - T_R^*}{\Delta T_R^*} = \Theta \frac{T_R^*}{\Delta T_R^*}$$

since often there is a well defined temperature difference ΔT_R^* which can be used as a reference quantity. Rewriting (3.5) with ϑ only changes $\tilde{E}c$ to $Ec = \tilde{E}c T_R^* / \Delta T_R^*$. The terms multiplied by Ec in the boundary layer equation (3.13) are the viscous dissipation and pressure work term, respectively. In most applications both can be neglected since $Ec = U_R^{*2} / c_P^* \Delta T_R^*$ is small. Only for high velocities and/or small heat transfer rates (i.e. small ΔT_R^*) can these effects not be neglected when compared to the effect of heat transfer by an imposed ΔT_R^*.

The scalings deduced so far all hold for forced convection, i.e. a heat transfer situation in which the flow is present independent of the heat transfer. The situation is quite different when heat transfer forces the flow as in natural convection which will be analysed next.

5.4 Natural convection boundary layers

Whereas thermal boundary layers are passive in character when they are superimposed on a forced convection boundary layer flow field, they actively influence the flow field in so-called natural convection flows. This has two immediate consequences:

(1) The reference velocity cannot be a given U_∞^* but in some way must include temperature effects if it is a characteristic velocity.

(2) The flow situation is necessarily one with variable property effects because only through variable properties (here: variable density) the temperature field may affect the flow field.

Both points in most textbooks about natural convection flows are not clearly accounted for but somehow are hidden in what is called *Boussinesq-approximation* which says: for natural convection flows all properties in the momentum equation can be assumed to be constant except the density in the buoyancy term. While this assumption sounds quite arbitrary asymptotic scaling considerations clearly show that it is not.

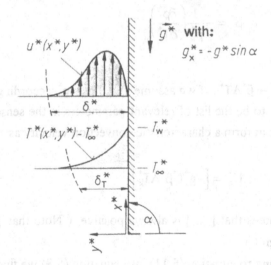

Fig. 5.2: Velocity and temperature boundary layers on a heated vertical flat plate

A typical natural convection flow is sketched in Fig. 5.2. The boundary layer scaling is deduced with exactly the same arguments as were used for forced convection flows. If we do not specify the characteristic convection velocity but instead name it U_c^* we can directly use equation (2.1) and (5.3) as scaling laws for the velocity and temperature boundary layers, respectively:

$$\frac{\delta^*}{L^*} \sim \left[\frac{v^*}{U_C^* L^*} \right]^{\frac{1}{2}} \quad ; \quad \frac{\delta_T^*}{L^*} \sim \left[\frac{a^*}{U_C^* L^*} \right]^{\frac{1}{2}} \tag{5.8}$$

The characteristic convection velocity U_C^* must be gained from the driving mechanism of the flow which is buoyancy.

Buoyancy forces occur due to near wall density changes $\Delta\rho_R^*$. These characteristic density changes $\Delta\rho_R^*$ are caused by characteristic temperature differences ΔT_R^*. Assuming small values of ΔT_R^* the density $\rho^*(T^*)$ can be expanded in a Taylor series which is truncated after the linear term, i.e.:

$$\rho^* = \rho_R^* + \left.\frac{\partial\rho^*}{\partial T^*}\right|_R (T^* - T_R^*) \quad\rightarrow\quad \Delta\rho_R^* = \left.\frac{\partial\rho^*}{\partial T^*}\right|_R \Delta T_R^* \tag{5.9}$$

With the *thermal expansion coefficient*

$$\beta^* = -\frac{1}{\rho_R^*}\left(\frac{\partial\rho^*}{\partial T^*}\right)_R \tag{5.10}$$

we thus have $\Delta\rho_R^* / \rho_R^* = -\beta^* \Delta T_R^*$. If we assume $\Delta\rho_R^* / \rho_R^*$, g_x^* according to Fig. 5.2 and a characteristic length L^* to be the list of relevant parameters in the sense of dimensional analysis we immediately can form a characteristic convection velocity as

$$U_C^* = \left[-g_x^* L^* \beta^* \Delta T_R^* \right]^{\frac{1}{2}} \tag{5.11}$$

The negative sign guarantees that [...] is always positive. (Note that g_x^* changes sign when ΔT_R^* changes its sign !).

Inserting U_C^* according to equation (5.11) into equation (5.8) we find:

$$\frac{\delta^*}{L^*} \sim \left[\frac{v^{*2}}{-g_x^* L^{*3} \beta^* \Delta T_R^*} \right]^{\frac{1}{4}} = Gr_\alpha^{-\frac{1}{4}} \quad \text{for} \quad Gr_\alpha \rightarrow \infty \tag{5.12}$$

$$\frac{\delta_T^*}{L^*} \sim \left[\frac{v^{*2}}{-g_x^* L^{*3} \beta^* \Delta T_R^*} \right]^{\frac{1}{4}} \left[\frac{a^*}{v^*} \right]^{\frac{1}{2}} = Gr_\alpha^{-\frac{1}{4}} Pr^{-\frac{1}{2}} \quad \text{for} \quad Gr_\alpha \to \infty \qquad (5.13)$$

with the definition of the Grashof number

$$Gr_\alpha = \frac{-g_x^* L^{*3} \beta^* \Delta T_R^*}{v^{*2}} \qquad (5.14)$$

The index α indicates that only the x-projection of the acceleration due to gravity is used in Gr_α, see Fig. 5.2.

For natural convection flows the Grashof number takes the role which the Reynolds number plays in forced convection flows. For example the boundary layer transformation is

$$N = \frac{y^*}{L^*} Gr_\alpha^{1/4} \quad ; \quad \bar{v} = \frac{v^*}{U_R^*} Gr_\alpha^{1/4} \qquad (5.15)$$

instead of equation (2.3) and equation (3.9) for forced convection flows.

Sometimes it is argued that only Re should be replaced by $Gr_\alpha^{1/2}$ to switch from forced to natural convection boundary layer flow. This, however, conceals the fundamental difference in the physics of both flows.

Now that the reference velocity is determined the second of the two points raised at the beginning of this chapter can be addressed. If natural convection flows are variable property flows the question arises which of the propertiy variations must be kept in the equations and which can be neglected, at least to the leading order in an asymptotic sense.

For this purpose we have to examine the basic equations in their original form in which all properties are variable. If, for example, we look at the x-momentum equation (3.2) it reads if we assume $\alpha = \pi / 2$ (vertical plate) for simplicity:

$$\rho \left(u \frac{\partial u}{\partial x} + v \frac{\partial u}{\partial y} \right) = -\frac{\partial p}{\partial x} + \frac{g^* L^*}{U_R^{*2}} (1 - \rho)$$

$$+ \frac{v^*}{U_R^* L^*} \left[\frac{\partial}{\partial x} \left(\eta \left(2 \frac{\partial u}{\partial x} - \frac{2}{3} \left(\frac{\partial u}{\partial x} + \frac{\partial v}{\partial y} \right) \right) \right) + \frac{\partial}{\partial y} \left(\eta \left(\frac{\partial u}{\partial y} + \frac{\partial v}{\partial x} \right) \right) \right] \qquad (5.16)$$

Different from (3.2) in (5.16) the nondimensional groups g^*L^* / U_R^{*2} and $v^* / \left(U_R^* L^*\right)$ are not replaced by Fr^{-2} and Re^{-1}, respectively, since these numbers typically represent forced convection flows with $U_R^* = U_\infty^*$.

Variable properties are $\rho = \rho^* / \rho_R^*$ and $\eta = \eta^* / \eta_R^*$. In order to systematically deduce how their temperature dependence must be accounted for they are expanded as Taylor series at a reference temperature T_R^*. Since this representation will be useful with respect to further aspects we here introduce the following formalism for a general property α^* which may stand for ρ^*, η^*, λ^*, c_p^* and D^*.

We expand $\alpha^* = \alpha^*(T^*)$:

$$\alpha^* = \alpha_R^* + \left.\frac{\partial \alpha^*}{\partial T^*}\right|_R \left(T^* - T_R^*\right) + \frac{1}{2}\left.\frac{\partial^2 \alpha^*}{\partial T^{*2}}\right|_R \left(T^* - T_R^*\right)^2 + \dots \tag{5.17}$$

which in nondimensional form reads:

$$\alpha = \frac{\alpha^*}{\alpha_R^*} = 1 + \varepsilon K_{\alpha 1}\vartheta + \frac{1}{2}\varepsilon^2 K_{\alpha 2}\vartheta^2 + O(\varepsilon^3) \tag{5.18}$$

with

$$\varepsilon = \frac{\Delta T_R^*}{T_R^*} \quad ; \quad \vartheta = \frac{T^* - T_R^*}{\Delta T_R^*} \tag{5.19}$$

$$K_{\alpha 1} = \left[\frac{\partial \alpha^*}{\partial T^*}\frac{T^*}{\alpha^*}\right]_R \quad ; \quad K_{\alpha 2} = \left[\frac{\partial^2 \alpha^*}{\partial T^{*2}}\frac{T^{*2}}{\alpha^*}\right]_R \tag{5.20}$$

Here $K_{\alpha 1}$ and $K_{\alpha 2}$ are properties of the fluid, ε is a small quantity which will be used as a perturbation parameter in heat transfer problems in chapter 6 below.

If ρ and η in equation (5.16) are replaced by their Taylor series representations $\rho = 1 + \varepsilon K_{\rho 1}\vartheta + \dots$, $\eta = 1 + \varepsilon K_{\eta 1}\vartheta + \dots$ the buoyancy term is of special interest. It reads:

$$\frac{g^*L^*}{U_R^{*2}}(1-\rho) = \frac{g^*L^*}{U_R^{*2}}\left(-\varepsilon K_{\rho 1}\vartheta + O(\varepsilon^2) \right) \tag{5.21}$$

With the reference velocity $U_R^* = U_C^*$ according to equation (5.11) the buoyancy term is

$$\frac{g^*L^*}{U_R^{*2}}(1-\rho) \;=\; \frac{g^*L^*}{-g_X^*L^*\beta^*\Delta T_R^*}\Big(-\varepsilon K_{\rho 1}\vartheta \,+\, O(\varepsilon^2)\Big) \;=\; \vartheta + O(\varepsilon) \qquad (5.22)$$

In (5.22) we applied $-g_X^* = -g^*$ (vertical wall), $\beta^* = -K_{\rho 1}/T_R^*$ according to (5.10) and (5.20) and $\varepsilon = \Delta T_R^* / T_R^*$ according to (5.19).

If we treat all variable properties in (5.16) in the same way it now takes the form:

$$\{1+O(\varepsilon)\}\left(u\frac{\partial u}{\partial x}+v\frac{\partial u}{\partial y}\right) \;=\; -\frac{\partial p}{\partial x} + \{\vartheta+O(\varepsilon)\}$$
$$+\, Gr^{-\frac{1}{2}}\left[\frac{\partial}{\partial x}\Big(\{1+O(\varepsilon)\}\,(\,\dots\,)\Big) + \frac{\partial}{\partial y}\Big(\{1+O(\varepsilon)\}\,(\,\dots\,)\Big)\right] \qquad (5.23)$$

If ε is a heat transfer perturbation parameter which accounts for deviations from $O(1)$-terms then the buoyancy term in (5.16) is an $O(1)$ - term with respect to this asymptotic expansion of the equations. This is the consequence out of choosing $U_R^* = U_c^*$ which asymptotically is of order $\varepsilon^{1/2}$. Setting $\varepsilon=0$ in (5.23) is what is called *Boussinesq approximation* in the literature. From an asymptotic point of view it is the leading order term of an asymptotic expansion of the basic equations with respect to a heat transfer parameter ε provided the reference velocity is of order $O(\varepsilon^{1/2})$ which it must be in order to be a characteristic velocity of the flow.

No systematic derivation of the Boussinesq approximation is possible if U_c^* would be chosen as $\sqrt{g^*L^*}$ which sometimes is used as reference velocity for natural convection flows.

The boundary layer equations that emerge from the basic equations for $Gr\rightarrow\infty$ after the boundary layer transformations (5.15) are applied, assuming the Boussinesq approximation (i.e. the leading term of an ε-expansion) to be sufficient, are:

$$\frac{\partial u}{\partial x} + \frac{\partial \bar{v}}{\partial N} = 0 \qquad (5.24)$$

$$u\frac{\partial u}{\partial x} + \bar{v}\frac{\partial u}{\partial N} = \frac{\partial^2 u}{\partial N^2} + \vartheta\sin\alpha \qquad (5.25)$$

$$u\frac{\partial \vartheta}{\partial x} + \bar{v}\frac{\partial \vartheta}{\partial N} = \frac{1}{Pr}\frac{\partial^2 \vartheta}{\partial N^2} + Ec\left(\frac{\partial u}{\partial N}\right)^2 \qquad (5.26)$$

They seem to be only slightly different from those for forced convection boundary layers, (3.10)-(3.13), if constant properties are assumed in these equations. However, the "small" difference, that $\partial p / \partial x$ in (3.11) is replaced by $\vartheta \sin\alpha$ in (5.25) reflects the very different physics of the two flows. From the mathematical point of view there is also an appreciable change: equations (5.25) and (5.26) are mutually coupled whereas (3.11) is independent of (3.13).

6. VARIABLE PROPERTY EFFECTS

6.1 Preliminary remark

In the previous chapter (see 5.4) we deduced the Boussinesq approximation for natural convection flows in a systematic way based on the Taylor series expansion of the physical properties involved. This procedure can be extended to all kinds of variable property flows and indeed is an ideal basis for a rational theory to account for variable property effects. Before we outline this method two empirical methods will be introduced that are widely used to account for variable property effects. There are, however, several shortcomings of these methods from an asymptotic point of view. Moreover, they can be identified as methods that are equivalent to the leading order part of the general asymptotic approach, as will be demonstrated at the end of this chapter.

6.2 Two empirical methods

There are two simple ideas how results that are gained under the assumption of constant properties can be corrected to take into account the effects due to variable properties:

(1) The final results for constant properties are multiplied by a correction factor of the general form

$$\left[\frac{\alpha_W^*}{\alpha_\infty^*}\right]^{m_\alpha} \qquad (6.1)$$

where α^* is a certain property of the fluid and m_α is an empirical exponent which either can be determined from experiment or from particular theoretical solutions that did not assume constant properties. The indexes w and ∞, indicate that α^* has to be taken at the temperature (and pressure) that prevails at the wall and far outside, respectively. If several properties are involved that are assumed to be equally important more than one factor may appear.

A complete ansatz to correct the skin friction result for constant properties, c_{f_0}, for example, would be

$$\frac{c_f}{c_{f_0}} = \left[\frac{\rho_w^*}{\rho_\infty^*}\right]^{m_\rho} \left[\frac{\eta_w^*}{\eta_\infty^*}\right]^{m_\eta} \left[\frac{\lambda_w^*}{\lambda_\infty^*}\right]^{m_\lambda} \left[\frac{c_{pw}^*}{c_{p\infty}^*}\right]^{m_c} \tag{6.2}$$

with 4 unknown exponents. This method is called *property ratio method*.

(2) The final result for constant properties in terms of $c_f = c_f(\text{Re})$, for example, is assumed to hold also for variable properties, but all properties that appear in the terms of the final results must be taken at an "a priori" unknown reference temperature

$$T_r^* = T_w^* + j\left(T_\infty^* - T_w^*\right) \tag{6.3}$$

with the common assumption: $0 \leq j \leq 1$.

The condition to determine j is that the formula for constant properties using T_r^* represents the physics for variable properties. This method is called *reference temperature method*.

The shortcomings of both methods will be discussed in chapter 6.5 below.

6.3 The expansion method

An analytical method to systematically account for the effects of temperature and pressure dependence of all properties can be based on the Taylor series expansion of $\alpha^* = \alpha^*(T^*, p^*)$, where α^* stands for ρ^*, η^*, ... It reads:

$$\alpha^* = \alpha_R^* + \left.\frac{\partial \alpha^*}{\partial T^*}\right|_R (T^* - T_R^*) + \left.\frac{\partial \alpha^*}{\partial p^*}\right|_R (p^* - p_R^*) + ... \tag{6.4}$$

or in a nondimensional form

$$\alpha = \frac{\alpha^*}{\alpha_R^*} = 1 + K_{\alpha 1}\Theta + \widetilde{K}_{\alpha 1}\widetilde{p} + \frac{1}{2}K_{\alpha 2}\Theta^2 + \frac{1}{2}\widetilde{K}_{\alpha 2}\widetilde{p}^2 + \overline{K}_{\alpha 2}\Theta\widetilde{p} + ... \tag{6.5}$$

with:

$$K_{\alpha 1} = \left[\frac{\partial \alpha^*}{\partial T^*}\frac{T^*}{\alpha^*}\right]_R \quad ; \quad \tilde{K}_{\alpha 1} = \left[\frac{\partial \alpha^*}{\partial p^*}\frac{\rho^* c_p^* T_R^*}{\alpha^*}\right]_R \tag{6.6}$$

$$K_{\alpha 2} = \left[\frac{\partial^2 \alpha^*}{\partial T^{*2}}\frac{T^{*2}}{\alpha^*}\right]_R \; ; \; \tilde{K}_{\alpha 2} = \left[\frac{\partial^2 \alpha^*}{\partial p^{*2}}\frac{\left(\rho^* c_p^* T_R^*\right)^2}{\alpha^*}\right]_R \; ; \; \overline{K}_{\alpha 2} = \left[\frac{\partial^2 \alpha^*}{\partial T^* \partial p^*}\frac{\rho^* c_p^* T_R^{*2}}{\alpha^*}\right]_R \tag{6.7}$$

Typical numbers of these nondimensional derivatives of α^* with respect to T^* and p^* are given in Table 6.1 for the viscosity of air and water. Since the pressure dependence is extremly small all effects in connection with $\tilde{K}_{\alpha i}$ can be neglected. The only exception is the pressure dependence of gas density which has an appreciable effect (compressibility).

In (6.5) Θ and $\tilde{p}$ in most cases are small numbers. If we rewrite them as

$$\Theta = \frac{T^* - T_R^*}{T_R^*} = \frac{T^* - T_R^*}{\Delta T_R^*}\cdot\frac{\Delta T_R^*}{T_R^*} \qquad ; \qquad \tilde{p} = \frac{p^* - p_R^*}{\rho^* c_p^* T_R^*} = \frac{p^* - p_R^*}{\rho^* U_R^{*2}}\cdot\frac{U_R^{*2}}{c_p^* T_R^*}$$

with $\dfrac{T^* - T_R^*}{\Delta T_R^*} = \vartheta$ and $\dfrac{p^* - p_R^*}{\rho^* U_R^{*2}} = p$ as O(1)-quantities and the two small parameters

$$\varepsilon = \frac{\Delta T_R^*}{T_R^*} \qquad ; \qquad \tilde{\varepsilon} = \frac{U_R^{*2}}{c_p^* T_R^*} \tag{6.8}$$

	$K_{\eta 1}$	$\tilde{K}_{\eta 1}$	$K_{\eta 2}$	$\tilde{K}_{\eta 2}$
AIR	0.8	0.0006	- 0.4	≈ 0
WATER	- 7.1	- 0.0003	80	≈ 0

Table 6.1: $K_{\eta i}$-values of air and water at $p_R = 1$ bar , $T_R = 293$ K

as heat transfer parameter and high speed parameter, respectively, the asymptotic representation of (6.5) is:

$$\alpha = 1 + \varepsilon K_{\alpha1}\vartheta + \widetilde{\varepsilon}\widetilde{K}_{\alpha1}p + \frac{1}{2}\varepsilon^2 K_{\alpha2}\vartheta^2 + \frac{1}{2}\widetilde{\varepsilon}^2\widetilde{K}_{\alpha2}p^2 + \varepsilon\widetilde{\varepsilon}\overline{K}_{\alpha2}p\vartheta + O(\varepsilon^3,\widetilde{\varepsilon}^3,\varepsilon^2\widetilde{\varepsilon},\varepsilon\widetilde{\varepsilon}^2) \qquad (6.9)$$

The expansion method to account for variable properties may be cast into four steps:

(S1) *Taylor series expansion of all physical properties involved in the problem*

In this step all properties are expanded according to equation (6.9) which introduces ε and $\widetilde{\varepsilon}$ as perturbation parameters into the problem. If, for example, only the temperature dependence of α is taken into account, the expansion is

$$\alpha = 1 + \varepsilon K_{\alpha1}\vartheta + \frac{1}{2}\varepsilon^2 K_{\alpha2}\vartheta^2 + O(\varepsilon^3) \qquad (6.10)$$

with the heat transfer parameter ε as the only perturbation parameter.

(S2) *Perturbation ansatz for all dependent variables*

If a represents all dependent variables of the problem, like u, $\bar{v}$, p, T, ... we assume the following expansion to be the asymptotic representation of a:

$$a = a_0 + \varepsilon\left[K_{\rho1}a_\rho + K_{\eta1}a_\eta + K_{\lambda1}a_\lambda + K_{c1}a_c\right] + \varepsilon^2\left[K_{\rho2}a_{\rho2} + K_{\eta2}a_{\eta2} + \dots\right.$$
$$\left. + K_{\rho1}^2 a_{\rho\rho} + K_{\eta1}^2 a_{\eta\eta} + \dots + K_\rho K_\eta a_{\rho\eta} + \dots\right] + O(\varepsilon^3) \qquad (6.11)$$

Here, again, we assume α to be only temperature dependent so that all variables a only must be expanded with respect to temperature effects.

In equation (6.11) four terms appear in the first order with respect to ε representing the effects of the four properties ρ, η, λ and c_p on a. In the second order there are 14 terms already due to the fact that four terms appear in connection with $K_{\alpha 2}$, four terms in connection with $K_{\alpha 1}^2$ and six terms due to combinations $K_{\alpha i}K_{\alpha j}$ which all are second order terms, i.e. of the asymptotic order $O(\varepsilon^2)$.

The number of terms is drastically reduced, however, if not all four properties are involved. If only the effects of one property α are strong enough to be accounted for, equation (6.11) is:

$$a = a_0 + \varepsilon K_{\alpha 1}a_\alpha + \varepsilon^2\left(K_{\alpha 2}a_{\alpha 2} + K_{\alpha 1}^2 a_{\alpha\alpha}\right) + O(\varepsilon^3) \tag{6.12}$$

(S3)*Insert the expansions for α and a into the complete basic equations and collect terms of equal asymptotic magnitude*

"Complete equation" means that of course the basic equations must include all variable property effects, i.e. the assumption of constant properties may be made in the basic equations only for those properties whose effects will not be accounted for by the expansion method.

After inserting the expansions into the basic equations, i.e. for example (6.10) and (6.12) if only the temperature effects of one property are important, terms of equal magnitude with respect to ε are collected. From this procedure we get zero, first, second, ... order equations which, due to the special form of the expansions, are free of ε and free of all constants $K_{\alpha i}$. As a consequence results emerging from these equations hold for arbitrary fluids (characterised by particular numbers for $K_{\alpha i}$) and arbitrary heat transfer rates ε as long as they are small enough to justify a series truncation after the $O(\varepsilon)$ or $O(\varepsilon^2)$ terms.

(S4)*Find the asymptotic results for a particular fluid*

Due to the expansion procedure all results will be of the same asymptotic form like the dependent variables, i.e. of the form (6.11) or (6.12) if only one α is accounted for.

For example, the skin friction coefficient and the Nußelt number will be of the form

$$\frac{c_f}{c_{f_0}} = 1 + \varepsilon K_{\eta 1}C_\eta + \varepsilon^2\left[K_{\eta 2}C_{\eta 2} + K_{\eta 1}^2 C_{\eta\eta}\right] + O(\varepsilon^3) \tag{6.13}$$

$$\frac{Nu}{Nu_0} = 1 + \varepsilon K_{\eta 1}N_\eta + \varepsilon^2\left[K_{\eta 2}N_{\eta 2} + K_{\eta 1}^2 N_{\eta\eta}\right] + O(\varepsilon^3) \tag{6.14}$$

if only the temperature dependence of viscosity is accounted for. The constants C_η, $C_{\eta 2}$ and $C_{\eta\eta}$ in (6.13) and N_η, $N_{\eta 2}$ and $N_{\eta\eta}$ are determined by solving the first and the second order equations. Solutions of the zero order equations are that for constant properties, i.e. they give c_{f_0} and Nu_0 in (6.13) and (6.14), respectively.

Examples of applying this expansion method to laminar boundary layer flows can be found in Gersten and Herwig [9], Herwig and Wickern [10] and Herwig, Wickern and Gersten [11].

The attractive part of the expansion method is that it is a rational theory that separates the effects of different properties clearly which allows a clear physical interpretation. The shortcoming of the method is its need for many sets of equations, at least when more than one property variation is involved and when higher order terms are sought. This problem is overcome by the so-called combined method described next.

6.4 The combined method

The simple idea of the combined method is the following: Instead of determining the constant $a_{\alpha i}$ in the final results (for example C_η, $C_{\eta 2}$, $C_{\eta\eta}$; N_η, $N_{\eta 2}$, $N_{\eta\eta}$ in (6.13) and (6.14) by solving the subsets of equations derived in the expansion method they are simply looked upon as coefficients of a Taylor series expansion of the final results with respect to $\varepsilon K_{\alpha 1}$, $\varepsilon^2 K_{\alpha 2}$ and $\varepsilon^2 K_{\alpha 1}^2$, (up to the order $O(\varepsilon^2)$).

For example C_η, $C_{\eta 2}$ and $C_{\eta\eta}$ in (6.13) are

$$C_\eta = \left[\frac{\partial \left(\dfrac{c_f}{c_{f_0}} \right)}{\partial \left(\varepsilon K_\eta \right)} \right]_0 \quad ; \quad C_{\eta 2} = \left[\frac{\partial \left(\dfrac{c_f}{c_{f_0}} \right)}{\partial \left(\varepsilon^2 K_{\eta 2} \right)} \right]_0 \quad ; \quad C_{\eta\eta} = \frac{1}{2} \left[\frac{\partial^2 \left(\dfrac{c_f}{c_{f_0}} \right)}{\partial \left(\varepsilon K_\eta \right)^2} \right]_0 \tag{6.15}$$

Next, these differentials are approximated by finite differences. For example C_η is approximated by

$$C_\eta = \lim_{h \to 0} \left\{ \frac{\left. \dfrac{c_f}{c_{f_0}} \right|_{\varepsilon K_\eta = h} - \left. \dfrac{c_f}{c_{f_0}} \right|_{\varepsilon K_\eta = -h}}{2h} \right\}_0 \tag{6.16}$$

Only two numerical solutions of the unexpanded basic equations are needed in which the viscosity η is taken as $\eta = 1 + h\vartheta$ and $\eta = 1 - h\vartheta$, respectively. Here, h serves just as a small number that is small enough to suppress higher order effects and large enough to be out of the range where numerical errors override the variable property effects.

By this method all constants in the final asymptotic results can be determined by a small number of numerical solutions of the basic unexpanded equations. Since this method combines highly occurate numerical solutions with asymptotic considerations it is called *combined method*. Details can be found in [12].

Figure 6.1 as an example shows that there is a wide range for the small number h which can be used to determine the constants $a_{\alpha i}$ with this method.

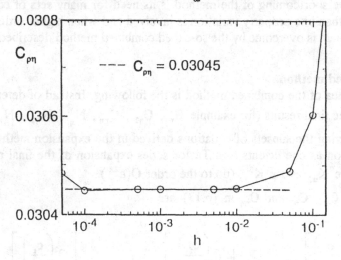

Fig. 6.1:   Determination of the constant $C_{\rho\eta} = \left[\dfrac{\partial^2\left(c_f/c_{f_0}\right)}{\partial\left(\varepsilon K_\rho\right)\partial\left(\varepsilon K_\eta\right)} \right]_0$ for a laminar

flat plate boundary layer with q_w = const ; Pr = 0.7

------ $C_{\rho\eta}$ according to the expansion method

6.5 Empirical methods in the light of the asymptotic results

The property ratio and the reference temperature method introduced in chapter 6.2 as empirical methods can be judged by comparison with the asymptotic methods.

If $\left(\alpha_w^*/\alpha_\infty^*\right)^{m_\alpha}$ of the property ratio method and $\alpha_r^*/\alpha_\infty^*$ in the reference temperature method are expanded into series with respect to ε the results of both methods can be compared directly:

(1) the exponents in the property ratio method can be identified with certain constants of the *linear* part of the asymptotic results.

(2) the factor j in the reference temperature method can also be identified with constants in the *linear* part of the asymptotic results.

It turns out that as long as m_α and j are constants (and not functions of ε, i.e. not functions of the strength of heat transfer) then they can only account for effects that are linear effects in an asymptotic sense.

Since m_α and j can be determined from the asymptotic results they are no longer empirical constants. Instead, the formulae of the two empirical methods may be used further on, but now in a different sense: not as an empirical correction to the constant property result but as a way to express the first order results of the analytical / numerical method to account for variable property effects.

References

1. Prandtl L.: Über Flüssigkeitsbewegung bei sehr kleiner Reibung, in: Verh. III Intern. Math. Kongr. (Ed. A. Krazer), Teubner, Leipzig 1904, 484-491.

2. Van Dyke, M.: Pertubation Methods in Fluid Mechanics, The Parabolic Press, Stanford 1975.

3. Schneider, W.: Mathematische Methoden in der Strömungsmechanik, Vieweg-Verlag, Braunschweig 1978.

4. Schlichting, H. and Gersten, K.: Grenzschichttheorie, Springer Verlag, Heidelberg 1996.

5. Gersten, K. and Herwig, H.: Strömungsmechanik, Vieweg-Verlag, Wiesbaden 1992.

6. Spurk, J. H.: Fluid Mechanics, Springer-Verlag, Berlin 1997.

7. Stewartson, K.: On the flow near the trailing edge of a flat plate, II. Mathematica, Vol. 16, 106-121, 1969.

8. Stewartson, K.: Multistructured boundary layers on flat plates and related bodies. Advances in Appl. Mech., Vol. 14, 145-239, 1974.

9. Gersten, K. and Herwig, H.: Impuls- und Wärmeübertragung bei variablen Stoffwerten für die laminare Plattenströmung, Wärme- Stoffübertragung, Bd. 16, 25-35, 1984.

10. Herwig, H. and Wickern, G.: The effect of variable properties on laminar boundary layer flow, Wärme- und Stoffübertragung, Bd. 20, 47-57, 1986.

11. Herwig, H.; Wickern, G. and Gersten, K.: Der Einfluß variabler Stoffwerte auf natürliche laminare Konvektionsströmungen, Wärme- und Stoffübertragung, Bd. 19, 19-30, 1985.

12. Herwig, H. and Schäfer, P.: A combined perturbation / Finite-difference procedure applied to temperature effects and stability in a laminar boundary layer, Archive of Applied Mechanics, Vol. 66, 264-272, 1996.

BOUNDARY-LAYER STABILITY - ASYMPTOTIC APPROACHES

R.J. Bodonyi
The Ohio State University, Columbus, OH, USA

1. Introduction

The stability of boundary-layer flows and their subsequent transition from a laminar state to a fully turbulent one has been the subject of intense study - theoretically, experimentally, and numerically - for many years. Indeed, hydrodynamic stability has generally been considered to be one of the central problems of fluid dynamics for over a century, and an explanation of boundary-layer stability and transition, in particular, has as yet to be described by a fully rational theory based on first principles, as noted by Reshotko [1], even though substantial progress has been made to this end during the past two decades.

Two types of transition are generally considered. The first type, bypass transition, is the process in which the external disturbances, e.g., sound, freestream turbulence, etc. or the internal disturbances, e.g., vibrations, roughness, etc. are strong enough that vortex stretching and other nonlinear mechanisms directly lead to turbulence without going through the known stability mechanisms. The second type is the so-called 'quiet' environment transition which evolves gradually, as has been experimentally documented, and can be followed by theory and numerical computations (see Reshotko [1] and Mack [2]). In this review we shall be concerned with this latter type of process.

The study of hydrodynamic stability and transition has a rich history, dating from the earliest studies by Helmholtz, Kelvin, Rayleigh, and Reynolds in the latter parts of the nineteenth century, and continuing to this day. Over the years, there have been many reviews written about the subject in general and about shear flows in particular, covering analytical, numerical, and experimental approaches to the subject over the entire speed range of interest. No attempt will be made here to review these 'classical' approaches to the study of shear flow instabilities. The interested reader is referred to one or more of the following reviews: Tani [3], Reshotko [1], Kozlov [4], Kachanov [5], and Reed, Saric & Arnal [6], and references quoted therein for additional information. Instead, this review will attempt to survey the rational, i.e., asymptotic, approaches which have been applied to shear flow stability problems. By rational, we mean those approaches by which linear and nonlinear effects can be taken into account without using empirical modeling or direct numerical solutions of the Navier-Stokes equations, but rather follow from asymptotic expansions applied to the complete Navier-Stokes equations when the Reynolds number of the flow is very large.

As discussed by Cowley & Wu [7] the classical *ad hoc* Orr-Sommerfeld (O-S) equation approach, along with other empirical models such as the e^n method, have been shown to provide useful tools for analyzing complex engineering problems involving boundary-layer stability and transition. However, from a mathematical point of view a stability analysis of shear flows at *finite* Reynolds numbers strictly requires a solution of the complete linearized Navier-Stokes equations. Indeed, even the more recent linear parabolized stability equation approach developed by Herbert [8] and coworkers is *ad hoc* in nature, even though this method has also proved quite successful in applications. However, there are other examples noted by Cowley & Wu for which these *ad hoc* approximations involving quasi-parallel flows have not been nearly as useful. Thus we must conclude that many of these linear theories, based on the Orr-Sommerfeld equation at finite Reynolds number, are not justifiable in any rigorous manner. Even though they have provided very useful tools for design and development in the past, many of these methods cannot be extended to problems involving nonlinear effects and hence cannot provide us with insight into these more complicated problems. A major goal of the asymptotic methods to be discussed in this review is to provide the designer with improved models for studying finite Reynolds number effects, much like that discussed by Smith, Papageorgiou & Elliott [9].

2. Multi-Structured Approach to Boundary-Layer Stability

An alternative approach to the stability problem for large Reynolds flows has been systematically developed by a number of researchers over the last two decades using the method of matched asymptotic expansions to construct self-consistent, rational approximations of the Navier-Stokes equations, i.e., approximations which become increasingly accurate as the perturbation parameter tends to zero (or infinity) and are such that one can improve upon the results by embedding them as a first step in a systematic scheme of successive approximations which can, in principle, be continued indefinitely (see Van Dyke [10]). The approach has as its basis the studies of Smith [11], [12] wherein he showed that the subsonic triple-deck structure of Stewartson [13] and Messiter [14] provides a rational description of the lower branch properties of the neutral stability curve for Blasius and other boundary-layer flows. Similarly, the upper branch neutral stability properties of the Blasius and other boundary-layer flows, including the effects of nonparallelism have been studied by Bodonyi & Smith [15] while the nonlinear critical layers associated with the upper branch stability structure have been considered by Smith & Bodonyi [16] and Bodonyi, Smith & Gajjar [17]. As these studies show, the upper branch structure is much more complicated than that of the lower branch, consisting of five asymptotically distinct zones as compared to the three zones appearing in the lower branch structure.

It is easily shown that the lower branch stability properties are governed by the triple-deck structure when the Reynolds number of the flow is large. Indeed, the essential scalings for the multi-structured theory can be found in Lin's [18] classical book on hydrodynamic stability. For example, consider the lower branch of the neutral stability curve for the Blasius boundary layer. In the asymptotic limit as the Reynolds number becomes large, Lin showed that the relationship between the wave number, α, and the Reynolds number, Re_δ, based on the boundary-layer thickness δ is given by

$$Re_\delta = \gamma_1(\alpha\delta)^{-4} \tag{1.1}$$

where γ_1 is an order one constant. Furthermore, Re_δ and δ are defined by

$$Re_\delta = U_\infty\delta/\nu, \quad \delta = \gamma_2 LRe^{-1/2}, \tag{1.2}$$

here U_∞ is a characteristic velocity, ν is the kinematic viscosity of the fluid, L is a characteristic length scale and γ_2 is another order one constant. The flow Reynolds number, Re, is given by

$$Re = U_\infty L/\nu. \tag{1.3}$$

By substituting (1.2) into (1.1) and rearranging, it is easily shown that the wavenumber scales as

$$\alpha L \propto Re^{3/8} \tag{1.4}$$

Or, equivalently, the wavelength of the disturbance is given by

$$(\alpha L)^{-1} \propto Re^{-3/8} \tag{1.5}$$

which is precisely the streamwise length scale of the triple-deck theory of viscous-inviscid interactions. It follows, therefore, that for Re $\gg$ 1 the linear disturbance quantities are governed by a triple-deck structure, on a streamwise length scale of $O(Re^{-3/8})$. A schematic of this structure is given in Figure 1a.

Similar arguments can be made concerning the upper branch stability characteristics. In this case, however, the properties of the critical layers are of most importance. Using the classical results of Lin [18] for the upper branch asymptotic properties, Bodonyi & Smith [15] showed that the stability structure for asymptotically large Reynolds numbers to be five-zoned with a streamwise length scale of $O(Re^{-9/20})$, as shown in Figure 1b.

Furthermore, if Δ denotes a typical disturbance size of the streamwise velocity relative to the freestream velocity, then for $\Delta = O(Re^{-1/8})$, a truly nonlinear response is provoked in the boundary layer, and it is controlled by the nonlinear triple-deck structure. For Δ just slightly less than this order, it has been found that weakly nonlinear theory applies and as Δ is decreased still further the effects of nonparallelism come into play until finally, when Δ is smaller than any power of $Re^{-1/8}$, the classical linear stability theory applies to leading order (see Smith [11], [12]).

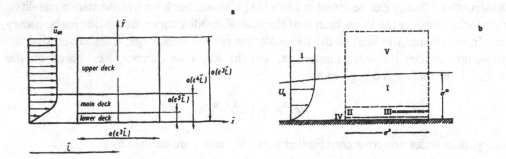

Fig. 1, Schematic of the Blasius boundary layer stability structure: (a) lower branch, $\epsilon = Re^{-1/8}$, (b) upper branch, $\sigma = Re^{-1/20}$.

There have been numerous studies in the literature, e.g., Bogdanova & Ryzhov [19], Terent'ev [20], Smith & Burggraf [21], Smith [22], and Ryzhov & Zhuk [23], to name a few, which show the link between triple-deck theory and hydrodynamic stability theory. These studies also indicate that triple-deck theory can describe not only the formation , but also the evolution of Tollmien-Schlichting (TS) waves. Thus the theory can provide a framework for a description of the earlier stages of some transition processes. Of particular importance for a study of the transition problem is the type of interaction being studied, e.g., two- or three-dimensional, small- or large-scale, viscous or inviscid, or wave/vortex interactions. When the concepts of the asymptotic theory are used to construct numerical models for the stability problem applicable at *finite* Reynolds numbers, as in the work of Smith, *et al.* [9], reasonable agreement with experiments and numerical solutions of the full unsteady Navier-Stokes equations underscore the possible practical value of the structured approach at the finite Reynolds numbers of practical concern. In some instances the formulas developed from the asymptotic theory are directly applicable to flows at finite Reynolds numbers. The accuracy of these formulas cannot be determined *a priori*. Each case must be checked against existing experimental and/or appropriate numerical computations to determine the range of applicability for which the formulas hold. Even in those cases where direct application of the results is not possible, valuable information concerning the flow field structure can be obtained which may be useful in developing approximate models for engineering applications.

2.1 Governing Equations

The starting point for a study of the stability problems to be considered here are the compressible Navier-Stokes equations. For simplicity, we shall restrict our attention to two-dimensional unsteady flows for the most part. Using the usual terminology, we first rewrite the governing equations in a nondimensional form using a characteristic length, L, and freestream reference conditions U_∞, T_∞, ρ_∞, $\rho_\infty U_\infty^2$, L/U_∞, and μ_∞, for the velocity, temperature, density, dynamic pressure, time scale, and the shear viscosity, respectively. Under such a change of variables the governing equations reduce to the usual nondimensional Navier-Stokes equations involving three basic parameters:

$$Re = \rho_\infty U_\infty L/\mu_\infty, \quad Pr = c_p \, \mu_\infty /k_\infty, \quad M_\infty = U_\infty /a_\infty,$$

where c_p, k_∞ are the fluid's specific heat and thermal conductivity while a_∞ is the fluid speed of sound in the freestream. These equations are naturally supplemented by the no-slip boundary condition at solid surfaces or other appropriate boundary conditions, depending on the problem being studied. In addition a condition on the thermal properties at the surface must be specified. For our purposes here we shall only need to note that the wall is either taken to be an order one constant, T_w, or else it is assumed that the wall is adiabatic.

Our interest in this review occurs for those physical situations where Re >> 1, Pr = O(1), and M_∞ = O(1). In such cases, as shown by the classical work of Prandtl [24], the flow can be divided into two asymptotically distinct regions, an outer inviscid region and a viscous boundary-layer region which is strongly affected by both viscosity and heat conduction near the surface. Most importantly, the pressure is found to be constant across the viscous boundary layer and is impressed on the flow by the inviscid pressure distribution found from a solution of the appropriate Euler equations. It is also important to note that Prandtl's theory assumes that the body surface is smooth without singularities in its slope and curvature.

Stewartson [13] and Messiter [14], in what are now classical papers, first showed how Prandtl's theory had to be modified to handle the discontinuity at the trailing edge of a flat plate aligned with the flow, and Stewartson & Williams [25] showed how to extend Prandtl's theory beyond the separation point at which the Goldstein singularity invalidated the classical theory. In both cases it was found that the classical theory failed in a small region of streamwise extent of $O(Re^{-3/8})$ around the point in question. In fact, this length scale is the same regardless of the particular disturbing agency in the flowfield. In particular, this is the same length scale typical of Tollmien-Schlichting waves, as noted above. Thus the triple-deck theory for viscous-inviscid interactions developed by Stewartson and others can be directly employed for investigating the stability properties of laminar boundary layers when the Reynolds number is sufficiently large.

A summary of triple-deck theory is given by Stewartson [26] and elsewhere in this volume. It suffices here to note that for subsonic or supersonic flows the fundamental problem can be reduced to a study of the lower-deck region in which the compressible Navier-Stokes equations reduce to the form

$$u_x + v_y = 0,$$
$$u_t + u u_x + v u_y = -p_x + u_{yy},$$

(2.1a,b)

subject to the boundary conditions

$$u = v = 0 \quad \text{on } y = 0,$$

$$u \sim y + A(x,t), \text{as } y \to \infty, \qquad (2.1\text{c-e})$$

$$u - y, v, A \to 0 \text{ as } x \to -\infty$$

These canonical lower-deck variables are related to nondimensional physical variables by the transformations:

$$\bar{x} = x^*/L^* - 1 = \epsilon^3 C^{3/8} \lambda^{-5/4} |M_\infty^2 - 1|^{-3/8} (T_w^*/T_\infty^*)^{3/2} x$$

$$\bar{y} = y^*/L^* = \epsilon^5 C^{5/8} \lambda^{-3/4} |M_\infty^2 - 1|^{-1/8} (T_w^*/T_\infty^*)^{3/2} y$$

$$\bar{t} = t^* U_\infty^*/L^* = \epsilon^2 C^{1/4} \lambda^{-3/2} |M_\infty^2 - 1|^{-1/4} (T_w^*/T_\infty^*) t$$

$$\bar{u} = u^*/U_\infty^* = \epsilon C^{1/8} \lambda^{1/4} |M_\infty^2 - 1|^{-1/8} (T_w^*/T_\infty^*)^{1/2} u \qquad (2.2\text{a-f})$$

$$\bar{v} = v^*/U_\infty^* = \epsilon^3 C^{3/8} \lambda^{3/4} |M_\infty^2 - 1|^{1/8} (T_w^*/T_\infty^*)^{1/2} v$$

$$\bar{p} = (p^* - p_\infty^*)/\rho^* U_\infty^{*2} = \epsilon^2 C^{1/4} \lambda^{1/2} |M_\infty^2 - 1|^{-1/4} p$$

$$\bar{A} = C^{5/8} \lambda^{-3/4} |M_\infty^2 - 1|^{-1/8} (T_w^*/T_\infty^*)^{3/2} A$$

In these equations, C refers to the Chapman Rubesin constant for the linear viscosity law μ^*/μ_∞^* = C T^*/T_∞^*, T_w^* is the constant wall temperature, λ is the scaled wall shear in the unperturbed boundary layer at x = 0 (λ = 0.3321 for Blasius flow), and T_∞^* is the freestream temperature. Finally, we note that A(x,t) can be interpreted as an unknown instantaneous displacement thickness. The mathematical formulation of the lower-deck problem is completed by an additional equation which relates A to the unknown pressure, p, viz,

$$p = \frac{1}{\pi} \int_{-\infty}^{\infty} \frac{\partial A/\partial \xi}{x - \xi} d\xi \quad \text{for } M_\infty < 1$$

$$p = -\frac{\partial A}{\partial x} \qquad\qquad \text{for } M_\infty > 1 \qquad (2.3\text{a,b})$$

Note that in terms of these scaled variables, the only difference between subsonic and supersonic flow is in the relationship between p and A. The above scalings are not appropriate for transonic flow conditions, as shown by Bodonyi & Kluwick [27] for steady viscous-inviscid interacting flows and Ryzhov [28] for the stability properties of transonic flows based on asymptotic methods.

3. Asymptotic Linear Theory

In an attempt to analyze the influence of boundary-layer growth on the stability of the incompressible Blasius boundary-layer flow Smith [11] used a rational, large Reynolds number, approach to study small disturbances of fixed frequency on the lower branch stability properties for Blasius flow. A schematic of the triple deck, large Reynolds number asymptotic scaling which is appropriate for the lower-branch Tollmien-Schlichting wave disturbances is shown in

Figure 1a. Smith's analysis yielded the classical parallel-flow solution to leading order and the non-parallel flow effects were deduced in a consistent manner from higher-order terms in the asymptotic expansions of the flow field variables. To this end, solutions to the lower-deck problem were sought in the form

$$\bar{u} = (U_1 + \epsilon U_2 + \epsilon^2 U_3 + \epsilon^3 \ln \epsilon U_{4L} + \epsilon^3 U_4 + ...)E$$
$$\bar{v} = (\epsilon^2 V_1 + \epsilon^3 V_2 + \epsilon^4 V_3 + \epsilon^5 \ln \epsilon V_{4L} + \epsilon^5 V_4 + ...)E \qquad (3.1\text{a-c})$$
$$\bar{p} = (\epsilon P_1 + \epsilon^2 P_2 + \epsilon^3 P_3 + \epsilon^4 \ln \epsilon P_{4L} + \epsilon^4 P_4 + ...)E$$

$$E = \exp[i(\theta(x) - \beta t)],$$

where

$$\beta = \beta_1 + \epsilon \beta_2 + \epsilon^2 \beta_3 + \epsilon^3 \ln \epsilon \beta_{4L} + \epsilon^3 \beta_4 + ... \qquad (3.2\text{a-c})$$

$$d\theta/dx = K_1(\bar{x}) + \epsilon K_2(\bar{x}) + \epsilon^2 K_3(\bar{x}) + \epsilon^3 \ln \epsilon K_{4L}(\bar{x}) + \epsilon^3 K_4(\bar{x}) + ...$$

Here β is the constant frequency of the disturbance, and θ is its wavenumber, taken to be a slowly varying function of $\bar{x}$. The solution for the first four terms in (3.2b,c) essentially yields the conventional parallel-flow results since these terms are uninfluenced by the slow spatial changes of the basic flow. To determine the major non-parallel flow contributions, however, the fifth terms in (3.2b,c) must be found. Furthermore, it should be noted that the non-parallel flow stability characteristics are dependent on the particular flow disturbance being considered, as has been noted by Bouthier [29]. The advantage of the asymptotic approach over the more heuristic methods is that the non-parallel flow effects emerge in a consistent fashion from the asymptotic expansions with the result that the non-parallel flow effects come into play much sooner than in the successive approximations used by others. Smith's results for the lower-branch neutral curve for the Blasius boundary layer along with the experimental results of Ross [30] and Gaster's [31] successive approximation results are given Figure 2a. The asymptotic results agree quite well with the other data over most of the Reynolds number range given. Furthermore, an important conclusion from these non-parallel flow studies of boundary-layer stability is that there is no universal criterion for determining the stability of real flows. Whether the boundary layer is stable or not, and to what extent, depends entirely on the particular disturbance quantity chosen to determine the stability characteristics of the flow.

A similar asymptotic analysis for the upper branch stability properties of the Blasius boundary layer was considered by Bodonyi & Smith [15]. Here, however, the scaling is quite different, resulting in a complicated five-zoned structure across the boundary layer as shown in Figure 1a, since now the critical layer is asymptotically distinct from the viscous sublayer which is not the case for the lower branch structure. Although the scalings are now different, the procedure is similar to that of the lower branch study of Smith [11]. The neutral frequency can be systematically determined along with the effects of non-parallelism. However, as shown in Figure 2b the results are not in good agreement with either the experimental data or the Orr-Sommerfeld parallel flow theory. With an origin shift of 300 in R_δ, which is allowable in the context of the asymptotic theory, the agreement between theory and experiment is improved somewhat. However, it does not appear that the asymptotic results provide a useful quantitative tool for predicting upper branch stability properties, even though the upper branch scalings have been quite useful in theoretical studies following the disturbances downstream and/or those involving nonlinear disturbances.

Although the above discussions have centered around the incompressible, flat plate Blasius problem, it should be noted that similar results also apply to subsonic compressible flows in principle since effects of compressibility, including the Mach number can be scaled out of the governing equations for the lower-deck problem as already noted above. Similar comments hold for the upper branch scaling laws. Smith [32] gives a comprehensive account of the effects of compressibililty on the lower branch stability characteristics of the flow to linear Tollmien-Schlichting modes for both two-and three-dimensional disturbances and the results are consistent with the more classical theories for compressible flows (see Reshotko [1] and Mack [2]) while Gajjar & Cole [33] consider the linear upper branch stability properties of compressible boundary-layer flows. An interesting question arises by considering what happens as the Mach number M_∞ approaches one from below, i.e., from a subsonic flow, or from above, i.e., from a supersonic flow. For a subsonic boundary layer, one can show that the dispersion relationships can lead to unstable oscillations which is in accord with classical theory and experiment. However, for a supersonic boundary layer, none of the dispersion relations allows for unstable disturbances. One must conclude, therefore that a supersonic boundary layer is stable in the framework of triple-deck theory and this result contradicts both theory and experiment.

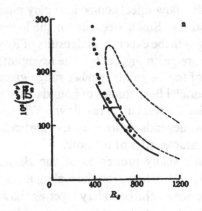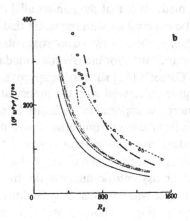

Fig. 2, (a) Comparison of O-S approximate neutral curve (- - -) and the non-parallel flow neutral curve of Smith [12] and to experimental measurements of Ross *et al.* [30] (from Smith [12]). (b) Comparison of upper branch neutral stability curve from Bodonyi & Smith [15]: ____, one term; __ __ __, two terms; __ ... __, three terms; ...., four terms; and - - - -, that from a numerical solution, according to parallel flow theory. ○, experimental data from Ross *et al.* [30]. — —.The curve obtained when an origin shift of 300 is applied in R_δ is applied to the curve .... (from Bodonyi & Smith [15]).

The above discussion does not hold for transonic flows where M_∞ is sufficiently close to one as can be seen in the scaling laws (2.2). Thus to consider transonic flows, one must rescale the governing equations as discussed by Bodonyi & Kluwick [27]. The most significant change now occurs in the upper-deck structure, where the inviscid flow is now governed by the nonlinear transonic small disturbance equation so that the flow can continuously pass through the sonic condition. The steady-state solutions for these transonic flows are thus complicated by the presence of the nonlinear behavior of the upper-deck structure as compared to the linear structures found in subsonic and supersonic flows. Similarly, the transonic stability problem becomes much more complicated and the question arises as to how to explain the problem noted above concerning the approach to sonic condition $M_\infty = 1$. A possible resolution to this dilemma has been considered by Ryzhov [28], and, according to Ryzhov, 'As a matter of principle these findings settle the main issue showing how the triple-deck theory of subsonic boundary-layer instability can be extended to provide a passage through the threshold value $M_\infty = 1$ and then cover a range of moderate supersonic velocities'. It is left to the individual reader to consider this result more completely. Further aspects of the transonic stability problem, have investigated by Bowles & Smith [34] but will not be reviewed here.

Although the above discussions have been limited to Blasius boundary-layer flows, it should be noted that according to the high Reynolds number asymptotic theory, the lower branch scaling laws also apply to boundary-layer flows with pressure gradients since the presence of a pressure gradient does not alter the scaling laws to leading order, including supersonic flows if one allows for three-dimensional disturbances (Smith [32]). This is not true of the upper branch analysis, however. In most two-dimensional boundary layers the upper branch stability structure has scalings different from those discussed above (see Reid [35], for instance), because of the action of the applied pressure gradient and, consequently, of a greater deviation from a uniform shear in the basic flow at the critical layer. For a further discussion of pressure gradient effects on linear stability problems see Cowley & Wu [7].

Finally, we note that it is well known that disturbances in a laminar boundary layer become quite three-dimensional as they propagate downstream. An attempt to elucidate the effects of nonparallelism on the three-dimensional linear stability properties of both two- and three-dimensional boundary layers has been undertaken by Stewart & Smith [36]. Their high frequency stability analysis shows that the effects of increasing nonparallelism are to accentuate three-dimensional instability and also to turn the direction of the maximum linear growth rate from the $0°$ orientation for two-dimensional TS waves towards a maximum angle of $64.68°$ to the freestream direction for three-dimensional waves. They also demonstrate that cross flow effects can have a significant influence on Tollmien-Schlichting instabilities, especially as the amount of cross flow increases.

4. Weakly Nonlinear Theory

As discussed above, the asymptotic approach to boundary-layer stability, centered around linearized triple-deck theory, has proved to be quite useful in delineating the basic mechanisms involved in the process for both parallel and nonparallel flows. However, to more fully understand the processes involved for nonparallel flows, it is imperative that a basis for continued study be that a large Reynolds number asymptotic theory be used to investigate the effects of nonparallelism and nonlinearity in a rationally consistent manner. Using this approach, new features in transition have been revealed as well as a deeper understanding of the underlying physical mechanisms involved in the transition process. A first step in analyzing the effects of nonlinearity on boundary-layer flows is to consider weakly nonlinear disturbances based on the assumption of small growth rates. By weak nonlinearity we mean a disturbance which corresponds to a small relative change in the basic flow from its undisturbed form. Smith [12] examined in detail how the size of a small two-dimensional disturbance affects its nonlinear stability properties by considering a nonlinear extension of the usual triple-deck scalings while Hall & Smith [37] considered three-dimensional disturbances in planar boundary-layer flows. Of particular interest were the stability characteristics arising near the lower branch of the non-parallel flow neutral curve of a boundary-layer flow. A fully nonlinear response is produced in the basic flow by a disturbance whose streamwise velocity amplitude is of $O(Re^{-1/8})$, i.e., it occurs in the lower deck of the triple-deck structure. For a weakly nonlinear response we then perturb the nonlinear flow by considering small disturbance amplitudes, h, such that

$$\beta = \beta_1 + O(h^2), \quad \alpha = \alpha_1 + \alpha_2 h^2 + \cdots, \quad E = \exp[i(\alpha_1 x - \beta_1 t)],$$
$$(4.1)$$
$$u = y + h(u_1 E + c.c.) + h^2 u_2 + h^3 u_3 + \cdots, \quad etc$$

with the multiple scaling:
$$\frac{\partial}{\partial x} \rightarrow \frac{\partial}{\partial x} + h^2 \frac{\partial}{\partial \bar{x}}$$

In applying these expansions, the equation for u_2 contains second harmonics $\propto E^{\pm 2}$ and a mean flow correction term $\propto E^0$ while u_3 contains terms $\propto E^{\pm 3}, E^{\pm 1}$, etc. After some algebra it turns out that the weakly nonlinear stability properties are controlled by a nonlinear Landahl amplitude equation

$$\frac{dA}{d\bar{x}} = i\alpha_2 A + a_1 |A|^2 A \qquad (4.2)$$

where A is the scaled amplitude of the leading order velocity perturbation. As discussed by Smith [12], a_1 is a constant such that $\Re(a_1)$ is negative and hence a supercritical bifurcation is found to occur with the result that for frequencies slightly greater than the neutral, stable nonlinear traveling waves are generated and an expression for the equilibrium amplitude, $|A^*|_e$ can be obtained. A comparison of Smith's results with those of Itoh [38] are given in Figure 3 for the special case $|A^*|_e = 0.138$, and, as shown, the agreement is encouraging, although some caution is necessary in comparing the results, as noted by Smith.

Hall & Smith [37] show that the while the equilibrium amplitude is stable for two-dimensional disturbances of the same frequency, it can be unstable to three-dimensional disturbances of a different frequency. They conclude that there are three main effects of nonparallelism of the basic flow on the nonlinear spatial evolution of three-dimensional disturbances: 1) nonparallelism makes any bifurcation from one stable form of motion to another a smooth process, 2) when more than one possible state can exist far downstream, the nonparallelism dictates exactly which state is achieved from the given initial conditions upstream, and 3) when no stable state is possible downstream, nonparallelism shows that the disturbance amplitude has a finite-distance singularity downstream of the initial state.

Further extensions of the two-dimensional analyses have also been considered by Smith & Walton [39] who included slow spanwise modulations consistent with lower-deck scalings. Sample computations show that solutions to the appropriate amplitude equation can terminate in a finite-distance singularity due to a focusing effect of spanwise modulations. To continue beyond this point, one must resort to a study of the fully nonlinear three-dimensional lower-deck equations.

Other weakly nonlinear problems have been studied in the literature, including both critical layers and vortex/wave interactions. For a good review of these studies the reader is referred to Cowley & Wu [7]. We also note that one of the major conclusions from the weakly nonlinear stability studies is that, for the most part, the growing wave amplitudes result from the solution of integral-differential equations which have quadratic and/or quartic type nonlinearities present, as discussed, for instance, by Goldstein [40], [41].

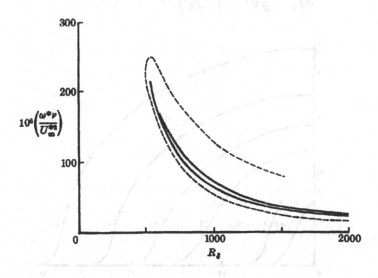

Fig. 3, The solid curves are curves of $|A_1^*|_e = 0.138$, the upper curve coming from Itoh's [38] calculations, while the lower curve is from (3.6) of Smith [12]. The dashed curve is the O-S approximation to a neutral curve. (from Smith [12]).

An important question concerning the two-dimensional weakly nonlinear theories is what happens downstream in either an initial-value problem or, equivalently, as $x > x_0$, in the corresponding steady-state problem, where x_0 is the neutral stability location at which the scaled frequency $\beta = \beta_0$ (≈ 2.29). These problems can be studied in terms of considering the solution as β increases beyond its neutral value β_0. Conlisk, Burggraf, & Smith [42] have undertaken numerical computations for traveling-wave solutions which continue on from (4.2) when β increases an O(1) amount from β_0, while Smith & Burggraf [21] have considered a large amplitude analysis of the relevant controlling equations which corresponds to high frequencies with $\beta = O(|\partial / \partial t|) \gg 1$ and suitably shortened streamwise length scales. Their study of this nonlinear TS regime reveals that an initially infinitesimally small TS disturbance passes through at least two distinct stages as it progresses downstream. These stages can be defined in terms of their representative pressure levels, e.g., $p = O(\beta^{1/2})$ (stage 1) and $p = O(\beta)$ (stage 2). When β is large any disturbance, although unstable, has only a relatively small growth rate and so, to leading order the corresponding wavenumber α is both real and large, i.e., $\alpha = \beta^{1/2} + O(\beta^{-1/2})$. Hence, wave-packet type solutions are appropriate in which their small growth rates are balanced by their nonlinearity. Following Smith [22] it is easily shown that for $\beta \gg 1$, using standard multiple scaling methods, the relevant governing equation, in terms of the disturbance pressure amplitude, P_0, reduces to a generalized cubic Schrödinger equation in slow variables X_1 and T_2 and in a frame of reference moving downstream with the group velocity of the form

$$\frac{\partial P_0}{\partial T_2} - i \frac{\partial^2 P_0}{\partial X_1^2} = \left(\frac{1-i}{\sqrt{2}} \right) P_0 - \frac{5i}{2} |P_0|^2 P_0 \qquad (4.3)$$

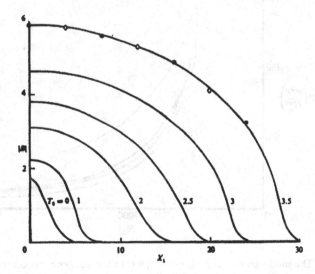

Fig. 4, The computed amplitude |B| versus X_1, at various times T_2. (from Smith [22]).

A numerical solution for |B|, where B is defined in terms of P_0 by the expression

$$P_0 = \sqrt{\frac{2}{5}} \exp\left(\frac{-iT}{\sqrt{2}} \right) B(X_1, T_2) \qquad (4.4)$$

is given in Figure 4. From both numerical and analytical studies it can be shown that the principal features of the solution of this equation are such that the disturbance broadens spatially in an exponentially fast form as T increases, and the amplitude of the motion also grows exponentially fast, being unaffected by nonlinear forces during this stage, with the amplitude solution taking on an elliptical form.   Only the effective phase of B is altered in a nonlinear fashion.

5. Nonlinear Theory

5.1 Lower-branch Asymptotics

The next stage in the nonlinear development of high frequency, shortened wavelength, disturbances occurs when the disturbance size increases to $O(\beta)$ and the fully nonlinear stage 2 is entered. Zhuk & Ryzhov [43] and, independently, Smith & Burggraf [21] showed that for asymptotically large Reynolds numbers, boundary layers can generate disturbances other than the usual Tollmien-Schlichting type wherein these new disturbances have much higher amplitudes and shorter length scales. Analysis indicates a four-deck structure for the disturbance flow which is governed by inviscid dynamics with the governing equation for the unknown displacement thickness, -A(X,T), given by the integro-differential equation known as the Benjamin-Ono equation

$$\frac{\partial A}{\partial T} + A \frac{\partial A}{\partial X} = -\frac{1}{\pi} \int_{-\infty}^{\infty} \frac{A_{\xi\xi}(\xi,T)}{(X-\xi)} d\xi \qquad (5.1)$$

This equation permits nonlinear traveling-wave and solitary-wave solutions as discussed by Smith & Burggraf [21] and Kachanov, Ryzhov & Smith [44]. Since (5.1) is based on inviscid dynamics, there must be a viscous sublayer of thickness $O(\beta^{-1/2})$ nearer to the surface, which can be shown to be of a classical unsteady boundary-layer form. However, it was shown by Van Dommelen & Shen [45] and Elliott, Cowley & Smith [46] that this sublayer is itself susceptible to breakdown via a finite time singularity. More recent numerical computations by Peridier, Smith & Walker [47] clearly show a singularity developing in the displacement thickness, δ^*, as can be seen in Figure 5, thus providing possible "eruptions of vorticity" or spikes into the outer inviscid layer where the Benjamin-Ono equation, (5.1), applies. Additional comments and comparisons between experiments and theory concerning the formation of solitons in transitional boundary layers and their possible role in K-type breakdown are given by Kachanov et al. [44]. The ultimate consequences of this sublayer eruption and precisely how these events can be incorporated into the inviscid analysis are not as yet resolved, although it seems obvious that new scales must be introduced. But it does appear likely that in the next stage of the evolution of nonlinear disturbances normal pressure gradient effects must be taken into account, and further discussion of these new effects are presented in the computations and analyses of Hoyle, Smith & Walker [48]. Additional comments on the effects of spikes and strongly nonlinear theory and comparisons with experimental data have recently been given by Smith [49].

Three-dimensional effects are also of importance in the study of nonlinear stability theory. Stewart & Smith [50] apply a small initial disturbance to a two-dimensional boundary-layer flow at large Reynolds number utilizing a type of vortex/wave interaction mechanism and study how three-dimensional nonlinear distortion effects come into play in the high frequency range.

Computations and analysis of the governing equations indicate a finite-time singularity occurring in the solution for both the wave and vortex amplitudes wherein the local flow quickly enters a strongly nonlinear three-dimensional stage in which the mean flow is significantly altered, and quantitative and/or order-of-magnitude comparisons are not inconsistent with available experimental data for K-type transition.

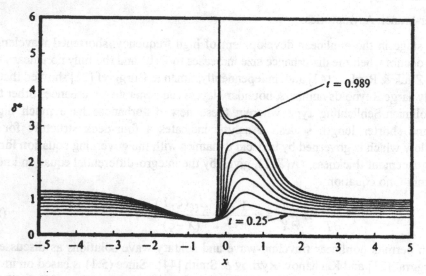

Fig. 5, Temporal development of the displacement thickness; plotted curves are at t = 0.25 (0.10) 0.95 and t_s = 0.989 (from Peridier *et al.* [47]).

The study of nonlinear three-dimensional vortex/wave interactions has become of increasing importance due to their possible relevance in boundary-layer transition. Hall & Smith [51] discuss nonlinear interactions between vortex flows and Tollmien-Schlichting waves, describing both analytical and computational properties of such interactions. Among the results discussed are the possibility, for spatially evolving interactions, of finite-distance breakdown with a singularity in the boundary-layer displacement thickness. Davis & Smith [52] investigated the effects of cross-flow on vortex/TS interactions. Their work centered on two lower-branch TS waves nonlinearly interacting in such a way that a longitudinal vortex flow is produced. This resulting vortex motion then gives rise to wave modulation via a wall shear forcing mechanism. Since the Reynolds number of the flow is assumed large, both the wave and vortex motions are principally governed by triple-deck theory. From analysis and computations, they found that there were three distinct responses which emerged in the nonlinear behavior of the downstream solution: an algebraic finite-distance singularity, far-downstream saturation, or far-downstream wave decay leaving only the vortex motion, depending on input conditions, wave angles involved, and the magnitude of the cross-flow. Further comments on nonlinear transition paths for vortex/wave interaction mechanisms are given by Smith [53] and Davis & Smith [54]. Finally, we note that the majority of the research efforts to date have concentrated on the stability characteristics of incompressible flow and as yet there is no completely acceptable transition theory for flows at arbitrary Mach number even given the large literature on the subject over the years. More recently, asymptotic multi-structured theories based on large Reynolds numbers have been advanced by several researchers. Most notably, the work of Smith [32], Smith & Walton [39], Bowles [55], Hall & Smith [51], Brown, Khorrami, Neish, & Smith [56], Bowles & Smith [34], and Davis & Smith [57] should be mentioned for the interested reader, as no detailed discussion of these compressible theories will be given in this review.

5.2 Upper branch Asymptotics

The correct asymptotic structure for large Reynolds number flows along the upper branch neutral stability curve was first proposed by Bodonyi & Smith [15] in their study of parallel and non-parallel flow stability of the Blasius boundary layer. In a related paper, Smith & Bodonyi [16] considered the effects of increasing the size of a disturbance relative to inverse powers of the asymptotically large Reynolds number, along the upper branch of the neutral stability curve for a general accelerating planar boundary layer on a fixed wall. As noted earlier, the main difference from the lower branch case arises from the well known distinction of the critical layer and the viscous wall layer in the balance of forces controlling stability along the upper branch. As shown in Figure 6 it is a five zoned structure similar to that shown by Bodonyi & Smith [15] which governs the upper branch parallel and non-parallel flow stability properties of the non-accelerating Blasius boundary layer. This structure provides the foundation for the study of nonlinear aspects necessarily induced as the disturbance size, δ, increases in size beyond the linear range. Smith & Bodonyi [16] found that when $\delta = O(Re^{-7/36})$, the stability structure is essentially unaltered from its linear form except that the critical layer becomes strongly influenced by nonlinear interactions and higher harmonics of the fundamental disturbance appear significantly everywhere in the flow field. At this stage the balance of forces in the critical layer then corresponds to the nonlinear critical layer problems addressed by Benny & Bergeron [58], Davis [59], Haberman [60], and Brown & Stewartson [61]. Smith & Bodonyi [16] fixed the scaling such that the viscous and inertia forces were of comparable size within the critical layer and determined the neutral instability wave amplitude by balancing the phase shift arising from the Stokes layer at the wall and the amplitude-dependent phase jump across the critical layer. They also deduced the changes that take place when the disturbance size is slightly increased, from $O(Re^{-7/36})$ to $O(Re^{-1/6})$, namely that the critical layer now moves out into the midst of the basic flow and the neutral wavelength becomes comparable with the characteristic lateral dimension of the basic flow. Thus the classical inviscid Rayleigh situation is recovered, and simultaneously, a significant lateral pressure gradient across the nonlinear critical layer is set up. Further studies of these larger amplitude waves have been given by Bodonyi, Smith & Gajjar [17] for the case where the critical layer is situated in the middle of the classical boundary layer, and by Smith, Doorly & Rothmayer [62] when the critical layer is situated at the outer edge of the boundary layer.

Gajjar & Smith [63], in an attempt to relax the restriction, present in a majority of studies on nonlinear critical layers up to this time, that disturbances are neutral traveling waves outside of the critical layer, studied the temporal-spatial growth and decay of disturbance amplitudes throughout the entire flowfield, not just inside the critical layer. Their results show that previous upper branch nonlinear studies based on quasi-neutral assumptions yield unstable subcritical threshold amplitudes, above which increasingly fast disturbance growth takes place globally. In a related paper, Goldstein & Durbin [64] conclude that the effect of nonlinear critical layers is to eliminate the classical upper branch of spatially growing TS waves predicted by linear theory, a result in agreement with the experimental observations of Bayliss *et al.* [65].

Further comments concerning spatial modulation of planar carrier waves under conditions which lead to nonlinear effects in the upper branch regime are discussed by Cowley & Wu [7] and references quoted therein, and will not be reviewed here. We note that they also give an extended discussion of Rayleigh waves in shear flows concentrating on those situations where the disturbances become nonlinear near a Rayleigh wave neutral point. Their discussion includes a review of two-dimensional regular and singular modes, a pair of oblique modes, and resonant-triad interactions.

Finally, we note that in a recent paper Wu, Lieb & Goldstein [66] studied the nonlinear

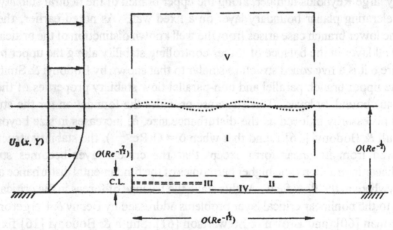

Fig. 6, Sketch of the relatively large five-zoned structure of a traveling wave disturbance of amplitude $\delta \leq O(Re^{-7/36})$ in an accelerating boundary layer with basic velocity profile $U_B(x,Y)$. (from Smith & Bodonyi [16]).

interaction and evolution of a pair of oblique TS waves using upper branch scalings so that a distinct critical layer appears. Analysis shows that the waves develop nonlinearly through four distinct stages. Of most interest is the presence of a new singularity in the third stage wherein the flow is governed by the fully nonlinear, three-dimensional inviscid triple-deck equations. The authors suggest that their analysis can characterize the 'so-called oblique breakdown' in a boundary layer.

The interaction of vortices with inviscid inflectional Rayleigh waves has also been the object of extensive study in recent years. Various types of interactions have been considered depending on the flow regime chosen for study. As noted by Brown & Smith [67], the common feature among these studies is that the large Reynolds number small-amplitude three-dimensional waves are nonlinearly coupled with the mean flow via its unknown longitudinal vortex component. Brown & Smith [67] extended the work of Smith, Brown & Brown [68] to study the effect of non-symmetry on vortex/wave interactions in the presence of inflectional disturbances for unequal input wave amplitudes and for small cross-flows. Of special interest in their study were the existence of bounded solutions to the governing equations and the possibility of providing a theory which could be used as a basis for studying flows with an enhanced non-symmetry in the input conditions or situations with stronger cross-flows present.

In a related paper, Allen, Brown & Smith [69] applied the ideas of Brown & Smith [67] to a study of an incident axisymmetric flow supplemented by a small amount of swirl. Although the flow structure is more complicated with the presence of swirl, it was found that the dominant interactions were much like those of Brown & Smith [67]. A principal difference between these papers is that in the latter paper numerical solutions of the full integro-differential equations were presented while only analytical studies of the bounded solutions were undertaken in the former study. More recently, Timoshin & Smith [70] analyzed the upper branch neutral modes for inviscid stability in boundary layers with significant longitudinal vortices present. The main results of their study are the demonstration of the presence of a non-inflectional logarithmically singular critical layer and the ensuing non-uniqueness present in the solutions. These new results are in direct contrast with previous linear and nonlinear studies used in vortex/wave interaction and secondary instability theories and should lead to a reconsideration of some of the previous results in this area of research.

6. Boundary-layer Receptivity

The term 'receptivity' was chosen by Morkovin [71] to comprise the processes by which external disturbances are ingested or converted into disturbances inside a boundary layer. These external disturbances may be a property of the flow field or may be caused by the boundaries themselves. For example, irrotational (pressure) disturbances are of the acoustic type which can consist of single-frequency tones all the way up to broad-band noise. Rotational disturbances comprise turbulent fluctuations and other two- or three-dimensional vorticity fluctuations in the freestream. Boundary disturbances are primarily caused by vibrations and roughness of the bounding surface itself. Receptivity, then, is the means by which a particular forced disturbance enters the boundary layer and initiates the transition process. If the initial disturbances are sufficiently large they can grow by forcing mechanisms to nonlinear levels and eventually lead to turbulence via a 'bypass' process. If the disturbances are 'small' they will tend to excite free disturbances in the boundary layer which are best known as Tollmien-Schlichting waves. These waves then go through various stages of development depending on the nature of the problem. It is this first stage that is termed the receptivity problem.

Receptivity is fundamentally different from the classical eigenvalue stability problem. It is by nature a boundary-value problem since it involves the response of the boundary layer to an externally imposed disturbance. However, naturally occurring freestream disturbances travel at much higher speeds than instability waves. Thus characteristic wavelengths of the freestream disturbances at a given frequency are much longer than the TS waves. Hence a wavelength conversion is required to transfer energy from these long waves to the much shorter TS waves. This wavelength conversion is the core of the receptivity problem (see Reshotko [1]). Murdock [72], Goldstein [73 - 75] and Goldstein, Leib & Cowley [76] have theoretically investigated the role that small freestream disturbances play in generating TS waves in boundary-layer flows in a variety of situations. In particular, we note that Ruban [77] and, independently, Goldstein [75] studied the effect that small variations in surface geometry have on scattering weak unsteady two-dimensional disturbances into TS waves. Using the triple-deck scalings of Stewartson [13], Goldstein concluded that relatively small surface variations which provoke correspondingly small pressure changes can produce a relatively large coupling between the TS waves and the imposed disturbance when these variations occur on the scale of a TS wave. However, Goldstein's analysis was limited in that he only considered a linearized steady flow solution for the interactive flow in the vicinity of a sharp corner in an incompressible flow. Thus he was not able to consider the effects of surface variations of sufficient size to provoke a nonlinear response in the steady boundary-layer flow.

Bodonyi, Welch, Duck & Tadjfar [78] considered a numerical study of the interaction of freestream disturbances and a small two-dimensional roughness element placed on an otherwise flat plate. For this study the two-dimensional nonlinear viscous-inviscid triple-deck equations were solved numerically to provide the basic steady motion, U and V. It was also shown therein that the unsteady motion is governed by the unsteady linearized triple-deck equations, in suitably scaled variables. Assuming the solution to be harmonic in time the appropriate equations to be solved for the disturbance quantities u,v,p, and a are

$$\frac{\partial u}{\partial X} + \frac{\partial v}{\partial Y} = 0$$

$$-i S_0 u + U \frac{\partial u}{\partial X} + u \frac{\partial U}{\partial X} + V \frac{\partial u}{\partial Y} + v \frac{\partial U}{\partial Y} = -\frac{\partial p}{\partial X} + \frac{\partial^2 u}{\partial Y^2}$$

$$(6.1)$$

along with boundary conditions

$$u = v = 0 \quad on \quad Y = F(X)$$

$$u \rightarrow 1 + a(X), \quad Y \rightarrow \infty, \quad all \; X,$$

$$u \sim 1 - e^{i^{\frac{3}{2}} \sqrt{S_0} Y}, \quad p \sim iXS_0, \quad for \; X \rightarrow -\infty$$

$$(6.2)$$

where $F(X)$ defines the hump shape, U,V are the velocity components of the related steady flow solution, and S_0 is a scaled Strouhal number, assumed to be of O(1). Finally, the perturbation displacement thickness, $a(X)$, and pressure, $p(X)$, are related by

$$\frac{d^2 a}{dX^2} = \frac{1}{\pi} \int_{-\infty}^{\infty} \frac{dp/d\xi - i S_0}{(\xi - X)} d\xi$$

$$(6.3)$$

Bodonyi et al. [78] undertook a detailed numerical study of this receptivity problem for a range of values of S_0, which represents the nature of the freestream disturbance, and a representative surface distortion taken to be $F(X) = h(1+X^2)^{-1}$, where h is an order-one factor giving the height of the distortion relative to the lower-deck scalings. Several values of h were chosen so that the corresponding steady flow solutions covered the range from linear theory up to nonlinear interactions including boundary-layer separation. Among their results, it was found that the amplitude of all disturbance quantities grows without bound, downstream of the hump, if $S_0 > 2.29...$ For values of $S_0 < 2.29...$, the disturbances eventually decay to zero amplitude and the flow remains stable. Thus the numerical solutions illustrate the growth or decay of the TS waves generated by the interaction between the freestream disturbance and the two-dimensional roughness element, depending on the value of the scaled Strouhal number. Figure 7 shows representative results for the disturbance wall shear $\tau_w(X,0)$ with h for two different peaks. For $h < 1$, the scaled disturbance amplitude depends approximately linearly on h, while for larger values of h an increasingly nonlinear enhancing effect occurs. They thus conclude that there is a linear dependence of the disturbance amplitude on h for sufficiently small h. For moderate h ($1 < h < 3$) they found an enhancement of the receptivity by the nonlinear effect of hump height. For h large, where local flow separation can occur in the steady flow, they found a possible short-wavelength instability in their time marching computations and a rapidly increasing enhanced receptivity.

In a related study, Tadjfar & Bodonyi [79] extended the work of Bodonyi et al. [78] to consider the receptivity of a boundary layer to the interaction of time harmonic freestream disturbances with a three-dimensional roughness element. They found that for supercritical values of $S_0 > 2$, the TS waves are amplified in a wedge-shaped region, 15° to 18° to the basic flow direction, extending downstream of the hump. Furthermore, they concluded that the amplification rate approaches a value somewhat higher than that of two-dimensional TS waves, as calculated by the linearized analysis, far downstream of the roughness element.

Recently Duck, Ruban & Zhikharev [80] considered the influence of freestream turbulence on boundary-layer receptivity by studying the effect of vorticity waves with a nondimensional frequency $\omega = O(Re^{1/4})$ on the generation of TS waves. Their results show that the mechanisms for generating TS waves by vortex waves, i.e. by freestream turbulence, is quite different than that caused by acoustic disturbances. The interaction of the vortex waves with a flat plate results in a 'vorticity deformation layer' of thickness $O(Re^{-1/4})$ along the plate surface. Using a combination of multi scale asymptotic techniques and numerical computations, they showed that a singularity forms in the flow at some distance downstream of the leading edge, indicating a possible abrupt transition to turbulence. Furthermore, they found that as the amplitude of these vortex waves is decreased the singular point moved further downstream and thus a suitably small surface irregularity could result in the generation of TS waves within the boundary layer at some distance downstream of the leading edge. Using triple-deck theory concepts, they were thus able to obtain an explicit formula for the amplitude of TS waves by the interaction of external vortex waves with stationary boundary-layer perturbations around the surface irregularity.

Additional comments on the receptivity of boundary layers to a variety of disturbance mechanisms, including vibrators, sound waves, small jets and vortices, are given by Kozlov & Ryzhov [81] who compare asymptotic analyses and corresponding experimental results where possible. Also, a more detailed review of boundary-layer receptivity to acoustic disturbances in a variety of physical settings is given by Goldstein & Hultgren [82] for the interested reader.

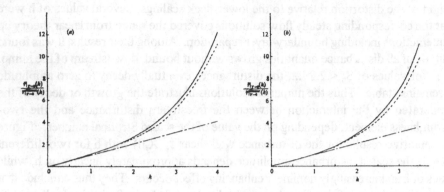

Fig. 7, (a) Disturbance amplitude versus h for $S_0 = 1$, normalized by the linearized results for h = 1. ___, minimum near X = 18.5; _ _ _, peak near X = 13; __ __ __, linear result. (b) $S_0 = 3$. ___, minimum near X = 9; _ _ _, peak near X = 7; __ __ __, linear result. (from Bodonyi et al. [78]).

7. Concluding Comments

This paper has attempted a partial review of the progress made since 1979 towards our understanding of boundary-layer instability mechanisms using the rational method of matched asymptotic expansions applied to the Navier-Stokes equations when the flow Reynolds number is asymptotically large. In the author's opinion, this approach is really the only viable way to systematically study the very complicated problem of boundary-layer stability and transition, especially when nonlinear effects are to be taken into account. A limiting feature of this approach is that the results of an asymptotic theory are not, in general, directly applicable to many of the problems of practical interest, since those problems always involve situations occurring at finite, albeit large, Reynolds numbers. This fact does not, however, diminish the value of the asymptotic approach in furthering our understanding of the stability characteristics of viscous fluids. Indeed, asymptotic results can be used as a guide in developing approximate theories for use at these more realistic Reynolds numbers. Also, they can help in increasing both our physical and theoretical understanding of the problems under study as well as providing estimates for more detailed numerical computations, as noted by Smith et al. [9].

By necessity, this review has been limited to a discussion of just some of the many possible topics that have been investigated over the last two decades. As a consequence, the contributions from many authors have not been reported here and the reader is left to consider other reviews on the subject which have been noted in this article for further details. These omissions do not imply in any way that their contributions are not of importance to the subject, but only that time and space were limited. Little has been said concerning the effects of both compressibility and three dimensionality, especially for supersonic/hypersonic flows even though there is a substantial literature available. In particular we note that a recent paper by Seddougui & Bassom [83] considers the linear stability properties for hypersonic flow over a slender cone with an attached shock within the framework of triple-deck theory. Their results suggest that curvature effects on the stability properties of hypersonic flow are important when the attached shock is properly accounted for in the theory. Their paper also provides the interested reader with additional recent references which pertain to effects that compressibility has on the stability of high speed flows in a variety of situations. Other problems, not reviewed here, involving nonlinear stability properties in both compressible and/or three-dimensional flows in the vicinity of both the upper and lower branches of the neutral stability curve are also of importance in enhancing our understanding of the transition problem and it is hoped that another review will be undertaken to summarize these effects for interested readers.

REFERENCES

1. Reshotko, E.: Boundary-Layer Stability and Transition, in *Annu. Rev. Fluid Mech.*, **8** (1976), 311-349.

2. Mack, L.: Boundary-layer Linear Stability Theory in *Special Course on Stability and Transition of Laminar Flows*, AGARD Report No. 709, (1984), 3.1 - 3.81.

3. Tani, I.: Boundary-Layer Transition, in *Annu. Rev. Fluid Mech.*, **1** (1969), 169-196.

4. Kozlov, V.V. (ed.): *Laminar-Turbulent Transition*, Springer-Verlag, Berlin 1985.

5. Kachanov, Y.S.: Physical Mechanisms of Laminar-Boundary-Layer Transition, in *Annu. Rev. Fluid Mech.*, **26** (1994), 411-482.

6. Reed, H.L., Saric, W.S. and Arnal, D.: Linear Stability Theory Applied to Boundary Layers, in *Annu. Rev. Fluid Mech.*, **28** (1996), 389-428.

7. Cowley, S.J. and Wu, X.: Asymptotic Approaches to Transition Modelling, in: AGARD-R-793 (Special course in progress in transition modelling), Von Kármán Inst. Rhode-Saint-Genése, Belg. (1993), 3.1-3.38.

8. Herbert, T.: Parabolized Stability Equations, in: AGARD-R-793 (Special course in progress in transition modelling), Von Kármán Inst. Rhode-Saint-Genése, Belg. (1993), 4.1-4.34.

9. Smith, F.T., Papageorgiou, D. and Elliott, J.W.: An alternate approach to linear and nonlinear stability calculations at finite Reynolds number, *J. Fluid Mech.*, **146** (1984), 313-330.

10. Van Dyke, M.: *Perturbation Methods in Fluid Mechanics*, Parabolic Press, Stanford 1975.

11. Smith, F.T.: On the nonparallel flow stability of the Blasius boundary layer, *Proc. Roy. Soc. Lond.*, **A366** (1979), 91-109.

12. Smith, F.T.: Nonlinear stability of boundary layers for disturbances of various sizes, *Proc. Roy. Soc. Lond.,* **A368** (1979), 573-589. (And corrections **A371** (1980), 439).

13. Stewartson, K.: On the flow near the trailing edge of a flat plate II, *Mathematika,* **16** (1969), 106-121.

14. Messiter, A.F.: Boundary-layer flow near the trailing edge of a flat plate, *SIAM J. Appl. Math.,* **18** (1970), 241-257.

15. Bodonyi, R.J. and Smith, F.T.: The upper branch stability of the Blasius boundary layer, including nonparallel flow effects, *Proc. Roy. Soc. Lond. A,* **375** (1981), 65-92.

16. Smith, F.T. and Bodonyi, R.J.: Nonlinear critical layers and their development in streaming-flow stability, *J. Fluid Mech.,* **118** (1982), 165-185.

17. Bodonyi, R.J., Smith, F.T. and Gajjar, J.: Amplitude-dependent stability of boundary-layer flow with a strongly nonlinear critical layer, *IMA J. Appl. Math.,* **30** (1983), 1-19.

18. Lin, C.C.: *The Theory of Hydrodynamic Stability,* Cambridge University Press, Cambridge 1955.

19. Bogdanova, E.V. and Ryzhov, O.S.: Free and induced oscillations in Poiseuille flow, *Q. J. Mech. Appl. Math.,* **36** (1983), 271-287.

20. Terent'ev, E.D.: On the nonstationary boundary layer with self-induced pressure near a vibrating wall in a supersonic flow, *Sov. Phys. Dokl.,* **23** (1978), 358.

21. Smith, F.T. and Burggraf, O.R.: On the development of large-sized short-scaled disturbances in boundary layers, *Proc. R. Soc. Lond.,* **A399** (1985), 22-55.

22. Smith, F.T.: Two-dimensional disturbance travel, growth, and spreading in boundary Layers, *J. Fluid Mech.,* **169** (1986), 353-377.

23. Ryzhov, O.S. and Zhuk, V.I.: in *Current Problems in Computational Fluid Dynamics.* Moscow : MIR Publications, (1986), 286.

24. Prandtl, L.: Über Flüssigkeitsbewegung bei sehr kleiner Reibung, Verhandl. d. III. Intern. Mathem. Kongresses, Heidelberg, (1904), 484-491.

25. Stewartson, K. and Williams, P.G.: Self-induced separation, *Proc. R. Soc. Lond.,* **A312** (1969), 181-206.

26. Stewartson, K.: Multistructured Boundary Layers on Flat Plates and Related Bodies,

Advances in Applied Mech., **14** (1974), 145-239.

27. Bodonyi, R.J. and Kluwick, A.: Freely interacting transonic boundary layers, *Phys. Fluids,* **20** (1977), 1432-1437.

28. Ryzhov, O.S.: An Asymptotic Approach to Separation and Stability Problems of a Transonic Boundary Layer, in *Transonic aerodynamics. Problems in Asymptotic Theory.* (Ed. L.P. Cook), SIAM 1993, 29-53.

29. Bouthier, M.: Stabilité linéaire des écoulements presque paralléles. La couche limite de Blasius, *J. de Mécanique,* **12** (1973), 75-95.

30. Ross, J.A., Barnes, F.H., Burns, J.G. and Ross, M.A.S.: The flat plate boundary layer. Part 3. Comparison of theory with experiment, *J. Fluid Mech.,* **43** (1970), 819-832.

31. Gaster, M.: On the effects of boundary-layer growth on flow stability, *J. Fluid Mech.,* **66** (1974), 465-480.

32. Smith, F.T.: On the first-mode instability in subsonic, supersonic or hypersonic boundary layers, *J. Fluid Mech.,* **198** (1989), 127-153.

33. Gajjar, J.S.B. and Cole, J.W.: The upper branch stability of compressible boundary layer flows, *Theo. Comput. Fluid Dyn.,* **1** (1989), 105.

34. Bowles, R.I.: Application of nonlinear viscous-inviscid interactions in liquid layer flows and transonic boundary layers transition. Ph.D. Thesis, University of London, (1990).

35. Reid, W.H.: The Stability of Parallel Flows., in *Basic Developments in Fluid Dynamics,* (ed. M. Holt), **1** (1965), 249-307.

36. Stewart, P.A. and Smith, F.T.: Three-dimensional instabilities in steady and unsteady nonparallel boundary layers, including effects of Tollmien-Schlichting disturbances and cross flow, *Proc. Roy. Soc. Lond. A,* **409** (1987), 229-248.

37. Hall, P. and Smith, F.T.: On the effects of nonparallelism, three-dimensionality, and mode interaction in nonlinear boundary-layer stability, *Stud. Appl. Math.,* **70** (1984), 91-120.

38. Itoh, N.: Spatial growth of finite wave disturbances in parallel and nearly parallel flows. Part 2. The numerical results for the flat plate boundary layer, *Trans. Japan Soc. Aero Space Sci.,* **17** (1974), 175-186.

39. Smith, F.T. and Walton, A.G.: Nonlinear interaction of near-planar TS waves and longitudinal vortices in boundary-layer transition, *Mathematika,* **36** (1989), 262-289.

40. Goldstein, M.E.: Nonlinear interactions between oblique instability waves on nearly parallel shear flows, *Phys. Fluids*, **6** (1994), 724-735.

41. Goldstein, M.E.: The role of nonlinear critical layers in boundary layer transition, *Phil. Trans. R. Soc. Lond. A*, **352** (1995), 425-442.

42. Conlisk, A.T., Burggraf, O.R. and Smith, F.T.: Nonlinear neutral modes in the Blasius boundary layer, *Forum on Unsteady Flow Separation,* **32** (1987), 119-121.

43. Zhuk, V.I. and Ryzhov, O.S.: Locally inviscid perturbations in a boundary layer with self-induced pressure, *Dokl. Akad. Nauk SSSR*, **263** (1982), 56-69 (in Russian). See also *Sov. Phys. Dokl.*, **27** (1982), 177-179 (in English).

44. Kachanov, Y.S., Ryzhov, O.S. and Smith, F.T.: Formation of solitons in transitional boundary layers: theory and experiment, *J. Fluid Mech.*, **251** (1993), 273-297.

45. Van Dommelen, L.L. and Shen, S.F.: The genesis of separation, in: *Numerical and Physical Aspects of Aerodynamic Flows* (ed. T. Cebeci), Springer, New York 1981, 293-311.

46. Elliott, J.W. Cowley, S.J. and Smith, F.T.: Breakdown of boundary layers: (i) on moving surfaces; (ii) in semi-similar unsteady flow; (iii) in fully unsteady flow, *Geophys. Astrophys. Fluid Dyn.*, **25** (1983), 77-138.

47. Peridier, V.J., Smith, F.T. and Walker, J.D.A.: Vortex induced boundary-layer separation. Part 1. The unsteady limit problem Re → ∞. Part 2. Unsteady interacting boundary-layer theory, *J. Fluid Mech.*, **232** (1991), 99-131, 133-165.

48. Hoyle, J.M., Smith, F.T. and Walker, J.D.A.: On sublayer eruption and vortex formation, *Comp. Phys. Comm.*, **65** (1991), 151-157.

49. Smith, F.T.: On spikes and spots: strongly nonlinear theory and experimental comparisons, *Phil. Trans. R. Soc. Lond. A*, **352** (1995), 405-424.

50. Stewart, P.A. and Smith, F.T.: Three-dimensional nonlinear blow-up from a nearly planar initial disturbance, in boundary-layer transition: theory and experimental comparisons, *J. Fluid Mech.*, **244** (1992), 79-100.

51. Hall, P. and Smith, F.T.: On strongly nonlinear vortex-wave interactions in boundary-layer transition, *J. Fluid Mech.*, **227** (1991), 641-666.

52. Davis, D.A.R. and Smith, F.T.: Influence of cross-flow on nonlinear Tollmien-Schlichting/vortex interaction, *Proc. R. Soc. Lond. A*, **446** (1994), 319-340.

53. Smith, F.T.: Nonlinear transition paths in boundary layers with cross-flow, in *IUTAM Symposium on Nonlinear Instability and Transition in Three-Dimensional Boundary Layers* (P.W. Duck and P. Hall eds.) (1996), 283-297.

54. Davis, D.A.R. and Smith, F.T.: Cross-flow influence on nonlinear Tollmien-Schlichting/vortex interaction, in *IUTAM Symposium on Nonlinear Instability and Transition in Three-Dimensional Boundary layers* (P.W. Duck and P. Hall, eds.) (1996), 309-316.

55. Bowles, R.I.: Application of nonlinear viscous-inviscid interactions in liquid layer flows and transonic boundary layers transition. Ph.D. Thesis, University of London, (1990).

56. Brown S.N., Khorrami, A.F., Neish, A. and Smith, F.T.: Hypersonic boundary-layer interactions and transition, *Trans. Roy. Soc.,* A **335** (1991), 139-152.

57. Davis, D.A.R. and Smith, F.T.: On subsonic, supersonic and hypersonic inflectional-wave/vortex interaction, *J. Engr. Maths.*, **30** (1996), 611-645.

58. Benney, D.J. and Bergeron, R.F.: A new class of nonlinear waves in parallel flows, *Stud. Appl. Math.*, **48** (1969), 181-204.

59. Davis, R.E.: On the high Reynolds number flow over a wavy wall, *J. Fluid Mech.*, **30** (1969), 337-346.

60. Haberman, R.: Critical layers in parallel flows, *Stud. Appl. Math.*, **51** (1972), 139-161.

61. Brown, S.N. and Stewartson, K.: The evolution of the critical layer of a Rossby wave. Part II, *Geophys. Astrophys. Fluid Dyn.*, **10** (1978), 1-24.

62. Smith, F.T., Doorly, D.J. and Rothmayer, A.P.: On displacement-thickness, wall-layer, and mid-flow scales in turbulent boundary layers, and slugs of vorticity in channel and pipe flows, *Proc. R. Soc. Lond. A,* **428** (1990), 255-281.

63. Gajjar, J. and Smith, F.T.: On the global instability of free disturbances with a time-dependent nonlinear viscous critical layer, *J. Fluid Mech.*, **157** (1985), 53-77.

64. Goldstein, M.E. and Durbin, P.A.: Nonlinear critical layers eliminate the upper branch of spatially growing Tollmien-Schlichting waves, *Phys. Fluids*, **29** (1986), 2344-2345.

65. Bayliss, A., Maestrello, L., Parrikh, P. and Turkel, E.: AIAA Paper 85-0565.

66. Wu, X., Lieb, S.J. and Goldstein, M.E.: On the nonlinear evolution of a pair of oblique Tollmien-Schlichting waves in boundary layers, *J. Fluid Mech.*, **340** (1997), 361-394.

67. Brown, S.N. and Smith, F.T.: On vortex/wave interactions. Part 1. Non-symmetrical input and cross-flow in boundary layers, *J. Fluid Mech.*, **307** (1996), 101-133.

68. Smith, F.T., Brown, S.N. and Brown, P.G.: Initiation of three-dimensional nonlinear transition paths from an inflectional profile, *Eur. J. Mech.*, **B12** (1993), 447-473.

69. Allen T., Brown, S.N. and Smith, F.T.: On vortex/wave interactions. Part 2. Originating from axisymmetric flow with swirl, *J. Fluid Mech.*, **325** (1996), 145-161.

70. Timoshin, S.N. and Smith, F.T.: Singular modes in Rayleigh instability of the three-dimensional streamwise-vortex flow, *J. Fluid Mech.*, **333** (1997), 139-160.

71. Morkovin, M.V.: Critical evaluation of transition from laminar to turbulent shear layers with emphasis on hypersonically traveling bodies, *Rep.* AFFDL-TR-68-149 (1969), Air Force Flight Dyn. Lab, Wright-Paterson AFB, Ohio.

72. Murdock, J.W.: The generation of a Tollmien-Schlichting wave by a sound wave, *Proc. R. Soc. Lond. A*, **372** (1980), 517-534.

73. Goldstein, M.E.: The evolution of Tollmien-Schlichting waves near a leading edge, *J. Fluid Mech.*, **127** (1983), 59-81.

74. Goldstein, M.E.: Generation of instability waves in flows separating from smooth surfaces, *J. Fluid Mech.*, **145** (1984), 71-94.

75. Goldstein, M.E.: Scattering of acoustic waves in Tollmien-Schlichting waves by small streamwise variations in surface geometry, *J. Fluid Mech.*, **154** (1985), 509-529.

76. Goldstein, M.E., Leib, S.J. and Cowley, S.J.: Generation of Tollmien-Schlichting waves on interactive marginally separated flows, *J. Fluid Mech.*, **181** (1987), 585-517.

77. Ruban, A.I.: On Tollmien-Schlichting wave generation by sound, *Izv. Akad. Nauk SSSR. MzhG*, **5** (1984), 44.

78. Bodonyi, R.J., Welch, W.J.C., Duck, P.W. and Tadjfar, M.: A numerical study of the interaction between unsteady freestream disturbances and localized variations in surface geometry, *J. Fluid Mech.*, **209** (1989), 285-308.

79. Tadjfar, M. and Bodonyi, R.J.: Receptivity of a laminar boundary layer to the interaction of a three-dimensional roughness element with time-harmonic freestream disturbances, *J. Fluid Mech.*, **242** (1992), 701-720.

80. Duck, P.W., Ruban, A.I. and Zhikharev, C.N.: The generation of Tollmien-Schlichting waves by freestream turbulence, *J. Fluid Mech.*, **312** (1996), 341-371.

81. Kozlov, V.V. and Ryzhov, O.S.: Receptivity of boundary layers: asymptotic theory and experiment, *Proc. R. Soc. Lond. A*, **429** (1990), 341-373.

82. Goldstein, M.E. and Hultgren, L.S.: Boundary layer receptivity to long-wave freestream disturbances, *Ann. Rev. Fluid Mech.*, **21** (1989), 137-166.

83. Seddougui, S.O. and Bassom, A.P.: Instability of hypersonic flow over a cone, *J. Fluid Mech.*, **345** (1997), 383-411.

THE EFFECT OF HEAT TRANSFER ON FLOW STABILITY

H. Herwig and J. Severin
Technical University of Chemnitz, Chemnitz, Germany

ABSTRACT

In the complex physics of the transition process from laminar to turbulent flow often distinct stages such as linear instability, nonlinear instability and chaotic behaviour can be identified. Various methods may be used to control this process by external means. One such method is heat transfer, which especially in the early stage of development (linear instability), can have considerable influence on the whole transition process.

The aim of this article is to provide a theoretical method by which the influence of heat transfer on flow stability can be accounted for in a general way. The basic idea is to use an asymptotic expansion based on the Taylor series expansion (with respect to temperature) of the properties involved.

This approach has been described in part 6 of the chapter "Laminar boundary layers" for the case of the undisturbed mean flow.

The following content will be covered with this asymptotic approach:

1. Introduction
2. The asymptotic approach
3. Boundary layers; linear stability theory
4. Other flows; linear stability theory
5. Nonlinear stability theory

1. INTRODUCTION

Among the studies that have investigated the stability of laminar boundary-layer flows, only a few have taken into account the effect of heat transfer, even though it can have a strong effect on the critical Reynolds number. For example, Wazzan et al. [1] investigated the boundary-layer stability of water under non-isothermal conditions. They found that the critical Reynolds number for a heated flat-plate boundary layer in water varies between 520 and nearly 16 000. Other studies of forced-convection stability which take into account heat transfer effects in a more or less systematic way are those by Hauptmann [2], Lee, Chen & Armaly [3] and Asfar, Masad & Nayfeh [4]. Natural-convection flows with variable property effects beyond that of the Boussinesq approximation were studied by Sabhapathy & Cheng [5] and Chen & Pearlstein [6], for example.

Experimental studies in this field are rather rare; typical examples are by Strazisar & Reshotko [7], Strazisar et al. [8] and Harrison et al. [9].

Our general method to account for heat transfer effects is based on a Taylor series expansion of the physical properties involved with a subsequent regular perturbation analysis for all dependent variables in the mean flow as well as in the disturbance equations. The zero order solutions are those for the isothermal case (constant properties), higher order solutions systematically account for temperature dependent fluid property effects.

2. THE ASYMPTOTIC APPROACH

Heat transfer affects flow stability through the temperature dependence of the properties involved. Therefore the first step in a systematic analysis is the Taylor series expansion of the properties with respect to temperature.

For the general property α^*, where α^* stands for ρ^*, η^*, λ^* and c_p^* it reads in nondimensional form (c.f. part 6 of the chapter "Laminar boundary layers"):

$$\alpha = \frac{\alpha^*}{\alpha_R^*} = 1 + K_{\alpha 1}\Theta + \frac{1}{2}K_{\alpha 2}\Theta^2 + ... \qquad (2.1)$$

with:

$$\Theta = \frac{T^* - T_R^*}{T_R^*} \qquad ; \qquad K_{\alpha 1} = \left[\frac{\partial \alpha^*}{\partial T^*}\frac{T^*}{\alpha^*}\right]_R \qquad ; \qquad K_{\alpha 2} = \left[\frac{\partial^2 \alpha^*}{\partial T^{*2}}\frac{T^{*2}}{\alpha^*}\right]_R \qquad (2.2)$$

or in the asymptotic form with:

$$\Theta = \varepsilon \vartheta \qquad ; \qquad \varepsilon = \frac{\Delta T_R^*}{T_R^*} \qquad ; \qquad \vartheta = \frac{T^* - T_R^*}{\Delta T_R^*} \qquad (2.3)$$

$$\boxed{\alpha = \frac{\alpha^*}{\alpha_R^*} = 1 + \varepsilon K_{\alpha 1} \vartheta + \frac{1}{2} \varepsilon^2 K_{\alpha 2} \vartheta^2 + O(\varepsilon^3)} \qquad (2.4)$$

If ´a´ represents all dependent mean flow variables of the problem, such as $\bar{u}$, $\bar{v}$, $\bar{p}$, $\bar{T}$, ... or its disturbances u', v', p', T', ... we assume the following expansion to be the asymptotic representation of a:

$$\boxed{\begin{aligned} a = {}& a_0 + \varepsilon\left[K_{\rho 1} a_\rho + K_{\eta 1} a_\eta + K_{\lambda 1} a_\lambda + K_{c1} a_c\right] + \varepsilon^2\left[K_{\rho 2} a_{\rho 2} + K_{\eta 2} a_{\eta 2} + \dots \right. \\ & \left. + K_{\rho 1}^2 a_{\rho\rho} + K_{\eta 1}^2 a_{\eta\eta} + \dots + K_\rho K_\eta a_{\rho\eta} + \dots\right] + O(\varepsilon^3) \end{aligned}} \qquad (2.5)$$

In equation (2.5) four terms appear in the first order with respect to ε representing the effects of the four properties ρ, η, λ and c_p on a. In the second order there are 14 terms already due to the fact that four terms appear in connection with $K_{\alpha 2}$, four terms in connection with $K_{\alpha 1}^2$ and six terms due to combinations $K_{\alpha i} K_{\alpha j}$ which all are second order terms, i.e. of the asymptotic order $O(\varepsilon^2)$.

The number of terms is drastically reduced, however, if not all four properties are involved. If only the effects of one property α are strong enough to be accounted for, equation (2.5) is:

$$a = a_0 + \varepsilon K_{\alpha 1} a_\alpha + \varepsilon^2\left(K_{\alpha 2} a_{\alpha 2} + K_{\alpha 1}^2 a_{\alpha\alpha}\right) + O(\varepsilon^3) \qquad (2.6)$$

With the two expansions, (2.4) for the properties α and (2.5) for all dependent variables ´a´ the asymptotic approach is straightforward.

As described in part 6 of the chapter "Laminar boundary layers" there are two methods to solve for the unknown functions a_α, $a_{\alpha 2}$, $a_{\alpha\alpha}$ etc. based on the expansions (2.4) and (2.5):

The expansion method:

After the expansions (2.4) and (2.5) are inserted in the mean flow as well as in the disturbance equations, terms of equal magnitude with respect to ε are collected. From this procedure one gets zero, first, second, ... order equations which due to the expansion procedure are free of ε. If in the higher order equations one also extracts the $K_{\alpha i}$ values, i.e. for example $K_{\rho 1}$ in the ε-order or $K_{\rho 2}$ in the ε^2 order, sets of higher order equations appear which now are also free of these $K_{\alpha i}$ values, i.e. their solutions hold for all fluids for which (2.4) is the representation of the property laws.

The shortcoming of this procedure is that in the general case we now have 4 equations in the first order and 14 in the second order as explained in connection with the expansion (2.5) already. The increasing amount of algebra can be circumvented however by the *combined method* described hereafter.

Once the solutions of the zero and higher order equations are known all final results can be cast into the same asymptotic form as was assumed for the dependent variables in (2.5). For example, the critical Reynolds number in its asymptotic representation is:

$$\frac{\text{Re}_c}{\text{Re}_{c0}} = 1 + \varepsilon \left[K_{\rho 1}A_\rho + K_{\eta 1}A_\eta + K_{\lambda 1}A_\lambda + K_{c1}A_c \right]$$

$$+ \varepsilon^2 \left[K_{\rho 2}A_{\rho 2} + K_{\eta 2}A_{\eta 2} + ... \right. \qquad (2.7)$$

$$+ K_{\rho 1}^2 A_{\rho\rho} + K_{\eta 1}^2 A_{\eta\eta} + ...$$

$$\left. + K_{\rho 1}K_{\eta 1}A_{\rho\eta} + ... \right] + O(\varepsilon^3)$$

Here the A_i, A_{ij} are coefficients in the final results which can be determined from the corresponding asymptotically derived equations.

For forced convection flows one knows beforehand that some of the coefficients in (2.7) must be zero. Since λ and c_p only appear in the energy and not in the momentum equation their influence on Re_c is shifted by one order of magnitude. This is due to the fact that they only act indirectly through a change in the temperature functions. Hence in (2.7) for forced convection flows $A_\lambda = A_c = 0$ holds. This decreased influence continues in the higher orders so that on the level of ε^2 we find $A_{\lambda 2} = A_{c2} = A_{\lambda\lambda} = A_{cc} = A_{\lambda c} = 0$ which reduces the number of nonzero coefficients to 9.

For natural convection flows all terms in results like (2.7) in the general case will be nonzero since the momentum and energy equations are equally affected by all physical properties.

The combined method:

The idea behind the combined method is the following: Instead of determining the coefficients A_i, A_{ij} in the final results like (2.7) by solving the subsets of equations derived in the expansion method they are looked upon as coefficients of a Taylor series expansion of a final result with respect to $\varepsilon K_{\alpha 1}$, $\varepsilon^2 K_{\alpha 2}$, ... For example A_η, $A_{\eta 2}$, and $A_{\eta\eta}$ in (2.7) are

$$A_\eta = \left[\frac{\partial\left(\dfrac{Re_c}{Re_{c0}}\right)}{\partial(\varepsilon K_{\eta 1})}\right]_0 \quad ; \quad A_{\eta 2} = \left[\frac{\partial\left(\dfrac{Re_c}{Re_{c0}}\right)}{\partial(\varepsilon^2 K_{\eta 2})}\right]_0 \quad ; \quad A_{\eta\eta} = \frac{1}{2}\left[\frac{\partial^2\left(\dfrac{Re_c}{Re_{c0}}\right)}{\partial(\varepsilon K_{\eta 1})^2}\right]_0 \quad (2.8)$$

Next, these differentials are approximated by finite differences. For example A_η is approximated by:

$$A_\eta = \lim_{h\to 0}\left\{\frac{\left.\dfrac{Re_c}{Re_{c0}}\right|_{\varepsilon K_{\eta 1}=h} - \left.\dfrac{Re_c}{Re_{c0}}\right|_{\varepsilon K_{\eta 1}=-h}}{2h}\right\} \quad (2.9)$$

Only two numerical solutions of the unexpanded equations are needed in which the viscosity η is taken as $\eta = 1 + h\vartheta$ and $\eta = 1 - h\vartheta$, respectively. Here, h serves just as a small number that is small enough to suppress higher order effects but large enough to be out of the range where numerical errors override the variable property effects.

By this method all coefficients in the final results can be determined by a small number of numerical solutions of the unexpanded equations for the mean and disturbance parts. Since this method combines highly accurate numerical solutions with asymptotic considerations it is called *combined method*, see [10] for more details.

3.   BOUNDARY LAYERS; LINEAR STABILITY THEORY

As an example for the asymptotic approach we first analyse the flat-plate flow (dp/dx=0) in chapter 3.1. Then, in 3.2 we show the influence of a nonzero pressure gradient.

3.1 Flat-plate flow

For the flat-plate flow we will analyse the influence of heat transfer on flow stability by the asymptotic method to the order $O(\varepsilon)$ with respect to the heat transfer parameter ε in chapter 3.1.2 and then by the combined method to the order $O(\varepsilon^2)$ in chapter 3.1.3. Since in both cases we need the corresponding mean flow results we will provide them first in the following chapter 3.1.1.

3.1.1 Flat-plate mean flow with heat transfer

The solution to the boundary layer equations for variable properties are only self similar for special thermal boundary conditions such as $T_W = $ const and $q_w = $ const.

In a systematic approach the partial boundary layer equations should be transformed according to the coordinate transformation.

$$(x,y) \; \to \; (\overline{x},\eta_s) \quad ; \quad \overline{x} = x \quad , \quad \eta_s = \frac{y^*}{\left(\dfrac{\eta_\infty^* x^*}{\rho_\infty^* U_\infty^*}\right)^{1/2}} \tag{3.1}$$

As a reference temperature difference

$$\Delta T^*(x) = T_W^*(x) - T_\infty^* = \left(T_W^* - T_\infty^*\right)_{L^*} \cdot x^e \tag{3.2}$$

can be used which assumes a power law dependence for the wall temperature. The temperature difference at a certain location $x^* = L^*$ downstream, i.e. $\left(T_W^* - T_\infty^*\right)_{L^*}$, is also used to define the perturbation parameter (c.p. = const properties):

$$\varepsilon = \frac{\left(T_W^* - T_\infty^*\right)_{L^*_{c.p.}}}{T_\infty^*} \tag{3.3}$$

With the similarity transformation and $\Delta T^*(x)$ given by (3.2), the thermal boundary conditions for which self similar solutions exist can easily be found.

The basic equations read (for details see [11]):

$$\left(\eta \left(\frac{f'}{\overline{\rho}}\right)^{\bullet}\right)' + \frac{1}{2}f\left(\frac{f'}{\overline{\rho}}\right)' = \overline{x}\left(f'\frac{\partial\left(f'/\overline{\rho}\right)}{\partial\overline{x}} - \left(\frac{f'}{\overline{\rho}}\right)'\frac{\partial f}{\partial\overline{x}}\right) \tag{3.4}$$

$$\left(\lambda\overline{\vartheta}'\right)' + \frac{1}{2}\mathrm{Pr}\,c_p\left(f\overline{\vartheta}' - 2ef'\overline{\vartheta}\right) = \mathrm{Pr}\,c_p\overline{x}\left(f'\frac{\partial\overline{\vartheta}}{\partial\overline{x}} - \overline{\vartheta}'\frac{\partial f}{\partial\overline{x}}\right) \tag{3.5}$$

with the boundary conditions

$$y = 0 : \quad f = f' = \overline{\vartheta} - 1 = 0 \tag{3.6}$$

$$y \to \infty : \quad f' - 1 = \overline{\vartheta} = 0 \tag{3.7}$$

Here f is the streamfunction with $\overline{\rho u} = f'$, $\overline{\vartheta}$ is the nondimensional mean flow temperature $\left(\overline{T}^* - T_\infty^*\right) / \Delta T^*(x)$ with $\Delta T^*(x)$ according to (3.2).

For the two standard thermal boundary conditions we find:

T_W = const.: $e = 0$ in (3.2); self similar velocity and temperature profiles

q_w = const.: $e = 1/2$ in (3.2); velocity and temperature profiles are quasi self similar since the x-dependence can be accounted for in the expansions. That for the streamfunction f, for example, reads:

$$f(\eta_s) = f_0(\eta_s) + \varepsilon \, x^e \left[K_{\rho 1} f_\rho(\eta_s) + K_{\eta 1} f_\eta(\eta_s) \right] + O(\varepsilon^2)$$

O(ε) analysis; asymptotic method

With the expansions (2.4) for the properties and (2.5) for all dependent variables we get the following sets of equations:

O(1):

$$f_0''' + \frac{1}{2} f_0 f_0'' = 0 \tag{3.8}$$

$$\overline{\vartheta}_0'' + \frac{1}{2} \Pr \left(f_0 \overline{\vartheta}_0' - f_0' \overline{\vartheta}_0 \right) = 0 \qquad ; \quad q_w = \text{const.} \tag{3.9}$$

$$\overline{\vartheta}_0'' + \frac{1}{2} \Pr f_0 \overline{\vartheta}_0' = 0 \qquad ; \quad T_w = \text{const.} \tag{3.10}$$

O(ε):

$$\begin{aligned}
f_\rho''' &+ \frac{1}{2} \left(f_0'' f_\rho + f_0 f_\rho'' \right) - \frac{1}{2} \left(f_0' f_\rho' - f_0'' f_\rho \right) \\
&= \left(\overline{\vartheta}_0' f_0' \right)' + \overline{\vartheta}_0' \left(f_0'' + \frac{1}{2} f_0 f_0' \right) - \frac{1}{2} \overline{\vartheta}_0 f_0'^2
\end{aligned} \qquad ; \quad q_w = \text{const.} \tag{3.11}$$

$$f_\rho''' + \frac{1}{2}\left(f_0''f_\rho + f_0f_\rho''\right) = \left(\overline{\vartheta}_0'f_0'\right)' + \overline{\vartheta}_0'\left(f_0'' + \frac{1}{2}f_0f_0'\right) \quad ; \quad T_w = \text{const.} \quad (3.12)$$

$$f_\eta''' + \frac{1}{2}\left(f_0''f_\eta + f_0f_\eta''\right) - \frac{1}{2}\left(f_0'f_\eta' - f_0''f_\eta\right) = -\left(\overline{\vartheta}_0f_0''\right)' \quad ; \quad q_w = \text{const.} \quad (3.13)$$

$$f_\eta''' + \frac{1}{2}\left(f_0''f_\eta + f_0f_\eta''\right) = -\left(\overline{\vartheta}_0f_0''\right)' \qquad ; \quad T_w = \text{const.} \quad (3.14)$$

with the corresponding boundary conditions

$$y = 0 \quad : \qquad f_0 = f_\rho = f_\eta = f_0' = f_\rho' = f_\eta' = \overline{\vartheta}_0 - 1 = 0$$

$$y = \infty \quad : \qquad f_0' - 1 = f_\rho' = f_\eta' = \overline{\vartheta}_0 = 0 \qquad\qquad (3.15)$$

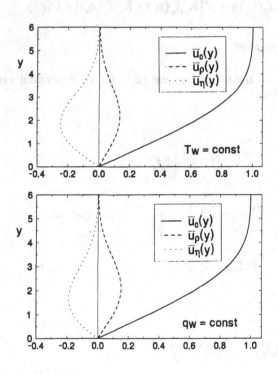

Figure 3.1: Velocity profiles according to the expansion method ; Pr=0.7

Note that the first order temperature function is not required to find the first order velocity functions because for forced convection flows the temperature effect is shifted by one order of magnitude in ε.

From the solutions of (3.8) - (3.15) we get $\bar{u}_0, \bar{u}_\rho$ and $\bar{u}_\eta$ in

$\bar{u} = \bar{u}_0 + \varepsilon \ x^e \left[K_{\rho 1} \bar{u}_\rho + K_{\eta 1} \bar{u}_\eta \right] + O(\varepsilon^2)$ as shown in Figure 3.1.

higher order analysis

$O(\varepsilon^2)$ or even higher order results could be obtained by continuing the process just described for the order $O(\varepsilon)$. Since higher order stability analysis will always be performed by applying the combined method there is no need for the explicit mean flow results of order $O(\varepsilon^2)$ or higher. In the combined method the mean flow will be determined from the non-expanded equations (3.4) and (3.5) with variable property effects to the appropriate order.

3.1.2 $O(\varepsilon)$ stability analysis; asymptotic method

For the isothermal case (constant properties, zero order with respect to the expansion in ε) the fundamental differential equation in the so-called method of small disturbances is the Orr-Sommerfeld (OS) equation, see for example Schlichting and Gersten [12]. In Squire [13] it was shown that it is sufficient to consider two-dimensional disturbances; the extension of this argument to variable properties can be found in Yih [14].

The present analysis presents an extended version of the OS equation which takes into account heat transfer effects.

In the method of small disturbances all quantities are decomposed into a mean value $\bar{a}^*$ and a superimposed disturbance $a^{*'}$. The common assumption is that an arbitrary two-dimensional disturbance can be expanded in a Fourier series; thus a single oscillation of the disturbance is assumed to be of the form (temporal stability)

$$a^{*'} = \hat{a}^*(y^*) \exp\left[i\alpha^*(x^* - \hat{c}^* t^*) \right] \qquad (3.16)$$

In (3.16) α^* is real with $2\pi/\alpha^*$ being the wavelength of the disturbance. The quantity $\hat{c}^*$ is complex, i.e. $\hat{c}^* = c_r^* + ic_i^*$. Here c_r^* denotes the phase velocity whereas c_i^* determines the degree of amplification or damping. The complex function $\hat{a}^*(y^*)$ is the shape function.

The shape function of the velocity disturbance can be obtained from the extended version of the OS equation. Since temperature effects are involved we also have to account for temperature disturbances. The corresponding shape function, $\hat{\vartheta}$, follows from the thermal energy equation for the disturbance.

The extended OS equation and the corresponding one for $\hat{\vartheta}$ are, for details see Schäfer et al. [16]:

$$\hat{\rho}(\bar{u}-\hat{c}) + \bar{\rho}\left[\hat{u}+\frac{\hat{v}'}{i\alpha}\right] + \frac{\overline{\rho}'\hat{v}}{i\alpha} = 0 \tag{3.17}$$

$$\overline{\rho}'\left[\hat{u}(\bar{u}-\hat{c})+\bar{u}'\frac{\hat{v}}{i\alpha}\right] + \bar{\rho}\left[\left(\hat{u}'+\alpha^2\frac{\hat{v}}{i\alpha}\right)(\bar{u}-\hat{c})+\bar{u}'\frac{\hat{v}'}{i\alpha}+\bar{u}''\frac{\hat{v}}{i\alpha}+\bar{u}'\hat{u}\right] +$$

$$\frac{i}{\alpha\,Re}\left[\overline{\eta}\left(\hat{u}'''-\alpha^2\hat{u}'+\alpha^2\frac{\hat{v}''}{i\alpha}-\alpha^4\frac{\hat{v}}{i\alpha}\right)-2\overline{\eta}'\left(\alpha^2\hat{u}-\hat{u}''\right)\right. \tag{3.18}$$

$$\left. + \overline{\eta}''\left(\hat{u}'-\alpha^2\frac{\hat{v}}{i\alpha}\right)+\hat{\eta}\left(\bar{u}'''+\alpha^2\bar{u}'\right)+2\hat{\eta}'\bar{u}''+\hat{\eta}''\bar{u}'\right] = 0$$

$$\overline{\rho}\overline{c}_p\left[\hat{\vartheta}(\bar{u}-\hat{c})+\overline{\vartheta}'\frac{\hat{v}}{i\alpha}\right] + \frac{i}{\alpha\,Re\,Pr}\left[\overline{\lambda}\left(\hat{\vartheta}''-\alpha^2\hat{\vartheta}\right)+\overline{\lambda}'\hat{\vartheta}'+\hat{\lambda}\overline{\vartheta}''+\hat{\lambda}'\overline{\vartheta}'\right] = 0 \tag{3.19}$$

For constant properties equations (3.17) and (3.18) can be combined by introducing a streamfunction $\hat{\varphi}$. Then, the classical OS equation emerges.

If now $\hat{u}$, $\hat{v}$, $\hat{\vartheta}$ and the parameter $\hat{c}$ are expanded according to (2.5) and the physical properties according to (2.4) we get the following sets of equations:

O(1):

$$\left(\bar{u}_0-\hat{c}_0\right)\left(\hat{\varphi}_0''-\alpha^2\hat{\varphi}_0\right) - \bar{u}_0''\hat{\varphi}_0 + \frac{i}{\alpha\,Re}\left(\hat{\varphi}_0''''-2\alpha^2\hat{\varphi}_0''+\alpha^4\hat{\varphi}_0\right) = 0 \tag{3.20}$$

$$\left(\bar{u}_0-\hat{c}_0\right)\hat{\vartheta}_0 - \overline{\vartheta}_0'\hat{\varphi}_0 + \frac{i}{\alpha\,Re\,Pr}\left(\hat{\vartheta}_0''-\alpha^2\hat{\vartheta}_0\right) = 0 \tag{3.21}$$

O(ε):

$$\hat{u}_\rho + \frac{\hat{v}_\rho'}{i\alpha} = -\hat{\vartheta}_0\left(\bar{u}_0-\hat{c}_0\right) + \overline{\vartheta}_0'\hat{\varphi}_0 \tag{3.22}$$

$$(\bar{u}_0 - \hat{c}_0)\left[\hat{u}'_\rho + \alpha^2 \frac{\hat{v}_\rho}{i\alpha}\right] + \bar{u}''_0 \frac{\hat{v}_\rho}{i\alpha} + \frac{i}{\alpha \, Re}\left[\hat{u}'''_\rho - 2\alpha^2 \hat{u}'_\rho - \alpha^4 \frac{\hat{v}_\rho}{i\alpha}\right] =$$

$$\frac{i\alpha}{Re}\left[\hat{\vartheta}'_0(\bar{u}_0 - \hat{c}_0) + \hat{\vartheta}_0 \bar{u}'_0 - \bar{\vartheta}''_0 \hat{\varphi}_0 - \bar{\vartheta}'_0 \hat{\varphi}'_0\right] \qquad (3.23)$$

$$- (\bar{u}_\rho - \hat{c}_\rho)\left[\hat{\varphi}''_0 - \alpha^2 \hat{\varphi}_0\right] + \bar{u}''_\rho \hat{\varphi}_0 + \hat{\vartheta}_0 \bar{u}''_0 \hat{\varphi}_0$$

$$+ (\bar{u}_0 - \hat{c}_0)\left[\bar{u}'_0 \hat{\vartheta}_0 - \bar{\vartheta}'_0 \hat{\varphi}'_0\right] - \hat{\vartheta}_0(\bar{u}_0 - \hat{c}_0)\left[\hat{\varphi}''_0 - \alpha^2 \hat{\varphi}_0\right]$$

$$(\bar{u}_0 - \hat{c}_0)\left[\hat{\varphi}''_\eta - \alpha^2 \hat{\varphi}_\eta\right] - \bar{u}''_0 \hat{\varphi}_\eta + \frac{i}{\alpha \, Re}\left[\hat{\varphi}''''_\eta - 2\alpha^2 \hat{\varphi}''_\eta + \alpha^4 \hat{\varphi}_\eta\right] =$$

$$\frac{-i}{\alpha \, Re}\left[\bar{\vartheta}_0\left[\hat{\varphi}''''_0 - 2\alpha^2 \hat{\varphi}''_0 + \alpha^4 \hat{\varphi}_0\right] + 2\bar{\vartheta}'_0\left[\hat{\varphi}'''_0 - \alpha^2 \hat{\varphi}'_0\right]\right. \qquad (3.24)$$

$$+ \bar{\vartheta}''_0\left[\hat{\varphi}''_0 + \alpha^2 \hat{\varphi}_0\right] + \hat{\vartheta}_0\left(\bar{u}'''_0 + \alpha^2 \bar{u}'_0\right) + 2\hat{\vartheta}'_0 \bar{u}''_0 + \hat{\vartheta}''_0 \bar{u}'_0\right]$$

$$- (\bar{u}_\eta - \hat{c}_\eta)\left[\hat{\varphi}''_0 - \alpha^2 \hat{\varphi}_0\right] + \bar{u}''_\eta \hat{\varphi}_0$$

with the associated boundary conditions

$$y = 0: \qquad \hat{\varphi}_0 = \hat{\varphi}'_0 = \hat{\vartheta}_0 = \hat{\varphi}_\eta = \hat{\varphi}'_\eta = \hat{u}_\rho = \hat{v}_\rho = 0 \qquad (3.25)$$

$$y \to \infty: \qquad \hat{\varphi}_0 = \hat{\varphi}'_0 = \hat{\vartheta}_0 = \hat{\varphi}_\eta = \hat{\varphi}'_\eta = \hat{u}_\rho = \hat{v}_\rho = 0 \qquad (3.26)$$

In (3.20), (3.21) and (3.24) streamfunctions could be introduced to combine $\hat{u}_0$, $\hat{v}_0$ and $\hat{u}_\eta$, $\hat{v}_\eta$; respectively. Note that this is not possible for the density effect $\hat{u}_\rho$, $\hat{v}_\rho$ due to the appearance of ρ in the continuity equation (3.17).

Expansion of the parameter $\hat{c} = c_r + ic_i$ is the crucial part of our asymptotic analysis. The effect of heat transfer on the amplification parameter c_i is given by its expansion

$$c_i = c_{i0} + \varepsilon\left[K_{\rho 1}c_{i\rho} + K_{\eta 1}c_{i\eta}\right] + O(\varepsilon^2) \qquad (3.27)$$

From c_i according to (3.27) we can directly infer how the critical Reynolds number is affected by heat transfer, i.e. we can determine the coefficients A_ρ and A_η in

$$\frac{\text{Re}_c}{\text{Re}_{c0}} = 1 + \varepsilon \left[K_{\rho 1} A_\rho + K_{\eta 1} A_\eta \right] + O(\varepsilon^2) \tag{3.28}$$

which is the $O(\varepsilon)$ version of (2.7). How this can be done will be demonstrated for A_ρ:
In (3.28), A_ρ can be interpreted as the Taylor series coefficient

$$A_\rho = \frac{1}{\text{Re}_{c0}} \left[\frac{\partial \text{Re}_c}{\partial (\varepsilon K_{\rho 1})} \right]_0 = \frac{1}{\text{Re}_{c0}} \lim_{\varepsilon K_{\rho 1} \to 0} \left[\frac{\Delta \text{Re}_c}{\varepsilon K_{\rho 1}} \right]_0 \tag{3.29}$$

of an expansion of Re_c with respect to $\varepsilon K_{\rho 1}$.
Here $\Delta \text{Re}_c = \text{Re}_c - \text{Re}_{c0}$ is the change in critical Reynolds number due to density/heat transfer effects.

The condition for the critical Reynolds number is $c_i = 0$ for just one value of the wave number α, i.e. the curve $c_i(\alpha)$ in a c_i-α diagram must be below the $c_i = 0$ line and touch it at one particular α, which then is the critical wave number α_c. Then, the parameter Re of the $c_i(\alpha)$ curve is the critical Reynolds number Re_c. Note that in this particular case

$$\frac{\partial c_i}{\partial \alpha} = 0 \quad \text{at} \quad \alpha = \alpha_c \tag{3.30}$$

When only density effects are accounted for, the condition $c_i = 0$ in its asymptotic form is:

$$c_{i0}(\alpha_c, \text{Re}_c) + \varepsilon K_{\rho 1} c_{i\rho}(\alpha_c, \text{Re}_c) = 0 \tag{3.31}$$

If the α- and Re-dependence in c_{i0} and $c_{i\rho}$ is accounted for by a Taylor series expansion with respect to α and Re at the reference point α_{c0}, Re_{c0} we get for (3.31):

$$c_{i0}(\alpha_{c0}, \text{Re}_{c0}) + \left[\frac{\partial c_{i0}}{\partial \alpha} \right]_0 \Delta \alpha_c + \left[\frac{\partial c_{i0}}{\partial \text{Re}} \right]_0 \Delta \text{Re}_c + ...$$

$$+ \varepsilon K_{\rho 1} \left\{ c_{i\rho}(\alpha_{c0}, \text{Re}_{c0}) + \left[\frac{\partial c_{i\rho}}{\partial \alpha} \right]_0 \Delta \alpha_c + \left[\frac{\partial c_{i\rho}}{\partial \text{Re}} \right]_0 \Delta \text{Re}_c + ... \right\} = 0 \tag{3.32}$$

with $\Delta \alpha_c = \alpha_c - \alpha_{c0}$ and $\Delta \text{Re}_c = \text{Re}_c - \text{Re}_{c0}$.

Since $c_{i0}(\alpha_{c0}, \text{Re}_{c0}) = \left[\partial c_{i0}/\partial \alpha \right]_0 = 0$ from (3.32) we get:

$$\frac{\Delta Re_c}{\varepsilon K_{\rho l}} = - \frac{c_{i\rho}(\alpha_{c0}, Re_{c0}) + \left[\frac{\partial c_{i\rho}}{\partial \alpha}\right]_0 \Delta \alpha_c}{\left[\frac{\partial c_{i0}}{\partial Re}\right]_0 + \varepsilon K_{\rho l}\left[\frac{\partial c_{i\rho}}{\partial Re}\right]_0} \tag{3.33}$$

which for $\varepsilon K_{\rho l} \to 0$ which also means $\Delta \alpha_c \to 0$ can be directly combined with (3.29) to give

$$A_\rho = - \frac{c_{i\rho}(\alpha_{c0}, Re_{c0})}{Re_{c0}\left[\frac{\partial c_{i0}}{\partial Re}\right]_0} \tag{3.34}$$

The corresponding coefficient A_η is

$$A_\eta = - \frac{c_{i\eta}(\alpha_{c0}, Re_{c0})}{Re_{c0}\left[\frac{\partial c_{i0}}{\partial Re}\right]_0} \tag{3.35}$$

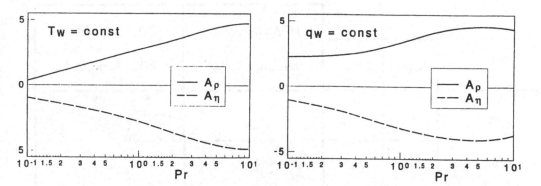

Figure 3.2: Coefficients A_ρ and A_η in equation (3.28)

The denominator in (3.34) and (3.35) can be determined from the isothermal (zero order) solution and for the flat plate flow is:

$$\text{Re}_{c0}\left[\frac{\partial c_{i0}}{\partial \text{Re}}\right]_0 = 0.0232 \tag{3.36}$$

The only information needed from the first order equation is $c_{i\rho}$ and $c_{i\eta}$. Since both are Prandtl number dependent we get A_ρ and A_η as functions of Pr.

Figure 3.2 shows the coefficients A_ρ and A_η for the Prandtl number range $0.1 \leq \text{Pr} \leq 10$.
An alternative though more tedious method to determine A_ρ and A_η is explained in Schäfer et al. [16].

A comparison of these results with experimental studies will be given in the next chapter which extends the analysis to $O(\varepsilon^2)$ effects.

In table 3.1 the qualitative effect of heating and cooling on the stability of different fluids based on the $O(\varepsilon)$ analysis of this chapter is summarized.

According to table 3.1 if water is heated and air is cooled then a stabilizing situation occurs (at least under the thermal boundary conditions T_w=const and q_w=const assumed here).

Numerical details of the solution procedure can be found in Herwig and Schäfer [15] where the Gram-Schmidt orthonormalization was applied to solve the systems of stiff ordinary differential equations (3.20)-(3.24) and in Severin [11] where the same is done by a spectral method.

	$K_{\rho 1}A_\rho$	$K_{\eta 1}A_\eta$	HEATING (ε>0) $\varepsilon(K_{\rho 1}A_\rho + K_{\eta 1}A_\eta)$	COOLING (ε<0) $\varepsilon(K_{\rho 1}A_\rho + K_{\eta 1}A_\eta)$
AIR	⊖	⊖	⊖	⊕
WATER	⊕	⊕	⊕	⊖

Table 3.1: Signs of the indicated terms in $\text{Re}_c/\text{Re}_{c0}$ according to equation (3.28)

⊕ = stabilizing effect

3.1.3 O(ε^2) stability analysis; combined method

All coefficients in the final results Re_c/Re_{c0} according to the expansion (2.7) can be directly determined by the combined method, briefly described in chapter 2 already and applied to the mean flow alone in part 6.4 of the chapter "Laminar boundary layers".

In the combined method the coefficients A_i and A_{ij} are looked upon as Taylor series coefficients of the appropriate expansions. The non-zero coefficients in (2.7) up to the order O(ε^2) are:

$$A_\rho = \left[\frac{\partial(Re_c/Re_{c0})}{\partial(\varepsilon K_{\rho 1})}\right]_0 \quad ; \quad A_\eta = \left[\frac{\partial(Re_c/Re_{c0})}{\partial(\varepsilon K_{\eta 1})}\right]_0 \tag{3.37}$$

$$A_{\rho 2} = \left[\frac{\partial(Re_c/Re_{c0})}{\partial(\varepsilon^2 K_{\rho 2})}\right]_0 \quad ; \quad A_{\eta 2} = \left[\frac{\partial(Re_c/Re_{c0})}{\partial(\varepsilon^2 K_{\eta 2})}\right]_0$$

$$A_{\rho\rho} = \left[\frac{\partial^2(Re_c/Re_{c0})}{\partial(\varepsilon K_{\rho 1})^2}\right]_0 \quad ; \quad A_{\eta\eta} = \left[\frac{\partial^2(Re_c/Re_{c0})}{\partial(\varepsilon K_{\eta 1})^2}\right]_0$$

$$A_{\rho\eta} = \left[\frac{\partial^2(Re_c/Re_{c0})}{\partial(\varepsilon K_{\rho 1})\partial(\varepsilon K_{\eta 1})}\right]_0 \quad ; \quad A_{\rho\lambda} = \left[\frac{\partial^2(Re_c/Re_{c0})}{\partial(\varepsilon K_{\rho 1})\partial(\varepsilon K_{\lambda 1})}\right]_0$$

$$A_{\rho c} = \left[\frac{\partial^2(Re_c/Re_{c0})}{\partial(\varepsilon K_{\rho 1})\partial(\varepsilon K_{c1})}\right]_0 \quad ; \quad A_{\eta\lambda} = \left[\frac{\partial^2(Re_c/Re_{c0})}{\partial(\varepsilon K_{\eta 1})\partial(\varepsilon K_{\lambda 1})}\right]_0$$

$$A_{\eta c} = \left[\frac{\partial^2(Re_c/Re_{c0})}{\partial(\varepsilon K_{\eta 1})\partial(\varepsilon K_{c1})}\right]_0 \tag{3.38}$$

The differentials are next approximated by finite differences. In equation (2.9) an example was given by the first order coefficient A_η. The second order coefficients are treated likewise. For example, $A_{\rho\eta}$ is approximated by:

$$A_{\rho\eta} = \left[\frac{\partial^2 (Re_c/Re_{c0})}{\partial(\varepsilon K_{\rho 1})\partial(\varepsilon K_{\eta 1})}\right] = \frac{\partial}{\partial(\varepsilon K_{\rho 1})}\left[\lim_{h\to 0}\frac{\left.\frac{Re_c}{Re_{c0}}\right|_{\varepsilon K_{\eta 1}=h} - \left.\frac{Re_c}{Re_{c0}}\right|_{\varepsilon K_{\eta 1}=-h}}{2h}\right]_0$$

$$= \lim_{h\to 0}\left\{\left[\frac{\left.\frac{Re_c}{Re_{c0}}\right|_{\varepsilon K_{\eta 1}=h} - \left.\frac{Re_c}{Re_{c0}}\right|_{\varepsilon K_{\eta 1}=-h}}{2h}\right|_{\varepsilon K_{\rho 1}=h} - \left.\frac{\left.\frac{Re_c}{Re_{c0}}\right|_{\varepsilon K_{\eta 1}=h} - \left.\frac{Re_c}{Re_{c0}}\right|_{\varepsilon K_{\eta 1}=-h}}{2h}\right|_{\varepsilon K_{\rho 1}=-h}\right]_0 / 2h\right\}$$

$$= \lim_{h\to 0}\frac{\left.\frac{Re_c}{Re_{c0}}\right|_{\varepsilon K_{\eta 1}=\varepsilon K_{\rho 1}=h} + \left.\frac{Re_c}{Re_{c0}}\right|_{\varepsilon K_{\eta 1}=\varepsilon K_{\rho 1}=-h} - \left.\frac{Re_c}{Re_{c0}}\right|_{\varepsilon K_{\eta 1}=-\varepsilon K_{\rho 1}=h} - \left.\frac{Re_c}{Re_{c0}}\right|_{-\varepsilon K_{\eta 1}=\varepsilon K_{\rho 1}=h}}{4h^2}$$

(3.39)

	Pr=0.7	Pr=7.0
A_ρ	2.39	4.62
A_η	-2.42	-4.78
$A_{\rho 2}$	1.93	1.85
$A_{\eta 2}$	-1.80	-1.63
$A_{\rho\rho}$	1.85	16.22
$A_{\eta\eta}$	5.20	20.81
$A_{\rho\eta}$	-7.12	-37.10
$A_{\rho\lambda}$	-0.95	0.49
$A_{\rho c}$	0.21	-0.29
$A_{\eta\lambda}$	0.79	-0.87
$A_{\eta c}$	-0.22	0.13

Table 3.2: Coefficients A_i and A_{ij} in equation (2.7), T_w = const.

To determine $A_{\rho\eta}$ according to equation (3.39) four different solutions of the unexpanded equations (3.17)-(3.19) are needed with property laws

$$\bar{\rho} = 1 + \varepsilon K_{\rho 1} \bar{\vartheta} \quad ; \quad \hat{\rho} = \varepsilon K_{\rho 1} \hat{\vartheta}$$

$$\bar{\eta} = 1 + \varepsilon K_{\eta 1} \bar{\vartheta} \quad ; \quad \hat{\eta} = \varepsilon K_{\eta 1} \hat{\vartheta} \tag{3.40}$$

and values of $\varepsilon K_{\rho 1}$ and $\varepsilon K_{\eta 1}$ as indicated by the suffixes in (3.39). Here again, h serves just as a small number, that is small enough to suppress higher order effects but also large enough to be out of the range where numerical errors override the variable property effects, for details see Herwig and Schäfer [10].

In table 3.2 all non-zero values of A_i and A_{ij} are listed for the Prandtl numbers Pr = 0.7 and Pr = 7.0 for the case T_w = const. Here A_ρ and A_η are exactly the same as those determined in the previous chapter by the expansion method, see Figure 3.2.

In Figure 3.3 the asymptotic results are compared to those of an experimental study by Harrison et al. [9].

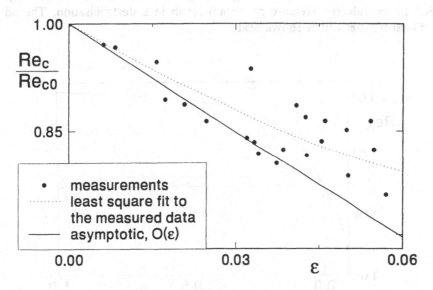

Figure 3.3: Comparison between experimental and asymptotic results; Pr = 0.7, q_w=const

3.2 Influence of a non-zero pressure gradient; O(ε) analysis
The only change in the analysis for a non-zero pressure gradient is that mean flow momentum equation must be supplemented by the pressure term. The form of the other equations, particulary the stability equations remain unchanged. Since, however, the mean

flow velocity profile has a strong influence on the stability behaviour of laminar boundary layers its changes due to pressure gradient effects strongly influence the stability of the whole flow.

The effect of variable properties will be demonstrated for the so-called Falkner-Skan flows which are self -similar boundary layer flows with a pressure gradient such that the external velocity is

$$u_e = x^m \qquad (3.41)$$

With $-0.0904 \leq m \leq 1.0$ they represent non-separating boundary layers over a wedge with a wedge angle $\overline{\beta} = 2\pi m/(m+1)$, see Herwig and Gersten [17], especially for $\overline{\beta} < 0$ when $m<0$.

The pressure gradient effect on the isothermal flow (zero order asymptotically) is shown in figure 3.4 where the critical Reynolds number Re_{c0} is plotted against the exponent m of the external flow.

Obviously there is a strong influence of m , i.e. of pressure gradient. Positive exponents (accelerated flows, favourable pressure gradient) stabilise the flow, whereas negative m (retarded flows, adverse pressure gradients) result in a destabilisation. The additional effect of heat transfer will be shown next.

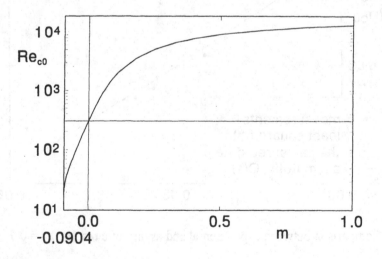

Figure 3.4: Critical Reynolds number for the isothermal case (zero order)

In figure 3.5 velocity profiles are shown for two different Prandtl numbers and four exponents m in $u_e = x^m$ for the case $T_w = $ const.

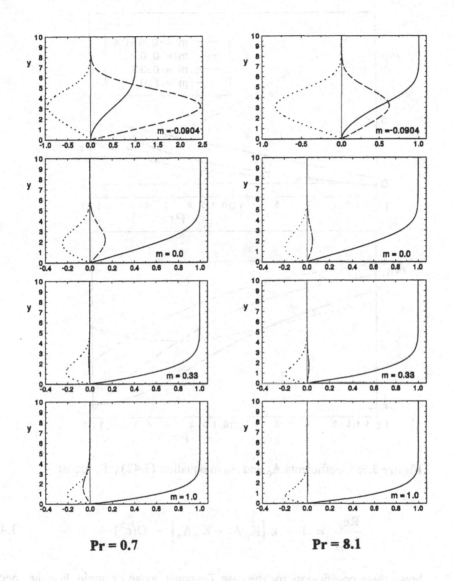

Figure 3.5: Velocity profiles $\bar{u}_0$, $\bar{u}_\rho$ and $\bar{u}_\eta$ for different Prandtl
numbers and different pressure gradients; T_w=const
—— $\bar{u}_0$; - - - $\bar{u}_\rho$; ... $\bar{u}_\eta$

There are considerable changes in the velocity profiles, especially in the first order profile $\bar{u}_\rho$. Based on these mean flow results the $O(\varepsilon)$ heat transfer effects can be obtained in terms of the coefficients A_ρ and A_η in

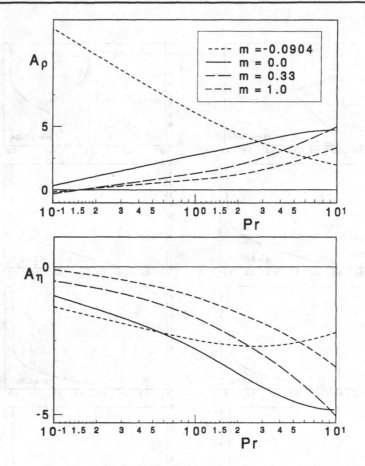

Figure 3.6: Coefficients A_ρ and A_η in equation (3.42) ; T_w=const.

$$\frac{Re_c}{Re_{c0}} = 1 + \varepsilon \left[K_{\rho 1} A_\rho + K_{\eta 1} A_\eta \right] + O(\varepsilon^2) \tag{3.42}$$

Figure 3.6 shows these coefficients for the case T_w=const, as an example. For the special case m=0 the values are repeated from figure 3.2.

4. OTHER FLOWS; LINEAR STABILITY THEORY

The method described here to account for heat transfer effects on flow stability is not of course restricted to boundary layer flows. It has been applied to a variety of other flows such as Poiseuille flow and Rayleigh Bénard natural convection flows. Some of these results will be shown in this chapter.

4.1 Plane Poiseuille flow; q_w=const.

The heat transfer situation is sketched in Figure 4.1. After a thermal adjustment zone there is a hydrodynamically and thermally fully developed region which, for constant properties, is characterized by a constant temperature difference $T_w^* - T_B^*$.

For this Poiseuille flow with constant wall heat flux and therefore a linearly increasing bulk temperature the choice of an adequate reference temperature is very important.

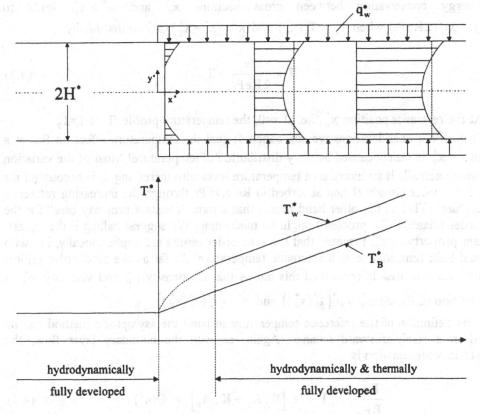

Figure 4.1: Development of the temperature profile

In the fully developed region the temperature can be split into two terms, one being x- and the other being y- dependent, i.e. $T^*(x^*,y^*) = T_B^*(x^*) + T_y^*(y^*)$

so that: $T_y^*(y^*) = T^*(x^*,y^*) - T_B^*(x^*)$ (4.1)

Nondimensionalising $T^* - T_R^*$ by $q_w^* H^*/\lambda^*$ gives:

$$\underbrace{\frac{T^* - T_R^*}{q_w^* H^*/\lambda^*}}_{T(x,y)} = \underbrace{\frac{T_B^* - T_R^*}{q_w^* H^*/\lambda^*}}_{T_B(x)} + \underbrace{\frac{T^* - T_B^*}{q_w^* H^*/\lambda^*}}_{T_y(y)} \tag{4.2}$$

where T_R^* is a yet to be determined reference temperature, q_w^* is the wall heat flux and H^* is half the channel width.

Energy conservation between cross sections x_R^* and $x^* > x_R^*$ leads to $T_B(x) = 3x/(2\,Re\,Pr)$ when $T_R^* = T_B^*(x_R^*)$ and $x = (x^* - x_R^*)/H^*$ so that finally

$$T = \frac{3x}{2\,Re\,Pr} + T_y \tag{4.3}$$

At the reference position x_R^*, i.e. at x=0, the temperature profile T is $T=T_y$.

From these considerations we can conclude that the temperature effect on Re_c at a certain $x^* = x_R^*$ is mainly caused by the y-distribution of temperature! Most of the variation in x, which actually is an increase of temperature level with increasing x, is accounted for in the zero order already (being absorbed in Re and Pr through the increasing reference temperature). This on the other hand means that a term "constant property case" for the zero order (classical OS problem) might be misleading. We suggest calling it the "quasi-constant property case". It means that the zero order results are applied locally, i.e. with the local bulk temperature as a reference temperature. As far as the zero order critical Reynolds number Re_{c0} is concerned this means that the density ρ_R^* and viscosity η_R^* in the definition of Re are $\rho_R^* = \rho^*\left(T_B^*(x_R^*)\right)$ and $\eta_R^* = \eta^*\left(T_B^*(x_R^*)\right)$.

With this definition of the reference temperature in mind the asymptotic method can be applied in a straightforward manner. Again, as with the boundary layer flow, the asymptotic representation is:

$$\frac{Re_c}{Re_{c0}} = 1 + \varepsilon \left[K_{\rho 1} A_\rho + K_{\eta 1} A_\eta\right] + O(\varepsilon^2) \tag{4.4}$$

with A_ρ and A_η according to (3.34) and (3.35), respectively. The denominator in these equations now is

$$Re_{c0}\left[\frac{\partial c_{i0}}{\partial Re}\right]_0 = 0.00971 \tag{4.5}$$

Figure 4.2 shows A_ρ and A_η in the Prandtl number range $0.1 \le Pr \le 10$. From this we can draw some general conclusions about the effect of $\rho(T)$ and $\eta(T)$ on flow stability.

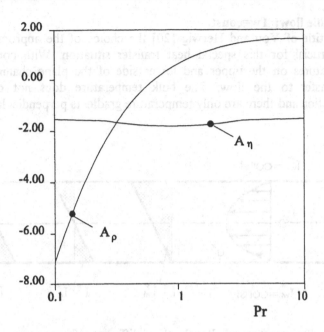

Figure 4.2: Coefficients A_ρ and A_η in equation (4.4)

$\rho(T)$ effect:

Since A_ρ changes its sign at $Pr \approx 0.6$ and for all fluids $K_{\rho 1} > 0$ holds (except the water anomaly) we find a stabilisation for
- a cooled fluid ($\varepsilon < 0$) with $Pr < 0.6$
- a heated fluid ($\varepsilon > 0$) with $Pr > 0.6$

$\eta(T)$ effect:

Since A_η is always negative the flow is stabilized for $\varepsilon K_{\eta 1} < 0$. The combination $\varepsilon K_{\eta 1}$ is negative for
- a heated fluid ($\varepsilon > 0$) with $K_{\eta 1} < 0$ (such as water with $K_{\eta 1} = -7.132$ at
  $T_R^* = 293K$, $p_R^* = 1bar$)
- a cooled fluid ($\varepsilon < 0$) with $K_{\eta 1} > 0$ (such as air with $K_{\eta 1} = 0.775$ at $T_R^* = 293K$,
  $p_R^* = 1bar$)

Details of the analysis and results can be found in Herwig and You [18]. For this flow it was also studied, how good the shape assumption with respect to the temperature disturbance amplitude function is, see Herwig and You [19].

4.2 Plane Poiseuille flow; Tw=const.

As stated in the title of You and Herwig [20] the choice of the appropriate reference temperature is crucial for this special heat transfer situation. With constant, though different, temperatures on the upper and lower side of the plane channel there is no overall heat transfer to the flow. The bulk temperature does not change in the downstream direction and there are only temperature gradients perpendicular to the wall.

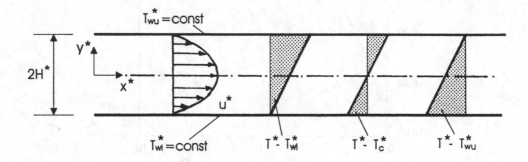

Figure 4.3: Temperature profiles for three different reference temperatures

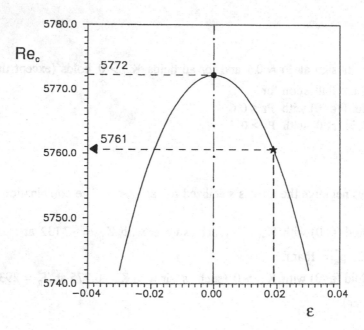

Figure 4.4: Critical Reynolds number for water, reference temperature $T_R^* = T_c^* = 293K$
----- refers to a special example with $\Delta T_w^* = 11.1°C$

In Figure 4.3 temperature profiles for three reference temperatures are shown. $T_R^* = T_{wl}^*$ and $T_R^* = T_{wu}^*$ suggest a heating and cooling situation, respectively, while $T_R^* = T_c^*$ visually reflects the fact that nothing happens in terms of bulk effects. Indeed it turns out that all reference temperatures other than $T_R^* = T_C^*$ are highly misleading, see You and Herwig [20] for details. There is no bulk effect of temperature. However, the gradients in lateral direction influence stability when, for example, viscosity is temperature dependent. Figure 4.4 shows how the critical Reynolds number of water (modeled as a fluid with $\eta(T)$, $K_{\eta 1} = -7.132$, $K_{\eta 2} = 78.12$) is affected by heat transfer (here: $\varepsilon = \Delta T_w^* / 2 T_c^*$).

4.3 Rayleigh Bénard natural convection

In Figure 4.5 a sketch of the Rayleigh Bénard problem shows that temperature differences (here between the upper and the lower wall) are a inherent feature of the problem. In Severin and Herwig [21] we determined the effect of a variable viscosity on the critical Rayleigh number by applying the asymptotic method.

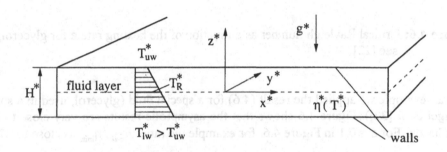

Figure 4.5: Sketch of the Rayleigh Bénard problem

Analogous to equation (2.7) for the critical Reynolds number we determined the ratio of the critical Rayleigh numbers, here up to the order $O(\varepsilon^4)$. It reads:

$$\frac{Ra_c}{Ra_{c0}} = 1 + \varepsilon^2 \left[-0.207 \ K_{\eta 1}^2 + 0.260 \ K_{\eta 2} \right]$$

$$+ \varepsilon^4 \left[-0.051 \ K_{\eta 1}^4 + 0.103 \ K_{\eta 1}^2 K_{\eta 2} - 0.014 \ K_{\eta 2}^2 - 0.056 \ K_{\eta 1} K_{\eta 3} \right. \quad (4.6)$$

$$\left. + 0.015 \ K_{\eta 4} \right] + O(\varepsilon^6)$$

with $Ra_{c0} = 1707.76$. Note that due to symmetry reasons all coefficients with odd exponents of ε are zero.

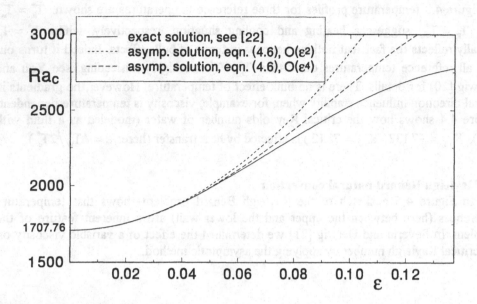

Figure 4.6: Critical Rayleigh number as a function of the heating rate ε for glycerol, see [22]

As an example we applied the result (4.6) for a special fluid (glycerol, used in a study by Stengel et al [22]). Figure 4.6 shows that the asymptotic results are very close to the exact solution. For ε = 0.1 in Figure 4.6, for example the ratio η_{max}/η_{min} is close to 270 !

5. NONLINEAR STABILITY THEORY

The asymptotic approach to account for heat transfer effects on flow stability is by no means restricted to the linear stability theory. It can be extended to nonlinear stability analysis in a straightforward manner. Since, however, there are several "routes" to nonlinear effects and finally to the occurence of real turbulence even for the isothermal case, the following can only be an example out of a large number of theoretical approaches to the nonlinear/turbulence problem.

In You and Herwig [23] we applied the weakly nonlinear theory of stability in a version that takes into account certain effects in the early stage of development and therefore is called "modified weakly nonlinear theory".

A crucial part of the analysis is a double asymptotic expansion of the disturbance stream function $\tilde{\psi}$ and the disturbance temperature $\tilde{T}$:

(1) Expansion of $\tilde{\psi}$ and $\tilde{T}$ with respect to an amplitude parameter $\hat{\varepsilon}$

A series expansion with respect to $\hat{\varepsilon}$ and truncation of the series after a finite number of terms is the basic idea of the so-called weakly nonlinear theory. With this expansion alone, we write:

$$\{\tilde{\psi}, \tilde{T}\} = \{\tilde{\psi}_0, \tilde{T}_0\} + \hat{\varepsilon}\{\tilde{\psi}_1, \tilde{T}_1\} + \hat{\varepsilon}^2\{\tilde{\psi}_2, \tilde{T}_2\} + O(\hat{\varepsilon}^3) \qquad (5.1)$$

(2) Expansion of $\tilde{\psi}$ and $\tilde{T}$ with respect to the heat transfer parameter ε

A series expansion with respect to ε accounts for variable property effects to the degree determined by the order of series truncation. With this expansion alone we write for the case of the variable viscosity:

$$\{\tilde{\psi}, \tilde{T}\} = \{\tilde{\psi}_0, \tilde{T}_0\} + \varepsilon K_{\eta 1}\{\tilde{\psi}_\eta, \tilde{T}_\eta\} + O(\varepsilon^2) \qquad (5.2)$$

If now the individual expansions (5.1) and (5.2) are combined it is necessary to asymptotically fix the ratio between the two perturbation parameters. They may be of the same or of different orders of magnitude. Each assumption about the ratio of $\hat{\varepsilon}$ and ε constitutes a particular asymptotic theory with this assumed ratio as a characteristic feature. None of these different theories is "right" or "wrong", they merely have different parameter ranges in which they may serve as a reasonable approximation to the physiscs. If, for example, a theory should cover the effects of small heat transfer rates superimposed on the (weakly) nonlinear development of flow disturbances, a ratio

$$\varepsilon = O(\hat{\varepsilon}^n) \quad \text{with} \quad n > 1 \qquad (5.3)$$

might be appropriate. Since it is exactly this case we have in mind we only have to choose the exponent n . With n = 2 we can account for heat transfer effects within a nonlinear theory with an $\hat{\varepsilon}^3$ order of series truncation.

Fixing a ratio between initially independent parameters in terms of asymptotic theory means to determine a distinguished limit. So the distinguished limit of our asymptotic theory is

$$\varepsilon = H \hat{\varepsilon}^2 \quad \text{with} \quad H = O(1) \qquad (5.4)$$

As a consequence of this choice the double expansion now reads:

$$\left\{\widetilde{\psi}, \widetilde{T}\right\} = \left\{\widetilde{\psi}_0, \widetilde{T}_0\right\} + \hat{\varepsilon}\left\{\widetilde{\psi}_1, \widetilde{T}_1\right\} + \hat{\varepsilon}^2\left\{\left(\widetilde{\psi}_2, \widetilde{T}_2\right) + H\,K_{\eta 1}\left(\widetilde{\psi}_\eta, \widetilde{T}_\eta\right)\right\} + O(\hat{\varepsilon}^3) \quad (5.5)$$

Within this theory heat transfer effects will enter the problem on the $\hat{\varepsilon}^2$-level of the nonlinear flow disturbance development.

Instead of (3.16) for the stream function disturbance we use the equivalent form

$$\hat{\varepsilon}\widetilde{\psi}_0 = b_0(t)\,\hat{\psi}_0(y)\,\exp\left[i(\alpha x - \theta)\right] + \text{c.c.} \quad (5.6)$$

with

$$b_0(t) = \hat{\varepsilon}\,\exp(\alpha c_{i_0}t)\,;\quad \theta = \alpha c_{r_0}t \quad (5.7)$$

We introduce the amplitude $b_0(t)$ which is asymptotically of order $O(\hat{\varepsilon})$. With $b_0(t) = \hat{\varepsilon}\,\exp(\alpha c_{i_0}t)$ we can also identify $\hat{\varepsilon}$, since for $t = 0$ the exponential function is equal to 1, so that

$$\hat{\varepsilon} = b_0(0) \quad (5.8)$$

i.e. $\hat{\varepsilon}$ is the initial amplitude of the fundamental wave. From $b_0(t)$ and $\theta(t)$ immediately follows

$$\frac{db_0}{dt} = \omega_{i0}b_0\,,\quad \frac{d\theta}{dt} = \omega_{r0}\quad \text{with}\quad \omega_{i0} = \alpha c_{i0}\,,\quad \omega_{r0} = \alpha c_{r0} \quad (5.9)$$

Within the linear stability theory, i.e. for $\varepsilon \to 0$, equation. (5.6) completely describes the flow disturbance. In the weakly nonlinear theory higher order modes will appear, triggered by $\hat{\varepsilon}\widetilde{\psi}_0$. That is why in the context of the nonlinear theory $\hat{\varepsilon}\widetilde{\psi}_0$ is called the fundamental wave. An assumption within the weakly nonlinear theory is that the shapes of the fundamental wave and all higher order modes are independent of time, i.e. for example $\hat{\psi}_0$ is $\hat{\psi}_0(y)$ and not $\hat{\psi}_0(y,t)$. In what we call *modified weakly nonlinear theory*, however, we only assume $\hat{\psi}_0$ to be independent of time. All other modes have shape functions $\hat{\psi}_i(y,t)$ that develop in time and only for large times will gain their final shape. With this assumption results are in close agreement to experimental as well as DNS data. Mathematically the modified weakly nonlinear theory is an initial value problem whereas the conventional weakly nonlinear theory is a boundary value one.

To account for temperature effects we also need $\widetilde{T}_0$, for which we assume again:

$$\hat{\varepsilon}\tilde{T}_0 = b_{0t}(t)\hat{T}_0(y,t)\exp\left[i(\alpha x - \theta)\right] + \text{c.c.} \tag{5.10}$$

This ansatz reflects the passive character of temperature disturbances. As mentioned already, we do not assume the shape of the amplitude function to be independent of time, i.e. we write $\hat{T}_0(y,t)$ instead of $\hat{T}_0(y)$.

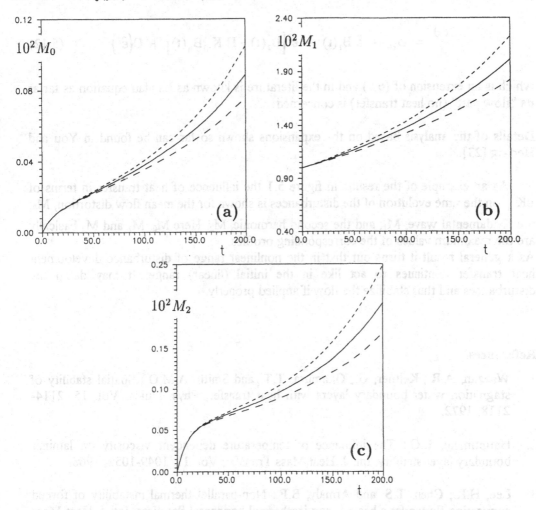

Figure 5.1: Influence of heat transfer, $\varepsilon K_{\eta l}$, on the time evolution of disturbances;

$$Re = 10^4, \quad \alpha = 1.04, \quad \hat{\varepsilon} = 5\times 10^{-3}, \quad Pr = 7.0$$

(a) M_0 ; mean flow distortion
(b) M_1 ; fundamental wave development
(c) M_2 ; second harmonic development
--- $\varepsilon K_{\eta l} = -0.0357$,   — $\varepsilon K_{\eta l} = 0.0$,   ---- $\varepsilon K_{\eta l} = 0.0357$

Due to the expansions of $\tilde{\psi}$ and $\tilde{T}$ in powers of $\hat{\varepsilon}$, c.f. (5.5), we expect $b_0(t)$ and $\theta(t)$ to be of the form

$$\frac{db_0}{dt} = \omega_{i0}b_0 + \hat{\varepsilon}\, A_1(t) + \hat{\varepsilon}^2\big[A_2(t) + H\, K_{\eta 1} A_\eta(t)\big] + O\big(\hat{\varepsilon}^3\big) \qquad (5.11)$$

$$\frac{d\theta}{dt} = \omega_{r0} + \hat{\varepsilon}\, B_1(t) + \hat{\varepsilon}^2\big[B_2(t) + H\, K_{\eta 1} B_\eta(t)\big] + O\big(\hat{\varepsilon}^3\big) \qquad (5.12)$$

which is an extension of (5.9) and in the literature is known as Landau equation as far as its "flow part" (no heat transfer) is concerned.

Details of the analysis based on the expansions shown so far can be found in You and Herwig [23].

As an example of the results in figure 5.1 the influence of heat transfer in terms of $\varepsilon K_{\eta 1}$ on the time evolution of the disturbances is shown for the mean flow distortion, M_0; the fundamental wave, M_1; and the second harmonic, M_2. Here M_0, M_1 and M_2 basically are the maximum values of the corresponding profiles.
As a general result it turns out that in the nonlinear range of disturbance development heat transfer continues to act like in the initial (linear) range. It may damp the disturbances and thus stabilize the flow if applied properly.

References

1. Wazzan, A.R.; Keltner, G.; Okamura, T.T. and Smith, A.M.O.: Spatial stability of stagnation water boundary layers with heat transfer, Phys. Fluids, Vol. 15, 2114-2118, 1972.

2. Hauptmann, E.G.: The influence of temperature dependent viscosity on laminar boundary layer stability, Int. J. Heat Mass Transfer, Vol. 11, 1049-1052, 1968.

3. Lee, H.R.; Chen, T.S. and Armaly, B.F.: Non-parallel thermal instability of forced convection flow over a heated, non-isothermal horizontal flat plate; Int. J. Heat Mass Transfer, Vol. 33, 2019-2028, 1990.

4. Asfar, O.R.; Masad, J.A. and Nayfeh, A.H.: A method for calculating the stability of boundary layers, Computers Fluids, Vol. 18, 305-315, 1990.

5. Sabhapathy, P. and Cheng, K.C.: The effect of temperature dependent viscosity and coefficient of thermal expansion on the stability of laminar, natural convective flow along an isothermal, critical surface, Int. J. Heat Mass Transfer, Vol. 29, 1521-1529, 1986.

6. Chen, Y.M. and Pearlstein, A.J.: Stability of free-convection flows of variable-viscosity fluids in vertical and inclined slots, J. Fluid Mech., Vol. 198, 513-541, 1989.

7. Strazisar, A.J. and Reshotko, E.: Stability of heated laminar boundary layers in water with nonuniform surface temperature, Phys. Fluids, Vol.21, 727-735, 1978.

8. Strazisar, A.J.; Reshotko, E. and Prahl, J.M.: Experimental study of the stability of heated laminar boundary layers in water, J. Fluid Mech., Vol. 83, 225-247, 1977.

9. Harrison, S.B.; Mee, D.J. and Jones, T.V.: Experiments of the influence of heating on boundary layer transition in favourable pressure gradients, Proc. of Eurotherm 25, 1-6, 1991.

10. Herwig, H. and Schäfer, P.: A combined perturbation/Finite-difference procedure applied to temperature effects and stability in a laminar boundary layer, Archive of Applied Mechanics, Vol. 66, 264-272, 1996.

11. Severin, J.: Der Einfluß der Wärmeübertragung auf die Stabilität von Strömungen, Dissertation TU Chemnitz, Germany, 1998.

12. Schlichting, H. and Gersten, K.: Grenzschichttheorie, Springer-Verlag, 1996.

13 Squire, H.B.: On the stability for three-dimensional disturbances of viscous fluid flow between parallel walls, Proc. R. Soc. London A, Vol. 142, 621-628, 1933.

14. Yih, C.-S.: Stability of two-dimensional parallel flows for three-dimensional disturbances, Q. Appl. Maths., Vol. 12, 434-435, 1955.

15. Herwig, H. and Schäfer, P.: Influence of variable properties on the stability of two-dimensional boundary layers, J. Fluid Mech., Vol. 243, 1-14, 1992.

16. Schäfer, P.; Severin, J. and Herwig, H.: The effect of heat transfer on the stability of laminar boundary layers, Int. J. Heat Mass Transfer, Vol. 38, 1855-1863, 1995.

17. Gersten, K. and Herwig, H.: Strömungsmechanik, Vieweg-Verlag, Braunschweig, Germany, 1992.

18. Herwig, H. and You, X.: Stability analysis of plane laminar Poiseuille flow with constant heat flux across the wall: asymptotic methods, submitted for publication, 1997.

19. Herwig, H. and You, X.: Thermal receptivity of unstable laminar flow with heat transfer, Int. J. Heat Mass Transfer, Vol. 40, 4095-4103, 1997.

20. You, X. and Herwig, H.: Linear stability of channel flow with heat transfer: the important role of reference temperature, Int. Comm. Heat Mass Transfer, Vol. 24, 485-496, 1997.

21. Severin, J. Herwig, H.: Onset of convection in the Rayleigh-Bénard flow with temperature dependent viscosity: an asymptotic approach, submitted for publication, 1997.

22. Stengel, K.C.; Oliver, D.S. and Brooker, J.R.: Onset of convection in a variable-viscosity fluid, J. Fluid Mech, Vol. 120, 411-431,1982.

23. You, X. and Herwig, H.: Nonlinear stability theory of channel flow with heat transfer; an asymptotic approach, accepted for publication in Heat and Mass Transfer, 1998.

TURBULENT BOUNDARY LAYERS I
FUNDAMENTALS

Ruhr University of Bochum, Bochum, Germany

1. INTRODUCTION

Quite often turbulent boundary layer theory is introduced by the statement, that turbulent boundary layers are governed by the same equations as the laminar boundary layers, except the kinematic viscosity ν and thermal diffusivity a have to be replaced by the so–called *effective* kinematic viscosity ν_{eff} and *effective* thermal diffusivity a_{eff}, respectively.

These values are given by

$$\nu_{eff} = \nu + \nu_t \quad , \quad a_{eff} = a + a_t, \tag{1.1}$$

where the momentum transfer due to molecular motion is characterized by ν and that of turbulent fluctuations characterized by $\nu_t = -\overline{u'v'}/(\partial u/\partial y)$. Similarly, a and $a_t = -\overline{T'v'}/(\partial T/\partial y)$ determine the heat transfer due to molecular motion and turbulent fluctuations, respectively. But ν_t and a_t are not physical properties of the fluid as ν and a, they are flow functions, which have to be connected with the flow field and the temperature field, respectively, by *model equations*. (Turbulence modelling)

It should be emphasized that there is no justification for introducing turbulent boundary layers this way. This is even more so, when the so–called boundary layer transformation, derived for laminar boundary layers to reach Reynolds–number independence, is also applied to turbulent boundary layers. The latter would not become independent of the Reynolds number after the boundary layer transformation mentioned above, because ν_t and a_t are implicitly functions of the Reynolds number.

The Reynolds–averaged Navier–Stokes equations are considered as the basis for deriving the equations for the turbulent boundary layers at high Reynolds numbers. For an asymptotically correct derivation it has to be taken into account that the turbulent boundary layer consists of two layers, the wall layer adjacent to the wall, where viscosity effects are important, and the fully turbulent layer, where viscosity effects are negligible. The flow in each of these two layers is governed by a different system of equations, which — to the leading order — are again independent of the Reynolds number, as it is shown in Section 2. Since the solution for the wall layer is universal, only the flow in the fully turbulent layer has to be calculated. This concept is called the *method of wall functions*. It will be shown that this method is also applicable to flows with separation (Sect.3) and whith heat transfer (Sect.4). The layer structure leads to conditions, which the model equations of the turbulence modelling have to satisfy in order to be asymptotically correct (Sect.5). To demonstrate that the asymptotic methods can also be applied to flows with massive separation, automobile aerodynamics is considered in Sect.6 as a typical example.

2. ATTACHED TURBULENT BOUNDARY LAYERS

2.1 LAYER STRUCTURE

At high Reynolds numbers the flow field consists of an inviscid outer flow and of the boundary layer. When the Reynolds number is high enough the flow in the boundary layer is turbulent. The turbulent boundary layer has two new features compared to the laminar one. First, it has a distinctive outer edge with the wall distance (or boundary layer thickness) δ as shown in Fig. 2.1. This outer edge separates the inviscid non–turbulent outer flow from the turbulent boundary–layer flow. Second, the turbulent boundary layer consists of two layers, the fully turbulent outer part, where viscosity effects are neglible and the viscous wall layer, where viscosity effects have to be taken into account. The thickness of the visous wall layer is $\delta_v(x)$. Both thicknesses $\delta(x)$ and $\delta_v(x)$ tend to zero for increasing Reynolds number, but $\delta_v(x)$ decreases much faster than $\delta(x)$. As it will be shown later the viscous wall layer is governed by universal laws which are determined mainly by the viscosity and the wall shear stress. From the asymptotic theory for high Reynolds numbers it can be concluded, that for a given pressure distribution only the flow in the fully turbulent outer part of the boundary layer has to be calculated. The flow equations valid for this fully turbulent layer become independent of the Reynolds number. The corresponding boundary conditions follow from the matching with the inviscid outer flow (at the outer edge $y = \delta$) and with the viscous layer (at the overlap layer $y \approx \delta_v$), respectively.

2.2 OVERLAP LAYER

Since the overlap layer must have the features of both adjacent layers, its flow field is (as part of the fully turbulent layer) independent of the viscosity and (as part of the viscous wall layer) independent of the wall layer thickness δ. Hence, the following relationship between the turbulent shear stress $\tau_t = -\rho\overline{u'v'}$ and the velocity gradient $\partial u/\partial y$ must be valid in the overlap layer

$$f\left(\frac{\partial u}{\partial y}, y, \frac{\tau_t}{\rho}\right) = 0 .$$ (2.1)

Applying the π-theorem of dimensional analysis leads to the *overlap law*:

$$\lim_{y\to 0}\frac{y}{\sqrt{\tau_t/\rho}}\frac{du}{dy} = \frac{1}{\kappa} ,$$ (2.2)

where κ is the *Karman constant*.

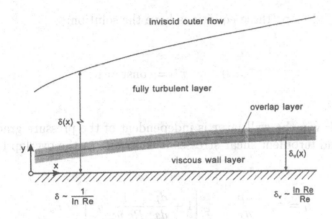

Figure 2.1 Layer structure of turbulent boundary layers

To describe properly the very thin viscous wall layer the following coordinate is used:

$$y^+ = \frac{u_\tau(x)y}{\nu} ,$$ (2.3)

where

$$u_\tau(x) = \sqrt{\frac{\tau_w(x)}{\rho}}$$ (2.4)

is the *friction velocity*.
When the variables are nondimensionalised by

$$x^* = \frac{x}{L}, \ u_\tau^* = \frac{u_\tau}{V}, \ Re = \frac{VL}{\nu},$$

$$p^* = \frac{p - p_\infty}{\rho V^2}, \ u^+ = \frac{u}{u_\tau}, \ v^+ = \frac{v}{u_\tau},$$ (2.5)

$$\tau^+ = (\mu\frac{\partial u}{\partial y} + \tau_t)/(\rho u_\tau^2)$$

the Reynolds–averaged flow equations reduce to

$$\frac{1}{Re\,u_\tau^*}\left(\frac{\partial u^+}{\partial x^*}+\frac{u^+}{u_\tau^*}\frac{du_\tau^*}{dx^*}+\frac{y^+}{u_\tau^*}\frac{\partial u^+}{\partial y^+}\frac{du_\tau^*}{dx^*}\right)+\frac{\partial v^+}{\partial y^+}=0\,,\tag{2.6}$$

$$\frac{1}{Re\,u_\tau^*}\left(u^+\frac{\partial u^+}{\partial x^*}+\frac{u^{+2}}{u_\tau^*}\frac{du_\tau^*}{dx^*}+\frac{u^+y^+}{u_\tau^*}\frac{du_\tau^*}{dx^*}\frac{\partial u^+}{\partial y^+}+\frac{1}{u_\tau^{*2}}\frac{dp^*}{dx^*}\right)+v^+\frac{\partial u^+}{\partial y^+}=\frac{\partial \tau^+}{\partial y^+}\,.\tag{2.7}$$

For the limit $Re\cdot u_\tau^*\to\infty$ these equations have the solution:

$$v^+=0\quad,\quad \tau^+=\text{const}=1\,.\tag{2.8}$$

The (attached) flow in the wall layer is independent of the pressure gradient, and the sum of viscous and turbulent shear stresses is constant. In the overlap layer we have

$$\tau_t=\tau_w+\frac{dp}{dx}=\tau_w\left[1+\frac{dp^*}{dx^*}\frac{1}{Re\,u_\tau^{*3}}y^+\right]\approx\tau_w\,.\tag{2.9}$$

and according to Eq.(2.2)

$$\lim_{y^+\to\infty}\frac{du^+}{dy^+}=\frac{1}{\kappa y^+}\,,\tag{2.10}$$

which leads to the well-known *universal logarithmic law*

$$\lim_{y^+\to\infty}u^+=\frac{1}{\kappa}\ln y^++C^+\tag{2.11}$$

with the integration constant C^+. This constant is $C^+=5.0$ for smooth walls and a well–known function of the dimensionless roughness parameter $k^+=ku_\tau/\nu$ for rough walls, respectively.

2.3 FULLY TURBULENT LAYER (DEFECT LAYER)

The flow field for high Reynolds numbers is calculated in two steps. In the first step the inviscid outer flow field is determined usually by methods of potential theory. The resulting pressure distribution (or velocity distribution $U(x)$) is input for calculating the fully turbulent layer within the boundary layer. For this layer the following form of the solution is assumed:

$$u(x,y) = U(x) - u_\tau(x)F'(x,\eta)$$
$$= U(x)\left[1 - \gamma(x)F'(x,\eta) + O(\gamma^2)\right]$$
$$v(x,y) = u_\tau\left[\frac{d\delta}{dx}(F - \eta F') + \frac{\delta}{u_\tau}\frac{du_\tau}{dx}F - \frac{\delta}{u_\tau}\frac{dU}{dx}\eta + \delta\frac{\partial F}{\partial x} + O(\gamma)\right] \qquad (2.12)$$
$$\tau_t(x,y) = -\rho\overline{u'v'} = \rho U^2\left[\gamma^2 S(x,\eta) + O(\gamma^3)\right]$$
$$p - p_\infty = \frac{\rho}{2}V^2\left[1 - \frac{U^2(x)}{V^2} + O(\gamma^2)\right] ,$$

where

$$\eta = \frac{y}{\delta(x)} , \quad \gamma(x) = \frac{u_\tau(x)}{U(x)} . \qquad (2.13)$$

The primes refer to derivatives with respect to η. The equations (2.12) can be considered as perturbations of the limiting solution $u(x,y) = U(x)$ for $Re = \infty$, where γ serves as perturbation parameter. Since $F'(x,\eta)$ is representing the velocity defect, this fully turbulent layer is also called *defect layer* (for attached flows, $\tau_w \neq 0$). As will be shown later, see Eq.(2.30), it is $\delta/L = O(\gamma) = O(1/\ln Re)$. Hence the limit $\gamma \to 0$ is equivalent to the limit $Re \to \infty$. When Eqs.(2.12) are introduced into the Reynolds–averaged Navier–Stokes equations the momentum equation for the x–direction is reduced — in the limit $Re \to \infty$, $\gamma \to 0$ — to the following defect–layer equation

$$A(x)\eta F'' + B(x)F' - S' = \frac{\partial F'}{\partial x}\Delta(x) , \qquad (2.14)$$

where

$$A(x) = \frac{d\Delta}{dx} + \frac{\Delta}{U}\frac{dU}{dx} \qquad (2.15)$$

$$B(x) = -2\frac{\Delta}{U}\frac{dU}{dx} \qquad (2.16)$$

$$\Delta(x) = \frac{U(x)\delta(x)}{u_\tau(x)} . \qquad (2.17)$$

The boundary conditions are:

$$\eta \to 0 : F = 0, \ F'' = -\frac{1}{\kappa\eta}, \ S = 1 \qquad (2.18)$$

$$\eta = 1 : F' = 0, \qquad\qquad S = 0 .$$

The conditions for $\eta \to 0$ are the matching conditions, Eq.(2.9) and (2.10), in the overlap layer.

A turbulence model is necessary to "close" the system of equations for the unknowns $F(x, \eta)$ and $S(x, \eta)$.

An example of a simple turbulence model is that of Michel et al.(1968):

$$S = \left(\frac{l}{\delta}\right)^2 F''^2, \quad \frac{l}{\delta} = c \tanh\left(\frac{\kappa}{c}\eta\right), \quad c = 0.078 \tag{2.19}$$

The system, Eqs.(2.14) to (2.19), is independent of the Reynolds number. Hence, only *one* calculation is necessary for all Reynolds numbers.

Integration of Eq.(2.14) over the defect layer yields

$$\frac{d(F_e\Delta)}{dx} + \frac{3}{U}\frac{dU}{dx}F_e\Delta = 1 \tag{2.20}$$

with the solution

$$F_e\Delta = \left(C + \int_0^x U^3 dx\right)/U^3 \tag{2.21}$$

where $F_e = F(x, \eta = 1)$.

When Eq.(2.14) is specialized for $\eta = 1$ (outer edge), it follows

$$A = \frac{d\Delta}{dx} + \frac{\Delta}{U}\frac{dU}{dx} = \frac{S_e'}{F_e''} \tag{2.22}$$

which can be considered as an equation determining the function $\Delta(x)$ for a given U(x).

From the solution $F(x, \eta)$ the following global values can be determined:

Boundary value:

$$F_e(x) = F(x, \eta = 1) = \frac{\delta_1}{\delta\sqrt{c_f/2}} \tag{2.23}$$

Wake parameter:

$$\Pi(x) = \frac{\kappa}{2}\lim_{\eta \to 0}\left[F'(x, \eta) + \frac{1}{\kappa}\ln\eta\right] \tag{2.24}$$

Shape parameter:

$$G(x) = \frac{\lim\limits_{y \to 0}\int_y^\delta (U - u)^2 dy}{u_\tau \lim\limits_{y \to 0}\int_y^\delta (U - u)dy} = \frac{1}{F_e}\lim_{\eta \to 0}\int_\eta^1 F'^2 d\eta . \tag{2.25}$$

All the global values are, as already mentioned, independent of Reynolds number as well as independent of the wall roughness k^+.

Eventually these parameters come into the picture, when the wall–shear–stress or the function $\gamma(x) = \sqrt{c_f(x)/2}$ have to be determined.

The matching condition in the overlap layer

$$\lim_{\eta \to 0} \frac{u(x,\eta)}{u_\tau(x)} = \lim_{y^+ \to \infty} u^+(y^+) \tag{2.26}$$

yields according to Eqs.(2.11) and (2.12)

$$\frac{U}{u_\tau} - \lim_{\eta \to 0} F'(x,\eta) = \frac{1}{\kappa} \ln \frac{y u_\tau}{\nu} + C^+ \tag{2.27}$$

and by use of Eq.(2.24) finally

$$\frac{1}{\gamma} = \frac{1}{\kappa} \ln(\gamma^2 Re) + C^+ + \tilde{C}(x), \tag{2.28}$$

where

$$\tilde{C}(x) = \frac{1}{\kappa} \left[2\Pi(x) + \ln \left\{ \frac{U(x)}{V} \frac{\Delta(x)}{L} \right\} \right]. \tag{2.29}$$

The implicit solution, Eq.(2.28), for the function $\gamma(x)$ can be written as an explicit *skin–friction law*

$$\gamma(x) = \frac{u_\tau(x)}{U(x)} = \sqrt{\frac{c_f(x)}{2}} = \frac{\kappa}{\ln Re} G(\Lambda; D) \tag{2.30}$$

where the function $G(\Lambda; D)$ is defined as

$$\frac{\Lambda}{G} + 2 \ln \frac{\Lambda}{G} - D = \Lambda. \tag{2.31}$$

In this case we have

$$\Lambda = \ln Re, \quad D(x) = 2 \ln \kappa + \kappa \left[C^+ + \tilde{C}(x) \right]. \tag{2.32}$$

Equation (2.30) shows that $\gamma = O(1/\ln Re)$ as mentioned earlier.

The first two terms on the right hand side of Eq.(2.28) have their origin in the viscous wall layer and hence contain the universal constants κ and C^+. Their effect on the skin friction is dominant. The term $\tilde{C}(x)$ characterizes the influence of the defect layer and is therefore dependent on the turbulence model. Its effect, however, is small and decreases with increasing Reynolds number.

2.4 EXAMPLES

a. Equilibrium boundary layers
Boundary layers whose defect layers have self–similar solutions are called *equilibrium boundary layers*. For these cases, Eq.(2.14) reduces to the ordinary differential equation

$$(1 + 2\beta)\eta F'' + 2\beta F' = F_e S' \ . \tag{2.33}$$

It can be shown easily, see Schlichting, Gersten (1997), that then the outer flows have power–law velocity distributions

$$U(x) \sim x^m \tag{2.34}$$

with

$$m = -\frac{\beta}{1 + 3\beta} \ , \tag{2.35}$$

where

$$\beta = \frac{\delta_1}{\tau_w} \frac{dp}{dx} \tag{2.36}$$

is the Rotta–Clauser–parameter. The turbulent boundary layer past the flat plate is one special case ($\beta = 0$).

b. Transition between two equilibrium boundary layers.
Figure 2.2 shows the results of flow calculations where the given velocity distribution $U(x)$ changes continuously from a constant value (flat plate) into a power–law distribution, see Gersten, Vieth (1995).

c. Friction drag of Joukowsky airfoils
The friction–drag coefficient of a symmetrical airfoil at zero angle of attack is given by

$$c_D = \frac{2D}{\rho V^2 bL} = 4 \int_0^1 \frac{\tau_w(X)}{\rho V^2} \, dX \tag{2.37}$$

where $X = x/L$. This yields according to Eq.(2.30) the drag law

$$c_D = 4 \left[\frac{\kappa}{\ln Re} G(\Lambda; D) \right]^2 \frac{U_R^2}{V^2} \int_0^1 \frac{\tau_w(X)}{\tau_{wR}} \, dX \tag{2.38}$$

where τ_{wR} is the wall shear stress at the reference point with the outer–flow velocity U_R. An airfoil with the thickness ratio $t/L = 0.15$ has the drag coefficient $c_D = 0.0058$ at $Re = 10^7$, which is lower than that of the flat plate ($c_D = 0.0062$) at the same Reynolds number, see Gersten, Herwig (1992).

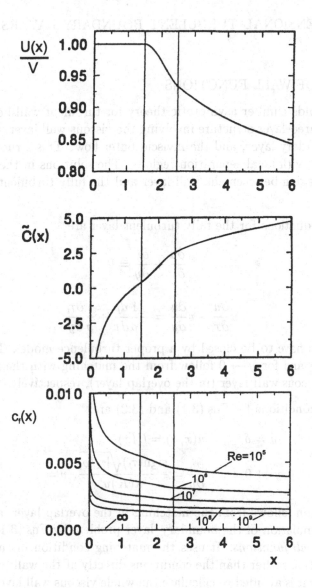

Figure 2.2 Results of the calculation of a turbulent boundary
layer, after Gersten, Vieth (1995)

The given velocity distribution is defined as follows:

$0 \leq X \leq 1.5 \quad U/V = 1$

$1.5 \leq X \leq 2.5 \quad U/V = 0.087X^3 - 0.547X^2 + 1.052X + 0.354$

$2.5 \leq X \qquad U/V = (X - 0.5)^{-0.1}$.

3. TWO–DIMENSIONAL TURBULENT BOUNDARY LAYERS WITH SEPARATION

3.1 METHOD OF WALL FUNCTIONS

The high Reynolds number asymptotic theory for turbulent wall-bounded flows leads to the typical three–layer structure involving the viscous wall layer, the fully turbulent part of the boundary layer, and the inviscid outer flow. This structure also holds for boundary layers with local separation regions. The solutions in the viscous layer and in the overlap region between the wall layer and the fully turbulent layer turn out to be universal.

The governing equations for the fully turbulent layer are

$$\frac{\partial u}{\partial x} + \frac{\partial v}{\partial y} = 0 \tag{3.1}$$

$$u\frac{\partial u}{\partial x} + v\frac{\partial u}{\partial y} = -\frac{1}{\rho}\frac{dp}{dx} + \frac{1}{\rho}\frac{\partial \tau_t}{\partial y} . \tag{3.2}$$

These equations have to be closed by a proper turbulence model. The boundary conditions at $y = \delta$ and for $y \to 0$ follow from the matching with the inviscid outer flow and with the viscous wall layer (in the overlap layer), respectively.

The boundary conditions for Eqs.(3.1) and (3.2) are:

$$y = \delta \quad : \quad u(x,y) = U(x) \tag{3.3}$$

$$y \to 0 \quad : \quad \frac{\partial u}{\partial y} = \frac{\text{sign}(\tau_t)\sqrt{|\tau_t/\rho|}}{\kappa(K)y}, \ v = 0 \quad . \tag{3.4}$$

The last condition follows from the matching in the overlap layer, as has been shown earlier. The formulation of the boundary–layer problem by Eqs.(3.1) to (3.4) is called the *method of wall functions*. It uses the matching condition in the overlap layer as boundary condition rather than the conditions directly at the wall (e.g.: no–slip condition). Therefore it is avoided to calculate the whole viscous wall layer which is anyhow universal and hence known a priori for a given wall shear stress and pressure gradient.

3.2 WALL FUNCTIONS

The turbulent shear stress

$$\tau_t = \tau_w + \frac{dp}{dx}y \tag{3.5}$$

can now be negative. Instead of the Karman constant a function $\kappa(K)$ appears in Eq.(3.4), where the new dimensionless parameter K characterizes the effect of the pressure gradient. At the point of vanishing wall shear stress (separation point) the turbulent shear stress is according to Eq.(3.5) proportional to the wall distance and the pressure gradient. For this case Eq.(3.4) yields

$$\lim_{y \to 0} \frac{\partial u}{\partial y} = \frac{\sqrt{(dp/dx)/\rho}}{\kappa_\infty \sqrt{y}} \qquad (\tau_w = 0) \tag{3.6}$$

which after integration leads to the square–root law written in proper coordinates for this particular wall layer

$$\lim_{y^\times \to \infty} u^\times(y^\times) = \frac{2}{\kappa_\infty} \sqrt{y^\times} + C^\times \tag{3.7}$$

where the following definitions have been used

$$u^\times = \frac{u}{u_s}, \tag{3.8}$$

$$y^\times = \frac{y u_s}{\nu}, \tag{3.9}$$

$$u_s = \left(\frac{\nu}{\rho} \frac{dp}{dx} \right)^{1/3}. \tag{3.10}$$

The square–root law according to Eq.(3.7) corresponds to the logarithmic law for attached boundary layers ($\tau_w \neq 0$). It will now be described how the logarithmic law changes over to the square–root law when an attached boundary layer approaches the separation point.

When the wall coordinate for attached boundary layers

$$y^+ = \frac{y u_\tau}{\nu} \tag{3.11}$$

is used, Eq.(3.5) reads

$$\frac{\tau_t}{\tau_w} = 1 + K y^+, \tag{3.12}$$

where the new dimensionless parameter K is defined by

$$K = \frac{\nu}{u_\tau \tau_w} \frac{dp}{dx} = \text{sign}(\tau_w) \left(\frac{u_s}{u_\tau} \right)^3. \tag{3.13}$$

The friction velocity

$$u_\tau = \sqrt{\frac{|\tau_w|}{\rho}} \tag{3.14}$$

is always positive, but K can be positive or negative depending on the sign of τ_w. This parameter characterizes the change–over from attached to separated boundary layers. It is the crucial parameter for flows with separation. For attached flows at high Reynolds numbers ($\nu \to 0$) we have $K \to 0$, whereas $K = \infty$ at the separation point ($\tau_w = 0$). At separation K changes sign and is usually negative in the separated region.

Equation (3.4), written in wall coordinates for attached flows, reads

$$\frac{\partial u^+}{\partial y^+} = \frac{\sqrt{1 + Ky^+}}{\kappa(K)y^+} . \tag{3.15}$$

Integration with respect to y^+ over the whole wall layer yields

$$u^+(y^+, K) = \frac{1}{\kappa(K)} \ln y^+ + \frac{2}{\kappa(K)} \ln \frac{2}{\sqrt{1 + Ky^+} + 1}$$
$$+ \frac{2}{\kappa(K)} \left(\sqrt{1 + Ky^+} - 1 \right) + C(K) . \tag{3.16}$$

This is the generalized wall function for the velocity $u^+(y^+, K)$. In the limit $K \to 0$ we get the logarithmic wall function

$$\lim_{K \to 0} u^+(y^+, 0) = \frac{1}{\kappa(0)} \ln y^+ + C(0) \tag{3.17}$$

where $\kappa(0) = \kappa_0 = 0.41$ and $C(0) = C^+ (= 5.0$ for smooth walls).

When the wall variables $y^\times$ and $u^\times$ according to Eqs.(3.8) and (3.9) are used the generalized wall function reads:

$$u^\times(y^\times, K) = \frac{1}{K^{1/3}} \frac{2}{\kappa(K)} \ln \left| \frac{\sqrt{y^\times + K^{-2/3}}}{\sqrt{y^\times}} \right| + \frac{1}{K^{1/3}} \frac{1}{\kappa(K)} \left(\ln \frac{4}{|K|} - 2 \right)$$
$$+ \frac{2}{\kappa(K)} \sqrt{y^\times + K^{-2/3}} + \frac{1}{K^{1/3}} C(K) . \tag{3.18}$$

This equation reduces to Eq.(3.7) in the limit $K \to \infty$, where

$$\kappa(\infty) = \kappa_\infty \approx 0.6, \ C(\infty) = K^{1/3} C^\times . \tag{3.19}$$

The wall functions, Eqs.(3.16) and (3.18), contain the two universal functions $\kappa(K)$ and $C(K)$. These functions have been determined by experiments, see Kiel (1995), Vieth (1996,1997), Vieth et al.(1998). The results are shown in Fig. 3.1. The largest uncertainties of these functions appear near the separation and reattachment points ($K \to \infty$). In Table 3.1 the values κ_∞ and $C^\times$ (see Eq.(3.7)) after various authors have been collected. It is obvious that κ_∞ is different from $\kappa_0 = 0.41$ and about $\kappa_\infty \approx 0.6$.

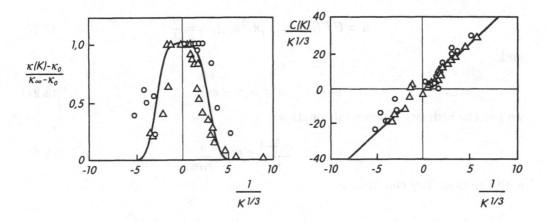

Figure 3.1 Experimental quantities $\kappa(K)$ and $C(K)$ of the generalized wall functions

author	kind of investigation	κ_∞	$C^\times$
Townsend (1960)	analytic	0.5 ± 0.05	/
Townsend (1961) and (1976)	analytic	0.48 ± 0.03	2.2
Mellor (1966)	analytic	0.41 (0.44)	1.33
Perry et al. (1966)	experimental	0.48	/
Szablewski (1972)	analytic	0.41	2.23
Kader/Yaglom (1978)	analytic	0.45	/
El Telbany/Reynolds (1980)	experimental	0.8	-3.2
Nakoyama/Koyama (1984)	analytic	0.5	0.0
Spalart/Leonhard (1987)	DNS	0.6	-3.36
Vieth (1996)	experimental	0.88	-3
present investigations	experimental	0.59	0.0

Table 3.1: Universal constants κ_∞ and $C^\times$ in the wall function at the separation point, Eq.(3.7), after various authors, see Kiel et al.(1997)

3.3 EXAMPLES

a. Stratford Flow

B.S. Stratford (1959) investigated experimentally the flow where the wall shear stress is equal to zero all along the wall. In this case the limiting solution is not the homogeneous velocity profile. The fully turbulent layer is not a defect layer, it has to satisfy the nonlinear boundary layer equation (3.2). It turns out that for this flow Eqs.(3.1) and (3.2) lead to a self–similar solution. Assuming

$$u = U f'(\eta), \ \tau_t = \rho U^2 s(\eta), \ \eta = \frac{y}{\delta} \tag{3.20}$$

and

$$U \sim x^m, \ \delta = \alpha(x - x_0) \tag{3.21}$$

we get the ordinary differential equation

$$f'^2 - 1 - \frac{m+1}{m} f f'' = \frac{s'}{\alpha m} \tag{3.22}$$

with the boundary conditions

$$\begin{aligned} \eta \to 0 \ &: \quad f = 0, \ f' = 0 \\ \eta = 1 \ &: \quad \quad \ f' = 1, \ f'' = 0 \ . \end{aligned} \tag{3.23}$$

Using the turbulence model (mixing length model) after Michel et al.(1968)

$$s(\eta) = \left(c_2 \tanh \frac{\kappa_\infty}{c_2} \eta \right)^2 f''^2, \ \kappa_\infty = 0.6; \ c_2 = 0.085 \tag{3.24}$$

the solution (for $C^\times = 0$) yields the following results, see Schlichting, Gersten (1997):

$$m = -0.219 \, , \ \alpha = 0.145 \ .$$

It is independent of the Reynolds number and can be considered as the limiting solution, which consists of the inviscid outer flow (potential flow) and a thin fully turbulent (inviscid!) layer (fixed thickness parameter α) adjacent to the wall. These flow characteristics are very similar to those of turbulent free shear layers, see Schneider (1991).

b. Turbulent Separation Bubble in a Diffuser

The flow in a two–dimensional diffuser with separation and reattachment has been investigated experimentally by Kiel (1995). In Fig. 3.2 the distributions of the displacement thickness $\delta_1(x)$ and skin friction coefficient $c_f(x)$ are shown. The curves have been calculated using the generalized wall functions, see Vieth (1996,1997). Since the prediction method had to be an inverse one, the distribution $\delta_1(x)$ was used as input. There is good agreement of the wall shear stress distribution between theory and experiment.

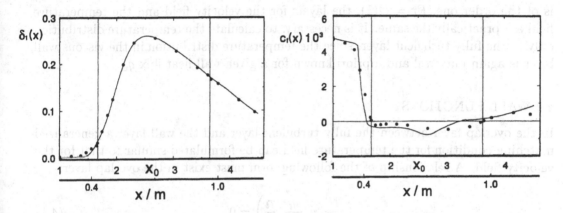

Figure 3.2 Distribution of displacement thickness $\delta_1(x)$ and skin friction
        coefficient $c_f(x)$ for a diffuser flow with a turbulent separation region
        o   Experiments after Kiel (1995)
        —— Theory after Vieth (1997).

Instead of using the method of wall functions low–Reynolds–number versions of various turbulence models are quite commonly in use. Their results (e.g. shear stress distributions) are basically the same as those of the method of wall functions. Their necessary numerical effort, however, is drastically higher, because the total boundary layer including the viscous wall layer is calculated, although the wall layer is universal and known a priori for given wall shear stress and pressure gradient. In other words, low–Reynolds–number versions of turbulence models for attached flows just determine the universal constant C^+, which of course has the value $C^+ = 5.0$ for smooth walls, see Wilcox (1993, page 142).

4. HEAT TRANSFER IN TURBULENT BOUNDARY LAYERS

4.1 LAYER STRUCTURE

The temperature field and in particular the heat transfer will be considered in turbulent
boundary layers at high Reynolds numbers. The flow is two–dimensional and steady.
The constant wall temperature T_w is only slightly different from the outer flow temperature T_∞, so that the physical properties such as density and viscosity are assumed to
be constant. At high Reynolds numbers the temperature field has also a layer structure: The inviscid outer flow has the constant temperature T_∞. The boundary layer
again consists of two layers: The fully turbulent layer free of viscosity effects and the
viscous wall layer. Under the assumption that the Prandtl number $Pr = c_p\mu/\lambda$, ($\lambda =$
thermal conductivity, $c_p =$ specific heat capacity of constant pressure, $\mu =$ viscosity)
is of the order one, $Pr = O(1)$, the layers for the velocity field and the temperature
field are practically the same. It is necessary, to calculate the temperature distribution
only in the fully turbulent layer, since the temperature distribution in the viscous wall
layer is again universal and a priori known for a given wall heat flux q_w.

4.2 WALL FUNCTIONS

In the overlap layer between the fully turbulent layer and the wall layer a generalized
matching condition for the temperature field can be formulated similar to that for the
velocity field. A relationship of the following form must exist in the overlap layer:

$$f\left(\frac{\partial T}{\partial y}, y, \frac{q_t}{\rho c_p}, \frac{\tau_t}{\rho}\right) = 0 \,, \tag{4.1}$$

where

$$q_t = \rho c_p \overline{v'T'} \,, \qquad \tau_t = -\rho \overline{u'v'} \tag{4.2}$$

are the turbulent heat flux and the turbulent shear stress, respectively. The dimensional
analysis reduces Eq.(4.1) to the *generalized matching condition*

$$\lim_{y \to 0} \frac{y\sqrt{|\tau_t/\rho|}}{-q_t/(\rho c_p)} \frac{\partial T}{\partial y} = \frac{1}{\kappa_\Theta(K)} \,, \tag{4.3}$$

where

$$K = \frac{\nu}{u_\tau \tau_w} \frac{dp}{dx} \tag{4.4}$$

characterizes the pressure–gradient effect. The function $\kappa_\Theta(K)$ corresponds to the
function $\kappa(K)$ for the velocity field.

Again, two limiting cases of Eq.(4.3) can be considered:

a.   Attached boundary layers ($\tau_w \neq 0$)

$$\tau_t = \tau_w \quad , \quad q_t = q_w \quad , \quad K = 0$$

$$\lim_{y \to 0} \frac{\partial \Theta^+}{\partial y} = \frac{1}{\kappa_{\Theta 0} y}, \quad \lim_{y^+ \to \infty} \frac{\partial \Theta^+}{\partial y^+} = \frac{1}{\kappa_{\Theta 0} y^+} \tag{4.5}$$

where

$$\Theta^+ = \frac{T - T_w}{T_\tau} \tag{4.6}$$

is analogous to $u^+ = u/u_\tau$.
The temperature difference $T - T_w$ is based on

$$T_\tau = -\frac{q_w}{\rho c_p u_\tau} \, , \tag{4.7}$$

which is called *friction temperature*. Integration of Eq.(4.5) over the wall layer leads to the logarithmic wall function

$$\lim_{y^+ \to \infty} \Theta^+(y^+, Pr) = \frac{1}{\kappa_{\Theta 0}} \ln y^+ + C_\Theta^+(Pr) \tag{4.8}$$

where

$$\kappa_{\Theta 0} = 0.47 \tag{4.9}$$

and

$$C_\Theta^+(Pr) = 13.7 \, Pr^{2/3} - 7.5 \quad (Pr > 0.5) \, . \tag{4.10}$$

b.   Separation point ($\tau_w = 0$)

$$\tau_t = \frac{dp}{dx} y \quad , \quad q_t = q_w \quad , \quad K = \infty$$

$$\lim_{y^\times \to \infty} \frac{\partial \Theta^\times}{\partial y^\times} = \frac{1}{\kappa_{\Theta \infty} (y^\times)^{3/2}} \tag{4.11}$$

where

$$\Theta^\times = \frac{T - T_w}{T_s} \, . \tag{4.12}$$

Here the temperature difference is based on

$$T_s = -\frac{q_w}{\rho c_p u_s} = -\frac{q_w}{\rho c_p} \bigg/ \left(\frac{\nu \, dp}{\rho \, dx} \right)^{1/3} \, . \tag{4.13}$$

Integration of Eq.(4.11) over the wall layer leads to the *reciprocal square-root law*

$$\lim_{y^\times \to \infty} \Theta^\times(y^\times, Pr) = \frac{-2}{\kappa_{\Theta \infty}} \frac{1}{\sqrt{y^\times}} + C_\Theta^\times(Pr) \, , \tag{4.14}$$

where

$$\kappa_{\Theta\infty} = 3.6 \quad , \quad C_\Theta^\times(Pr = 0.72) = 1.8 \,. \tag{4.15}$$

The values of $C_\Theta^\times(Pr)$ for other Prandtl numbers are not known yet. When a general flow with heat transfer is considered, we have

$$\tau_t = \tau_w + \frac{dp}{dx} y \quad , \quad q_t = q_w \tag{4.16}$$

and hence Eq.(4.3) in wall–layer coordinates (y^+, u^+, Θ^+) as

$$\lim_{y^+ \to \infty} \frac{d\Theta^+}{dy^+} = \frac{1}{\kappa_\Theta(K)y^+\sqrt{1 + Ky^+}} \,. \tag{4.17}$$

Integration over the entire wall layer leads to the *generalized matching condition*:

$$\Theta^+(y^+, K) = \frac{1}{\kappa_\Theta(K)} \ln y^+ + \frac{2}{\kappa_\Theta(K)} \ln \frac{2}{\sqrt{1 + Ky^+} + 1} + C_\Theta(K, Pr) \,. \tag{4.18}$$

If one uses the wall-layer coordinates $(y^\times, u^\times, \Theta^\times)$ for the separation point the generalized matching condition reads:

$$\Theta^\times(y^\times, K) = 2\frac{K^{1/3}}{\kappa_\Theta(K)} \ln \left| \frac{\sqrt{y^\times + K^{-2/3}} - K^{-1/3}}{\sqrt{y^\times}} \frac{2}{|K|^{1/2}} \right| + K^{1/3}C_\Theta(K, Pr) \tag{4.19}$$

with Eq.(4.14) as the limiting case $K \to \infty$.

The *wall functions*, Eqs.(4.18) and (4.19), contain the two universal functions $\kappa_\Theta(K)$ and $C_\Theta(K, Pr)$. These functions have been determined experimentally by Kiel (1995). The results are shown in Fig. 4.1.

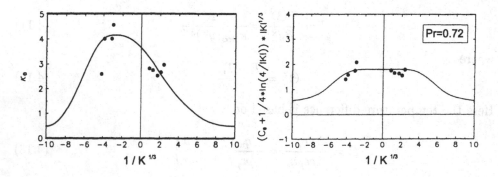

Figure 4.1 Universal functions $\kappa_\Theta(K)$ and $C_\Theta(K, Pr = 0.72)$

4.3 BOUNDARY LAYER CALCULATION

The governing equation for the temperature in the fully turbulent layer is

$$u\frac{\partial T}{\partial x} + v\frac{\partial T}{\partial y} = -\frac{1}{\rho c_p}\frac{\partial q_t}{\partial y} \,. \tag{4.20}$$

The boundary conditions are

$$y = \delta: \quad T(x,y) = T_\infty \tag{4.21}$$

$$y \to 0: \quad \frac{\partial T}{\partial y} = -\frac{q_t/(\rho c_p)}{\kappa_\Theta(K)y\sqrt{|\tau_t/\rho|}} \,. \tag{4.22}$$

They follow from the matching of the fully turbulent layer with the outer flow and with the viscous wall layer (overlap layer), respectively. Equation (4.20) needs a proper turbulence model for the turbulent heat flux q_t in order to close the system of equations.

4.4 EXAMPLES

a. Equilibrium boundary layers

In these boundary layers the velocity profiles in the fully turbulent layer are self–similar. It turns out that the temperature profiles are also self–similar for the two standard boundary conditions of the temperature field at the wall T_w =const and q_w =const, respectively.

The *defect formulation* for the temperature

$$T(x,y)=T_\infty - T_\tau(x)G'(\eta) \tag{4.23}$$

$$q_t(x,y)=-\rho c_p u_\tau(x)T_\tau(x)S_\Theta(\eta) \tag{4.24}$$

$$\eta=\frac{y}{\delta(x)} \tag{4.25}$$

leads to the following ordinary differential equations:

$$T_w = \text{const} \quad (1+2\beta)\eta G'' = S'_\Theta \tag{4.26}$$

$$q_w = \text{const} \quad (1+2\beta)\eta G'' - \beta G' = S'_\Theta \,, \tag{4.27}$$

$$\text{where} \quad \beta = -\frac{m}{1+3m} \tag{4.28}$$

$$\text{and} \quad U(x) \sim x^m \,. \tag{4.29}$$

The boundary conditions are

$$\eta = 1: \quad G' = 0, \ S_\Theta = 0$$

$$\eta \to 0: \quad G'' = -\frac{1}{\kappa_{\Theta 0}\eta}, \ S_\Theta = 1 \tag{4.30}$$

The matching of the temperature in the overlap layer

$$\lim_{\eta \to 0} \Theta^+(\eta) = \lim_{y^+ \to \infty} \Theta^+(y^+) \tag{4.31}$$

or

$$\lim_{\eta \to 0} \frac{T - T_w}{T_\tau} = \frac{1}{\kappa_{\Theta 0}} \ln y^+ + C_\Theta^+(Pr) \tag{4.32}$$

yields the heat–transfer law

$$\frac{T_\infty - T_w}{T_\tau} = \frac{1}{\kappa_{\Theta 0}} \ln \frac{u_\tau \delta}{\nu} + C_\Theta^+(Pr) + \bar{C}_\Theta \tag{4.33}$$

where

$$\bar{C}_\Theta = \lim_{\eta \to 0} \left[G'(\eta) + \frac{1}{\kappa_{\Theta 0}} \ln \eta \right] . \tag{4.34}$$

Combining the heat transfer law, Eq.(4.33), and the skin–friction law

$$\frac{U}{u_\tau} = \sqrt{\frac{2}{c_f}} = \frac{1}{\kappa_0} \ln \left(\frac{u_\tau \rho}{\nu} \right) + C^+ + \bar{C} , \tag{4.35}$$

gives an explicit formula for the Stanton number or the Nusselt number:

$$St = \frac{q_w}{\rho c_p (T_w - T_\infty) U} = \frac{Nu}{Re \ Pr} = \frac{c_f / 2}{\frac{\kappa_0}{\kappa_{\Theta 0}} + \sqrt{\frac{c_f}{2}} D_\Theta(Pr)} \tag{4.36}$$

where

$$D_\Theta(Pr) = C_\Theta^+(Pr) + \overline{C}_\Theta - \frac{\kappa_0}{\kappa_{\Theta 0}}(C^+ + \overline{C}) . \tag{4.37}$$

The equations (4.26) and (4.27) have been solved by using the turbulence model of constant turbulent Prandtl number

$$Pr_t = \frac{\nu_t}{a_t} = -\tau_t c_p \frac{\partial T}{\partial y} \Big/ \left(q_t \frac{\partial u}{\partial y} \right) = \frac{\kappa_0}{\kappa_{\Theta 0}} = 0.87 , \tag{4.38}$$

see Gersten, Herwig (1992). The dependence of the temperature field on the velocity field is realised only via this turbulence model. The resulting values $\overline{C}$ and $\overline{C}_\Theta$ are given in Table 4.1.

The ratio $\kappa_0/\kappa_{\Theta 0}$ in Eq.(4.36) is universal. Only the values $\overline{C}$ and $\overline{C}_\Theta$ follow from the solution for the fully turbulent layer and depend on the turbulence model. Their influence on the heat transfer is small (a few percent) and decreases for increasing Reynolds numbers.

β	m	$(\overline{C}_\Theta - 0.87\overline{C})_{T_w=\text{const}}$	$(\overline{C}_\Theta - 0.87\overline{C})_{q_w=\text{const}}$
0	0	0.91	0.91
0.5	-0.200	-4.19	-5.87
1	-0.250	-7.37	-9.31
2	-0.286	-11.92	-13.60
10	-0.323	-30.94	-29.49

Table 4.1 Constants in the heat–transfer law, Eqs.(4.36) and (4.37), for equilibrium boundary layers.

b.   Heat transfer in separation region of a diffuser.

R. Kiel (1995) carried out experiments on heat transfer in a diffuser flow, where a separated region of finite length occurred. Figure 4.2 shows the distribution of the Stanton number. The theoretical curves have been calculated using different model constants for the q_t–equation. The tendency of the experimental distribution is simulated by the theory, but refinements of the turbulence modelling are still necessary.

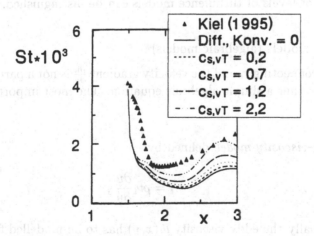

Figure 4.2 Stanton number in a separated region of a diffuser flow, after Kiel (1995) and Vieth (1997).

5. TURBULENCE MODELLING FROM AN ASYMPTOTIC POINT OF VIEW

5.1 CLOSURE OF EQUATION SYSTEM

The Reynolds–averaged Navier–Stokes equations reduce to the following boundary-layer equations for steady incompressible flow:

$$\frac{\partial u}{\partial x} + \frac{\partial v}{\partial y} = 0 \qquad (5.1)$$

$$\rho \left(u \frac{\partial u}{\partial x} + v \frac{\partial u}{\partial y} \right) = -\frac{dp}{dx} + \frac{\partial}{\partial y} \left(\mu \frac{\partial u}{\partial y} + \tau_t \right) \qquad (5.2)$$

$$\rho c_p \left(u \frac{\partial T}{\partial x} + v \frac{\partial T}{\partial y} \right) = -\frac{\partial}{\partial y} \left(-\lambda \frac{\partial T}{\partial y} + q_t \right) . \qquad (5.3)$$

Since $\frac{dp}{dx}$ is a given function derived from calculating the inviscid outer flow, Eqs.(5.1) and (5.2) are formally two equations for the two unknown velocity components $u(x, y)$ and $v(x, y)$. Unfortunately, the *turbulent shear stress* $\tau_t(x, y) = -\rho \overline{u'v'}$ in Eq.(5.2) is an additional unknown function. In order to close the system of equations turbulence modelling has to develop equations for $\tau_t(x, y)$, by which it is connected with the velocity field. In a similar way the *turbulent heat–flux rate* $q_t = \rho c_p \overline{v'T'}$ must be connected with the velocity and temperature field by model equations.

5.2 TURBULENCE MODELS

Different categories or levels of turbulence models can be distinguished.

a.   Zero–equation models (algebraic models)

The equation connecting τ_t and the velocity gradient $\frac{\partial u}{\partial y}$ is not a partial differential equation, but rather a simple algebraic equation. The most important models of this type are:

–   The *eddy–viscosity model*, defined by

$$\tau_t = \rho \nu_t \frac{\partial u}{\partial y} , \qquad (5.4)$$

where usually the eddy viscosity $\nu_t(x, y)$ has to be modelled further.

–   The *mixing–length model*, defined by

$$\tau_t = \rho l^2 \left| \frac{\partial u}{\partial y} \right| \frac{\partial u}{\partial y} , \qquad (5.5)$$

where the mixing length $l(x, y)$ has to be modelled, which is usually much simpler than the modelling of $\tau_t(x, y)$.

Example (model after R. Michel et al.(1968))

$$\frac{l}{\delta} = c \, \tanh \frac{\kappa y}{c\delta} \, , \quad c = 0.085 \, . \tag{5.6}$$

b.   One–equation models

These models use the boundary–layer equation for the kinetic energy $k(x, y)$ of the turbulent fluctuation:

$$\underbrace{u\frac{\partial k}{\partial x} + v\frac{\partial k}{\partial y}}_{\text{convection}} = \underbrace{\nu\frac{\partial^2 k}{\partial y^2}}_{\text{viscous}} + \underbrace{\frac{\partial B}{\partial y}}_{\text{turbulent}} + \underbrace{\frac{\tau_t}{\rho}\frac{\partial u}{\partial y}}_{\text{production}} - \underbrace{\epsilon}_{\text{dissipation}} , \tag{5.7}$$

$$\underbrace{\phantom{\nu\frac{\partial^2 k}{\partial y^2} + \frac{\partial B}{\partial y}}}_{\text{diffusion}}$$

where

$$k = \frac{1}{2}\overline{q^2} \, , \tag{5.8}$$

$$B = -\overline{\left(\frac{1}{2}q^2 + \frac{p'}{\rho}\right)v'} \, , \tag{5.9}$$

$$q^2 = u'^2 + v'^2 + w'^2 \, . \tag{5.10}$$

When the turbulent diffusion dB/dy is modelled properly, Eq.(5.7) can be considered as an equation with three unknowns $k(x, y)$, the dissipation $\epsilon(x, y)$ and $\tau_t(x, y)$. The general relationship

$$\nu_t = f(k, \epsilon) \tag{5.11}$$

yields via dimensional analysis

$$\nu_t = c_\mu \frac{k^2}{\epsilon} \, , \quad c_\mu = \text{const} \, . \tag{5.12}$$

Some models use the *dissipation length*

$$l_\epsilon = c_\epsilon \frac{k^{3/2}}{\epsilon} \tag{5.13}$$

instead of the dissipation ϵ.

When $l_\epsilon(x, y)$ is properly modelled by an algebraic equation similar to Eq.(5.6) the Eqs.(5.1), (5.2), (5.4), (5.7), (5.12) and (5.13) form a closed system of equations for $u(x, y)$, $v(x, y)$, $\tau_t(x, y)$, $\nu_t(x, y)$, $k(x, y)$ and $\epsilon(x, y)$. Such a model is called one–equation model, because *one* partial differential equation, Eq.(5.7), is used in the turbulence modelling.

c.   Two–equation models

These models use a second balance equation as partial differential equation in addition to Eq.(5.7). Examples are:

- $k - l$ model by J.C. Rotta
  This model uses an equation for l_ϵ (see Eq.(5.13)) based on a balance equation for the two–point correlation function

$$R_{ii} = \overline{u_i(x, y, z)\, u_i(x, y + r_y, z)}\,. \tag{5.14}$$

- $k - \epsilon$ model by W.R. Jones and B.E. Launder.
  This model uses an equation for $\epsilon(x, y)$, which was formulated heuristically in analogy to the k–equation (5.7):

$$u\frac{\partial \epsilon}{\partial x} + v\frac{\partial \epsilon}{\partial y} = \frac{\partial}{\partial y}\left(\frac{\nu_t}{Pr_\epsilon}\frac{\partial \epsilon}{\partial y}\right) + c_{\epsilon 1}\frac{\epsilon}{k}\frac{\tau_t}{\rho}\frac{\partial u}{\partial y} - c_{\epsilon 2}\frac{\epsilon^2}{k}\,. \tag{5.15}$$

Here the viscous terms are already neglected.

- $k - \omega$ model by D.C. Wilcox
  This model uses an heuristically developed equation for the *specific dissipation rate*

$$\omega = \epsilon/(c_\mu k)\,. \tag{5.16}$$

d.   Reynolds stress models (second moment closure models)
These models use the balance equations for the Reynolds stresses $\overline{u_i u_j}$ ($i, j = 1, 2, 3$):

$$u\frac{\partial \overline{u_i u_j}}{\partial x} + v\frac{\partial \overline{u_i u_j}}{\partial y} = D_{i,j} + P_{i,j} - \epsilon_{i,j} + \Phi_{ij} \tag{5.17}$$

where D_{ij} : diffusion
             P_{ij} : production
             ϵ_{ij} : dissipation
             Φ_{ij} : pressure–strain correlation

The Reynolds stress models have the following advantages compared to the lower–order models:

- The simple algebraic equation (5.4) for τ_t is replaced by a partial differential equation for τ_t. According to Eq.(5.4) the velocity has to have a maximum or minimum at locations where $\tau_t = 0$, which is, however, not the case in many flows, e.g. wall jets, separated flows, natural convection etc.

- In flows along highly curved walls the turbulence modelling depends strongly on the wall curvature. Since the Reynolds stress models can distinguish between stresses parallel and normal to the centrifugal forces, the curvature effects can be taken into account easily, which is not the case for the lower–order models.

The Reynolds stress models, however, have mainly the following problems. They still use the heuristic ϵ–equation (5.15) and there are difficulties to model the pressure–strain correlation Φ_{ij}, for which experimental data are not available.

e. Models for the heat–flux rate

For $q_t(x, y)$ mainly two types of models are available:

- Concept of *turbulent Prandtl number*

$$Pr_t = \frac{\nu_t}{a_t} = -\tau_t c_p \frac{\partial T}{\partial y} \bigg/ \left(q_t \frac{\partial u}{\partial y} \right) , \qquad (5.18)$$

where a_t is defined by

$$q_t = -\rho c_p a_t \frac{\partial T}{\partial y} . \qquad (5.19)$$

In most cases it is assumed $Pr_t =$const., but functions $Pr_t(y/\delta)$ are also used.

- Heat–flux rate balance equation

An equation for q_t similar to Eq.(5.17) for $\overline{u_1 u_2} = -\tau_t/\rho$ can be formulated where three terms have to be modelled.

5.3 CONDITIONS FOR TURBULENCE MODELS

According to the asymptotic theory for high Reynolds–number flows the boundary–layer calculation is restricted to the fully turbulent layer, where the viscosity effects are negligible. The matching of the solution in this layer with the inviscid outer flow and with the wall layer leads to boundary conditions which the solution has to satisfy independently of the turbulence model. Hence, the turbulence models have to satisfy certain conditions in the so–called overlap layers to guarantee the proper matching. These conditions will be discussed first for the matching with the *viscous wall layer* and second for the matching with the *viscous superlayer*, which is located between the fully turbulent layer and the inviscid outer flow as shown in Fig. 5.1.

5.3.1 MATCHING WITH THE VISCOUS WALL LAYER

The flow field in the overlap layer between the fully turbulent layer and the viscous wall layer is characterized by the dimensionless pressure–gradient parameter

$$K = \frac{\nu}{u_\tau \tau_w} \frac{dp}{dx} . \qquad (5.20)$$

Each balance equation leads to an a–priori information in the overlap layer described by the limit $y \to 0$. The momentum equation, Eq.(5.2), leads to

$$\lim_{y \to 0} \frac{y}{(\tau_t/\rho)^{1/2}} \frac{\partial u}{\partial y} = \frac{1}{\kappa(K)} . \qquad (5.21)$$

The k–equation, Eq.(5.7), leads to

$$\epsilon = f\left(y, \frac{\tau_t}{\rho}, k, \frac{\partial B}{\partial y}\right) \tag{5.22}$$

which by dimensional analysis results in the formula

$$C_\epsilon = F(C_k, C_B) , \tag{5.23}$$

where

$$C_\epsilon(K) = \lim_{y \to 0} \frac{y\epsilon}{(\tau_t/\rho)^{3/2}} \tag{5.24}$$

$$C_k(K) = \lim_{y \to 0} \frac{\tau_t}{\rho k} \tag{5.25}$$

$$C_B(K) = \lim_{y \to 0} \frac{y}{(\tau_t/\rho)^{3/2}} \frac{\partial B}{\partial y} . \tag{5.26}$$

The value C_k is called *structural parameter*. It follows that k is proportional to τ_t in the overlap layer.

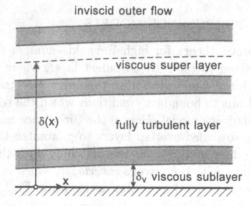

Figure 5.1: Various layers in turbulent boundary layers

It can be shown by intergrating the k–equation, Eq.(5.7), over the viscous wall layer that for $K \to 0$ the diffusion vanishes in the overlap layer and hence the following equilibrium exists:

$$\text{production} = \text{dissipation (overlap layer, } K \to 0) . \tag{5.27}$$

Therefore, in this case ($K \to 0$) the overlap layer is called *equilibrium layer*. The k–equation, Eq.(5.7), reduces to

$$\epsilon = \frac{\tau_t}{\rho} \frac{\partial u}{\partial y} + \frac{\partial B}{\partial y} \tag{5.28}$$

or

$$C_\epsilon(K) = \frac{1}{\kappa(K)} + C_B(K) .$$ (5.29)

At separation $(K \to \infty)$ the diffusion does not vanish in the overlap layer $(C_B(\infty) \neq 0)$.

In Table 5.1 the wall functions are given for the two limiting cases $K \to 0$ $K \to \infty$.

$K \to 0$	$K \to \infty$
$\frac{\tau_t}{\rho} = \frac{\tau_w}{\rho} = u_\tau^2$	$\frac{\tau_t}{\rho} = \frac{1}{\rho}\frac{dp}{dx}y = u_p^2\frac{y}{\delta}$
$\frac{\partial u}{\partial y} = \frac{u_\tau}{\kappa_0 y}$	$\frac{\partial u}{\partial y} = \frac{u_p}{\kappa_\infty \delta}\sqrt{\frac{\delta}{y}}$
$\kappa_0 = 0.41$	$\kappa_\infty = 0.6$
$k = \frac{1}{C_k(0)}\frac{\tau_w}{\rho}$	$k = \frac{u_p^2}{C_k(\infty)}\frac{y}{\delta}$
$C_k(0) = \sqrt{c_\mu} = 0.3$	$C_k(\infty) = 0.2$
$\epsilon = \frac{u_\tau^3}{\kappa_0 y}$	$\epsilon = C_\epsilon(\infty)\frac{u_p^3}{\delta}\sqrt{\frac{y}{\delta}}$
$\frac{\partial B}{\partial y} = 0$	$\frac{\partial B}{\partial y} = \left(C_\epsilon(\infty) - \frac{1}{\kappa_\infty}\right)\frac{u_p^3}{\rho}\sqrt{\frac{y}{\delta}}$
$\nu_t = \kappa_0 u_\tau y$	$\nu_t = \kappa_\infty u_p \delta \left(\frac{y}{\delta}\right)^{3/2}$
$l = \kappa_0 y$	$l = \kappa_\infty y$
$q_t = q_w = -\rho c_p u_\tau T_\tau$	$q_t = q_w = -\rho c_p u_p T_p$
$\frac{\partial T}{\partial y} = \frac{T_\tau}{\kappa_{\Theta 0} y}$	$\frac{\partial T}{\partial y} = \frac{T_p}{\kappa_{\Theta\infty}\delta}\left(\frac{y}{\delta}\right)^{-3/2}$
$a_t = \kappa_{\Theta 0} u_\tau y$	$a_t = \kappa_{\Theta\infty} u_p \delta \left(\frac{y}{\delta}\right)^{3/2}$
$Pr_t = \frac{\kappa_0}{\kappa_{\Theta 0}}$	$Pr_t = \frac{\kappa_\infty}{\kappa_{\Theta\infty}}$

Table 5.1: Functions in the overlap layer $(y \to 0)$ for $K \to 0$ (attached) and $K \to \infty$ (separation)

There exist two types of functions, those which change their dependence on y when K changes from $K \to 0$ to $K \to \infty$, and those which do not change.

The latter are the *structural function* $a(x,y) = \tau_t/(\rho k)$ with $\lim\limits_{y\to 0} a(x,y) = C_k(K)$, $l(x,y)$ and Pr_t. Hence, turbulence models using these three functions should be preferred. All models using for instance the eddy viscosity concept have the disadvantage that $\nu_t(x,y)$ changes the functional dependence on y when flow separation occurs. An a–l–model would be better in this respect than the existing two–equation models. The functions $\kappa(K)$, $\kappa_\Theta(K)$, $C_\epsilon(K)$, $\dot{C}_k(K)$ and $C_B(K)$ are universal and have been determined experimentally by Kiel (1995). All turbulence models have to be consistent with the wall functions in the overlap layer.

The diffusion term in the k–equation is usually modelled by a gradient concept

$$B = \frac{\nu_t}{Pr_k}\frac{\partial k}{\partial y} \ , \qquad (5.30)$$

where Pr_k is a model constant. Since it is connected with $C_B(K)$ via Eq.(5.26), it has to be a function of K. For $K \to \infty$ we have

$$Pr_k(\infty) = \frac{3}{2}\frac{\kappa_\infty}{C_B(\infty)C_\epsilon(\infty)} \ . \qquad (5.31)$$

In the same way the model constants in Eq.(5.15), Pr_ϵ, $c_{\epsilon 1}$, and $c_{\epsilon 2}$, must be functions of K, as can bee seen, when Eq.(5.15) is specialized for the overlap layer:

$$K \to 0: \quad Pr_\epsilon(c_{\epsilon 2} - c_{\epsilon 1}) = \frac{\kappa_0^2}{C_k(0)} \qquad (5.32)$$

$$K \to \infty: \quad Pr_\epsilon[\kappa_\infty C_\epsilon(\infty)c_{\epsilon 2} - c_{\epsilon 1}] = \frac{\kappa_\infty^2}{2C_k(\infty)} \ . \qquad (5.33)$$

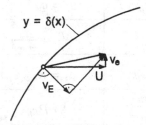

Figure 5.2: Entrainment velocity v_E

5.3.2. MATCHING WITH THE VISCOUS SUPERLAYER

The matching of the fully turbulent layer with the inviscid outer flow results in the boundary conditions at $y = \delta$:

$$y = \delta : \ U - u = 0, \ \tau_t = 0, \ k = 0, \ \epsilon = 0 \ \text{etc.} \qquad (5.34)$$

It can be shown that all these variables tend to zero linearly, see Jeken (1992).

A characteristic parameter for the superlayer is the *entrainment velocity* v_E, which follows from the integration of the continuity equation, see Fig. 5.2:

$$v_E = \frac{d}{dx} \int_0^{\delta(x)} u\, dy = U \frac{d\delta_1}{dx} - v_e \, . \tag{5.35}$$

This value is positive for an increasing boundary layer thickness. When the new superlayer coordinate

$$n = \delta(x) - y \tag{5.36}$$

is introduced, the balance equations reduce for $n = 0$ to the following relations, which turbulence models have to satisfy:

a.   momentum equation

$$\rho v_E \left(\frac{\partial u}{\partial n} \right)_0 = - \left(\frac{\partial \tau_t}{\partial n} \right)_0 \, . \tag{5.37}$$

For the eddy viscosity model

$$\tau_t = -\rho \nu_t \frac{\partial u}{\partial n} \tag{5.38}$$

follows

$$\left(\frac{\partial \nu_t}{\partial n} \right)_0 = v_E \tag{5.39}$$

b. k–equation

There is an equilibrium of convection and diffusion in the superlayer:

$$v_E = \frac{1}{Pr_k} \left(\frac{\partial \nu_t}{\partial n} \right)_0 \tag{5.40}$$

which means by comparing Eqs.(5.39) and (5.40), that it must be

$$Pr_k = 1 \, .$$

In the same way it can be shown, that it also follows in the ϵ–equation $Pr_\epsilon = 1$.

c. τ_t–equation

For flows in which the turbulent shear stress changes sign in the flow field (e.g. reversed flow, wall jet, natural convection) the eddy viscosity concept according to Eq.(5.4) fails. The following balance equation for τ_t can be used instead:

$$u \frac{\partial \tau_t}{\partial x} + v \frac{\partial \tau_t}{\partial y} = \frac{\partial}{\partial y} \left(c_S \frac{k}{\epsilon} \overline{v'^2} \frac{\partial \tau_t}{\partial y} \right) + \rho \overline{v'^2} \frac{\partial u}{\partial y} - c_1 \epsilon \frac{\tau_t}{k} - c_2 \rho \overline{v'^2} \frac{\partial u}{\partial y}$$

$$+ \left(-\frac{3}{2} c_{1w} \epsilon \frac{\tau_t}{k} + \frac{3}{2} c_{2w} \rho \overline{v'^2} \frac{\partial u}{\partial y} \right) \frac{k^{3/2}}{c_L \epsilon y} \, . \tag{5.41}$$

The matching of the fully turbulent layer with the viscous superlayer leads to

$$c_1 = \frac{1}{c_S}(1 - c_2) \quad , \tag{5.42}$$

see Jeken (1992). This relation between the three constants c_1, c_2, and c_S is illustrated in Fig.5.3. For a fixed value c_S Eq.(5.42) corresponds to a straight line in this figure. As an example the widely used value $c_S = 0.22$ has been chosen. All the various sets of model constants proposed by different authors are very close to the straight line. But none of the model–constants sets — except one — satisfies the relation Eq.(5.42) exactly.

More conditions for the model constants in Reynolds–stress models can be found in Jeken (1992).

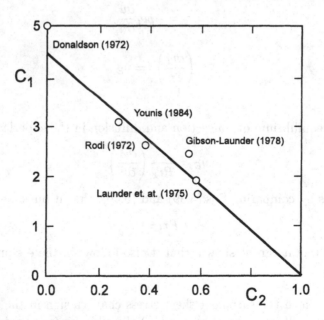

Figure 5.3 Relation between the model constants c_1 and c_2
according to Eq.(5.42), after Jeken (1992). The straight
line corresponds to $c_S = 0.22$.

6. AUTOMOBILE AERODYNAMICS – AN EXAMPLE FOR THE APPLICATION OF ASYMPTOTIC METHODS

6.1 ZONAL METHOD

The general objective of automobile aerodynamics is to determine the flow–induced forces on the car. The distributions of the pressure and the shear stress at the body wall as well as the aerodynamic drag and lift are of particular interest. According to the asymptotic theory for high Reynolds numbers the entire flow field is devided into various zones, as shown schematically in Fig.6.1a. These zones are the inviscid outer flow, the boundary layer attached to the body and the turbulent wake flow, including the dead water region close to the backside of the body. Since the various zones are governed by different equations and have to be solved separately and matched with each other afterwards, the asymptotic concept is sometimes called *zonal method*, see Kline (1982). The outer inviscid flow will lead to the pressure distribution, and the boundary layer attached to the body yields the wall shear stress distribution. To get these results a detailed calculation of the wake zone is not necessary. It is sufficient to take into account the global effect of the wake, i.e. its displacement effect on the inviscid outer flow.

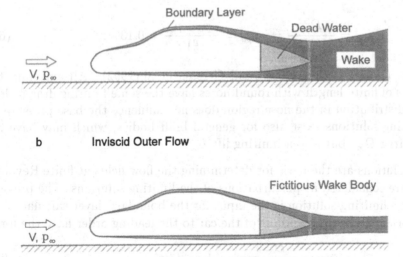

Figure 6.1 Flow zones around an automobile. Shaded areas
         are turbulent or viscous zones
        a   Flow zones
        b   Fictitious body for calculating the outer flow.

6.2 CONCEPT OF THE FICTITIOUS BODY

The key element of the zonal method is the principle of the *fictitious body*. This means that the geometry of the given body is modified such that the displacement effect of the viscous zones is reproduced by placing the fictitious body in inviscid flow. In the region of the attached boundary layer the fictitious body is formed by the original body (car) which is enlarged by the boundary–layer displacement thickness. The contour of this fictitious body continues into the wake region representing the wake displacement, see Fig.6.1b. But a precise knowledge of the contour far downstream is not needed, since the far–field displacement effects on the pressure distribution at the body surface decreases very rapidly with distance. The fictitious body has a finite cross–section at infinity (half body), the area of which is directly related with the aerodynamic drag of the car.

6.3 LIMITING SOLUTION

While the thickness of the boundary layer attached to the car surface tends to zero for $Re \to \infty$, the fictitious body representing the wake tends to a finite size in the limit $Re \to \infty$. According to this limiting solution the car with a bluff–body shape has a finite drag. As a simple example, Fig.6.2 shows a scetch of the limiting flow past a semi-infinite bluff circular cylinder, the limiting drag of which is given by

$$c_{D\infty} = \frac{D_\infty}{\frac{\varrho}{2}V^2 \frac{\pi}{4} d^2} = -c_{pb} = -\frac{p_b - p_\infty}{\frac{\varrho}{2}V^2} = 0.13 \quad , \tag{6.1}$$

see VanWagenen (1968), Becker (1994) and Gersten et al.(1992). All axisymmetric circular cylinders of finite length with round noses have the same limiting drag as long as the pressure distribution in the nose region does not influence the base pressure p_b. Equivalent limiting solutions exist also for general bluff bodies, which may have not only a limiting drag D_∞, but also a limiting lift L_∞.

These limiting solutions are the basis for determining the flow fields at finite Reynolds number, which are considered as perturbations of the limiting solutions. The pressure distribution of the limiting solution is the input for the boundary–layer calculation on the car surface. Hence, the drag and lift of the car to the leading order have the form

$$c_D = \frac{D}{\frac{\varrho}{2}V^2 A} = c_{D\infty} + c_{Df}(Re) \tag{6.2}$$

$$c_L = \frac{L}{\frac{\varrho}{2}V^2 A} = c_{L\infty} \tag{6.3}$$

where A is the frontal area. The friction drag coefficient is given by

$$c_{Df}(Re) \sim \frac{1}{\ln^2 Re} \tag{6.4}$$

with the Reynolds number $Re = VL/\nu$.

The shape of the fictitious wake body can be determined because it must reproduce certain universal properties of the static pressure distribution in the wake. For more details see Papenfuß (1997), Dilgen (1995), Papenfuß, Dilgen (1993) and Gersten, Papenfuß (1992). It is worth mentioning that the fictitious wake body should be free of transverse forces since such forces cannot be supported due to the absence of rigid walls.

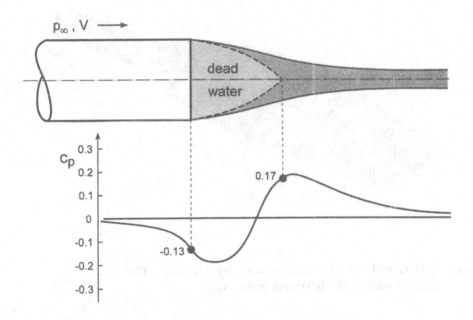

Figure 6.2 Limiting solution for the flow past a semi–infinite
cicular cylinder at infinite Reynolds number
a   cylinder and fictitious wake body
b   pressure distribution at the fictitious wake body
contour $c_p = 2(p - p_\infty)/(\rho V^2)$

6.4 EXAMPLES

Fig.6.3 shows a typical fictitious body of a compact car (squareback). It is assumed, that the pressure at the blunt base is constant in agreement with many experimental results. Fig.6.4 shows the geometry of a model (scale 1:8) of a car for windtunnel experiments. In the lower part of the figure the pressure distribution in the symmetry plane is shown. Excellent agreement can be seen between the the theoretical curves found by the zonal method and the data from the windtunnel experiments. The comparison of the theoretical and experimental results for the pressure distribution in six different cross–sections (Fig.6.5) is also very good. The calculated values of the drag and lift coefficients are given in the small table of Fig.6.5. The theoretical base pressure coefficient is in complete agreement with the experiment. The rather low drag coefficient of $c_D = 0.216$ is plausible because the model has no external parts or wheels.

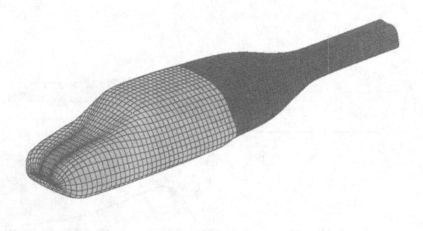

Figure 6.3 Fictitious body of a compact car (squareback). The shaded part is the fictitious wake body.

6.5 AERODYNAMIC DESIGN

It is well known that separation of the three–dimensional boundary layer on the car surface — in particular on the rear part of fastback cars — generates longitudinal vortices of high energy, which increase the aerodynamic drag of the car drastically. Therefore, the concept of the aerodynamic design of automobiles should be to avoid the separation of the boundary layer on the car surface, but fix the flow separation by sharp edges of the properly chosen vertical base plane. The zonal method just described is a perfect tool to find the optimal shape of the car which satisfies these aerodynamic–design conditions, see Gersten, Papenfuß (1993). It has the great advantage that it requires only a small fraction of computer capacity compared to the Navies–Stokes codes which automobile industry is presently trying to apply.

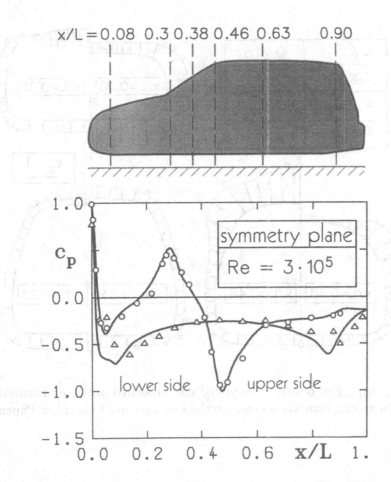

Figure 6.4 Comparison between calculated and measured pressure distribution
in the symmetry plane of a compact car, after Papenfuß (1997)

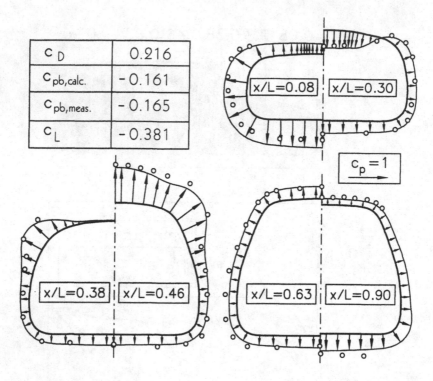

c_D	0.216
$c_{pb,calc.}$	−0.161
$c_{pb,meas.}$	−0.165
c_L	−0.381

Figure 6.5 Comparison between calculated and measured pressure distribution in several transverse cross–sections of a compact car, after Papenfuß (1997)

7 CONCLUSIONS

The theory of turbulent boundary layers is an asymptotic theory of the Reynolds-averaged Navier–Stokes equations for high Reynolds numbers. The whole flow field is devided into the inviscid outer flow and the turbulent boundary layer. The latter consists of two layers: the fully turbulent layer and the viscous wall layer. The calculation can be restricted to the fully turbulent layer, since the solutions of the wall layer are universal. The procedure is called *method of wall functions*. This method can be applied to attached and separated flows as well as to heat transfer problems.

The matching conditions lead to conditions which asymptotically correct turbulence models have to satisfy. Many examples confirm these results. In particular the example of the automobile aerodynamics shows that the asymptotic methods can also applied to flows with massive separation.

8 REFERENCES

Becker, A. (1994): Messungen im Nachlauf von Körpern mit stumpfem Heck mittels der Laser–Doppler–Anemometrie. VDI–Fortschrittsberichte, Reihe 8: Meß–, Steuerungs– und Regelungstechnik, No. 402, VDI–Verlag, Düsseldorf.

Dilgen, P. (1995): Berechnung der abgelösten Strömung um Kraftfahrzeuge: Simulation des Nachlaufs mit einem inversen Panel–Verfahren. VDI–Fortschrittsberichte, Serie 7, Strömungstechnik, No. 258, Düsseldorf, VDI–Verlag.

Gersten, K. (1995): What can asymptotic theory do for the turbulence modellers? Proceedings of Symposium on Developments in Fluid Dynamics and Aerospace Engineering, 9–10 Dec. 1993 (60th Birthday of Prof. R. Narasimha), Indian Institute of Science, Bangalore, pp 122-143.

Gersten, K.; Becker, A.; Demmer, T. (1993): Experimental investigation of the wake past bluff bodies. In: K. Gersten (Ed.): Physics of Separated Flows –Numerical, Experimental, and Theoretical Aspects. Notes on Numerical Fluid Mechanics, Vol. 40, Vieweg–Verlag, Braunschweig, 233–240.

Gersten, K.; Herwig, H. (1992): Strömungsmechanik. Grundlagen der Impuls– Wärme– und Stoffübertragung aus asymptotischer Sicht. Vieweg–Verlag, Braunschweig, Wiesbaden.

Gersten, K.; Klauer, J.; Vieth, D. (1993): Asymptotic analysis of two–dimensional turbulent separating flows. In: K. Gersten (Ed.): Physics of separated flows. Notes on Numerical Fluid Mechanics, Vol.40, Vieweg–Verlag, Braunschweig, Wiesbaden, pp. 125–132.

Gersten, K.; Papenfuß, H.–D. (1992): Separated flows behind bluff bodies at low speeds including ground effects. In: H. Ramkissoon (Ed.): Proceedings 2nd Caribbean Conference of Fluid Dynamics, University of West Indies, St. Augustine, Trinidad, 115–122.

Gersten, K.; Papenfuß, H.–D. (1993): Aerodynamischer Entwurf beim Automobil. RUBIN, Wissenschaftsmagazin der Ruhr-Universität Bochum, Vol.1, 36–41.

Gersten, K.; Vieth, D.(1995): Berechnung anliegender Grenzschichten bei hohen Reynolds– Zahlen. Festschrift anläßlich des 70. Geburtstages von Prof. Dr. J. Siekmann, Universität–GH Essen.

Jeken, B. (1992): Asymptotische Analyse ebener turbulenter Strömungen an gekrümmten Wänden bei hohen Reynolds–Zahlen mit einem Reynolds–Spannungs–Modell. VDI: Reihe 7, Nr.215, VDI–Verlag, Düsseldorf.

Kaplun, S. (1957): Low Reynolds number flow past a circular cylinder. J. Math. Mech.,Vol. 6, 595–603

Kaplun, S.; Lagerstrom, P.A. (1957): Asymptotic expansions of Navier–Stokes solutions for small Reynolds numbers. J. Math. Mech., Vol. 6, 585–593.

Kiel, R. (1995): Experimentelle Untersuchung einer Strömung mit beheiztem lokalen Ablösewirbel an einer geraden Wand. VDI–Fortschritt–Berichte, Reihe 7, Nr. 281, VDI– Verlag, Düsseldorf.

Kline, S.J. (1982): Universal or zonal modelling — the road ahead. In: S.J. Kline, B.J. Cantwell, G.M. Lilley (Eds.): Proceedings of the 1980–81 AFOSR/HTTM Stanford Conference on Complex Turbulent Flows. Stanford University, Stanford, Cal. Vol. II: Comparison of Computation and Experiment, 991–1015.

Messiter, A.F. (1970): Boundary–layer flow near the trailing edge of a flat plate. SIAM J. Appl. Math., Vol. 18, 241–257.

Michel, R.; Quémard, C.; Durant, R. (1968): Hypotheses on the mixing length and application to the calculation of the turbulent boundary layers. In: Kline, S.J. et al. (Eds.): Proceedings Computation of Turbulent Boundary Layers–1968. AFOSR–IFP–Stanford Conference, Vol. I, 195–207.

Neiland, V.Y. (1969): Theory of laminar boundary layer separation in supersonic flow. Izv. Akad. Nauk SSSR Mekh. Zhid. i Gaza, Vol. 4, 53–57. Engl. Übersetzung: Fluid Dynamics, Vol. 4, 1969, 33-35.

Papenfuß, H.-D. (1997): Theoretische Kraftfahrzeug–Aerodynamik – die Struktur des Strömungsfeldes bestimmt das Konzept. ATZ Automobiltechnische Zeitschrift, Bd. 99, 100–107.

Papenfuß, H.-D.; Dilgen, P. (1993): Three–dimensional separated flow around automobiles with different read profiles: Application of the zonal method. In: K. Gersten (Ed.): Physics of Separated Flows –Numerical, Experimental, and Theoretical Aspects. Notes on Numerical Fluid Mechanics, Vol. 40, Vieweg–Verlag, Braunschweig, 241–248.

Prandtl, L. (1904): Über Füssigkeitsbewegung bei sehr kleiner Reibung, In: Krazer, A. (Hrsg.): Verh. III Intern. Math. Kongr., Heidelberg, 484–491, Teubner, Leipzig. 1905– Repr. Nendeln 1967.

Schlichting, H.; Gersten, K. (1997): Grenzschicht–Theorie, 9th Edition, Springer–Verlag, Berlin, Heidelberg.

Schneider, W. (1991): Boundary–layer theory of free turbulent shear flows. Z. f. Flugwiss. u. Weltraumforsch., Bd.15, 143–158.

Stewartson, K. (1969): On the flow near the trailing edge of a flat plate. II. Mathematika, Vol. 16, 106–121.

Stewartson, K. (1974): Multistructured boundary layers on flat plates and related bodies. Advances in Applied Mechanics, Vol. 14, 145–239.

Stewartson, K.; Smith, F.T.; Kaups, K. (1982): Marginal separation. Studies in Appl. Math., Vol. 67, 45–61.

Stratford, B.S. (1959): An experimental flow with zero skin friction throughout its region of pressure rise. J. Fluid Mech., Vol.5, 17–35.

Van Dyke, M. (1975): Perturbation Methods in Fluid Mechanics. The Parabolic Press, Stanford, California.

Vieth, D. (1996): Heat and momentum transfer in turbulent boundary layers in the presence of strong adverse pressure gradients. In: K. Gersten (Ed.): Asymptotic Methods for Turbulent Shear Flows at High Reynolds Numbers. Kluwer Academic Publishers, Dordrecht, 155–168.

Vieth, D. (1997): Berechnung der Impuls– und Wärmeübertragung in ebenen turbulenten Strömungen mit Ablösung bei hohen Reynolds–Zahlen. VDI–Fortschritt–Berichte, Reihe 7, Nr.311, VDI–Verlag, Düsseldorf.

Vieth, D.; Kiel, R. (1995): Experimental and theoretical investigations of the near–wall region in a turbulent separated and reattached flow. Exp. Thermal and Fluid Sci., Vol.II, 243-254.

Vieth, D.; Kiel, R.; Gersten, K. (1998): Two–dimensional turbulent boundary layers with separation and reattachment including heat transfer. In: M. Fiebig (Ed.): Monographien der Forschergruppe "Wirbel und Wärmeübertragung". Notes on Numerical Fluid Mechanics, Vieweg–Verlag, Braunschweig, Wiesbaden.

VanWagenen, R.G. (1968): A study of axially–symmetric subsonic base flow. Ph.D. Thesis, Washington University.

Young, A.D. (1989): Boundary Layers. BSP Professional Books, Oxford.

TURBULENT BOUNDARY LAYERS II
FURTHER DEVELOPMENTS

J.D.A. Walker
Lehigh University, Bethlehem, PA, USA

1. Introduction

In this chapter, asymptotic analysis is used to consider various aspects of turbulent boundary layers in the limit of large Reynolds number. Turbulent wall-bounded shear flows are common in engineering practice and, although such motions can be very complex, generic trends are exhibited over a wide range of Reynolds numbers. In such circumstances, asymptotic theory is an essential tool for revealing the critical aspects of boundary-layer structure, as well as the dominant physical processes in the turbulence. In the following six sections, specific issues related to both the prediction and physics of turbulent shear flows near walls will be addressed. In §2, some of the classical results for two-dimensional incompressible flows will be reviewed and extended; special emphasis is placed on the minimum information and the numerical algorithms that are required to structure a prediction scheme for such flows; this chapter forms a basis for the more complicated types of boundary layers considered in subsequent sections. In §3, a model for the mean flow profile in the near-wall region of the boundary layer is described; this model is based on the observed coherent structure of the wall-layer flow and provides a simple alternative to conventional mixing-length formulations. In §4, the case of incompressible two-dimensional flow with heat transfer at the wall is addressed; the asymptotic theory constrains the types of models that can be used in the energy equation and provides an effective way to determine the heat transfer at the surface in a prediction method.

In §5, the dynamics of the time-dependent flow in the turbulence are considered. It is argued that the main features of the turbulent boundary layer, including regeneration, can be explained in terms of how hairpin vortices interact with one another and their environment.

Attention is returned to the mean flow problem in §6 through consideration of pressure-driven three-dimensional flows. It is shown via asymptotic theories how the streamwise and cross-stream velocities must behave in both the outer part and the wall layer of the boundary layer. Embedded function algorithms are described wherein accurate numerical predictions can be produced by matching an outer-layer numerical solution to a set of embedded functions in the wall layer. In this manner, it is not necessary to calculate the flow all the way to the wall, complicated inner-region models are not needed, and savings of up to 50% of the total mesh points may be achieved (as compared to a conventional calculation).

Lastly, in §7, the issue of compressibility is taken up for mainstream Mach numbers approaching and throughout the supersonic range. Here again asymptotic theory is invaluable in both constraining the types of turbulence models that are feasible and also in revealing the structure of the boundary layer in the limit of large Reynolds number. Modifications of existing turbulence models for compressibility are suggested, as well as a self-consistent general formulation in terms of the Howarth-Dorodnitsyn variable; a set of self-similar profiles is discussed and compared with data. The extension of the embedded function methodology to the prediction of high-speed compressible flow is discussed.

2. Two-Dimensional Turbulent Boundary Layers

2.1 Introduction

The two-dimensional, nominally steady turbulent boundary layer has been the subject of countless theoretical and experimental studies, many of which are often vague and sometimes contradictory. Such flows occur at high Reynolds numbers and involve many subtle issues. In order to make lasting progress, it is desirable to adopt a rational approach based on modern asymptotic methods. However, the study of the mean turbulence equations is complicated by the closure problem; that is, the governing differential equations contain unknown functions associated with time-averaging of the turbulence and are therefore not precise. Unfortunately, this deficiency has led to a number of diverse approaches to the problem, even on a rational basis. In order to set the stage for the results that will be described in the subsequent chapters, as well as to appreciate the limitations of the theory and how it may be extended into more complex situations, it is useful to review some features of the constant property turbulent boundary layer and the underlying assumptions concerning some central results.

Defining dimensionless variables in terms of a representative length L and flow speed U_o, a dimensionless friction velocity u_τ can be defined in terms of the wall shear stress by

$$u_\tau^2(x, Re) = \frac{1}{Re} \frac{\partial u}{\partial y}\bigg|_{y=0}, \tag{2.1}$$

where (x,y) are Cartesian coordinates measuring distance along and normal to the wall, respectively; the corresponding mean velocity components are (u, v) and the Reynolds number is defined by $Re = U_o L/\nu$, where ν is the kinematic velocity. In a laminar flow at high Re, the mainstream velocity is adjusted to relative rest on the wall through a layer having thickness $O(Re^{-1/2})$ within which u is $O(1)$; thus from equation (2.1), u_τ is $O(Re^{-1/4})$. By contrast, a turbulent boundary layer is a composite double layer, consisting of an effectively inviscid outer layer and a thin viscous wall layer. As subsequently shown, $u_\tau = O(1/\log Re)$ for a turbulent boundary layer, and the thickness of the outer layer is $O(u_\tau)$. Thus the skin friction and boundary-layer thickness are much greater in a turbulent flow than for the laminar counterpart.

For an attached turbulent boundary layer, experimental measurements [1], [2], [3] for situations with and without pressure gradients show that the Reynolds stress $-u'v'$ and normal stresses $\overline{u'^2}$ and $\overline{v'^2}$ are all $O(u_\tau^2)$. Substitution in the Reynolds-averaged Navier-Stokes equations leads to the special form in a thin boundary layer of

$$ u\frac{\partial u}{\partial x} + v\frac{\partial u}{\partial y} = U_e\frac{dU_e}{dx} + \frac{\partial \tau}{\partial y}, \qquad \frac{\partial u}{\partial x} + \frac{\partial v}{\partial y}, \qquad (2.2)$$

where τ denotes a total stress in the x direction defined by

$$ \tau = \frac{1}{Re}\frac{\partial u}{\partial y} + \sigma, \qquad \sigma = -\overline{u'v'}. \qquad (2.3)$$

The normal momentum equation indicates that the pressure does not vary in y across the boundary layer to leading order (to $O(u_\tau^2)$), and the relative error in the first of equations (2.2) is $O(u_\tau \partial u_\tau /\partial x)$.

It is well known that the turbulent boundary layer is double-structured consisting of : (1) a relatively thick outer layer where the turbulent stress σ is dominant, (2) a thin wall layer adjacent to the surface, where the viscous stress must come into play (see, for example [1] – [3]). For attached turbulent boundary layers, all modern experimental and theoretical research supports this concept; various features of how these two regions interact in the time-dependent turbulence will be subsequently described in §5. In the overlap region between the two layers, the mean profile u is widely believed to be logarithmic in y. This is essentially an empirical observation, and although many arguments have been put forth over the years to justify the logarithmic behavior, none are entirely satisfactory as a proof, including Millikan's argument and developments based on dimensional analysis. At present, there is no sound reasoning that elucidates what precise sequence of events leads to a logarithmic behavior in u near the wall. Indeed, in recent times, a number of authors have suggested that the mean profile in turbulent pipe and boundary-layer flows may be algebraic in y near the wall, with an exponent that depends on Re [4].

In the present development, the commonly accepted logarithmic behavior will be assumed in order to develop a set of detailed results. A derivation of the leading-order results from first principles is given elsewhere [5], [6]; note, however, that to one extent or another, the leading-order results are all based on the empirical logarithmic law.

2.2 Leading-Order Theory

Let $\Delta_o(x,Re)$ be a scaling quantity characteristic of the outer (and hence the boundary-layer) thickness and write the streamwise velocity in a defect law of the form

$$u = U_e(x) + u_\tau(x, Re)\frac{\partial F_1}{\partial \eta} + ..., \qquad (2.4)$$

where $\eta = y/\Delta_o$ is the scaled outer variable and $U_e(x)$ is the mainstream speed. Near the wall, the defect function $\partial F_1/\partial \eta$ is taken to be logarithmic according to

$$\frac{\partial F_1}{\partial \eta} \sim \frac{1}{\kappa}\log\eta + C_o(x) + ... \quad \text{as} \quad \eta \to 0, \qquad (2.5)$$

where κ is the von Kármán constant, which is often assumed to have a universal value of 0.41; the function $C_o(x)$ is to be found, usually from a numerical solution for $\partial F_1/\partial \eta$, and its ultimate value is dependent on the choice of outer-layer turbulence model. In the wall layer, the velocity is written in terms of the conventional inner variable as

$$u = u_\tau(x, Re)U^+(y^+) + ..., \quad y^+ = Re\,u_\tau y. \qquad (2.6)$$

It may be verified by substitution in the Navier-Stokes equations that a possible dependence on x in U^+ is at most parametric and again a logarithmic behavior must occur at the wall-layer edge of the form

$$U^+(y^+) \sim \frac{1}{\kappa}\log y^+ + C_i \quad \text{as} \quad y^+ \to \infty. \qquad (2.7)$$

This relation is the classical law of the wall, and it is normally believed that C_i is a universal constant having a typical value of 5.0. It may readily be verified that the velocities (2.4) and (2.6) match provided

$$\frac{U_e}{u_\tau} = \frac{1}{\kappa}\log\{Re\,u_\tau\Delta_o\} + C_i - C_o. \qquad (2.8)$$

This relation, called the match condition (or skin friction relation), relates u_τ to Re and Δ_o and establishes that u_τ/U_e is $O(\kappa/\log Re)$ as $Re \to \infty$.

Upon substitution of these expansions into equation (2.2), it may be verified that the primary balance in the wall layer occurs between the Reynolds and the viscous stresses, with the pressure gradient and convective terms being of relatively lower order. It is easily shown that the system (2.2) reduces to $\partial\tau/\partial y^+ = 0$ and, as a consequence, the total stress is constant across the wall layer and equal to the value at the wall, viz.,

$$\tau = u_\tau^2.$$ (2.9)

However, since the viscous stress is expected to become small at the outer edge of the wall layer ($y^+ \to \infty$), equation (2.9) also fixes the order of magnitude of the Reynolds stress and writing

$$\sigma = u_\tau^2 \, \sigma_1(x,y^+) + ...,$$ (2.10)

it follows from the first of equations (2.3) that the leading order wall-layer equation is

$$\frac{\partial U^+}{\partial y^+} + \sigma_1 = 1.$$ (2.11)

Consequently if σ_1 is specified, the velocity profile U^+ is known and vice versa; either one of U^+ or σ_1 must be modeled in order to resolve the closure problem.

It also follows that the total stress in the outer layer should be expanded according to

$$\tau = u_\tau^2 \, T_1(x,\eta) + ...,$$ (2.12)

where $T_1 \to 1$ as $\eta \to 0$ in order to match the total stress in the wall layer. If the expansions (2.4) and (2.12) are substituted in the momentum equation, it is easily confirmed that $\Delta_o = O(u_\tau)$ in order that the total stress balance the convective terms. Furthermore, defining

$$\alpha = \frac{U_e \Delta_o'}{u_\tau}, \quad \beta = -\frac{\Delta_o U_e'}{u_\tau}, \quad \sigma = \frac{\Delta_o U_e}{u_\tau},$$ (2.13)

where the primes denote differentiation with respect to x, it may be deduced by differentiation of the match condition (2.8) that

$$u_\tau' = \frac{u_\tau U_e'}{U_e} + \frac{u_\tau^2}{U_e^2}(\beta - \alpha + \kappa\sigma C_o')\frac{U_e}{\kappa\sigma} + O\left(\frac{u_\tau}{U_e}\right)^3.$$ (2.14)

It is then readily shown that the leading-order outer-layer equation is

$$\frac{\partial T_1}{\partial \eta} - (\beta - \alpha)\eta\frac{\partial^2 F_1}{\partial \eta^2} + 2\beta\frac{\partial F_1}{\partial \eta} = \sigma\frac{\partial^2 F_1}{\partial x \partial \eta}.$$ (2.15)

Once a turbulence model for T_1 is specified, along with an initial profile $\partial F_1 / \partial \eta$ at some station $x = x_o$, equation (2.15) can be integrated in the downstream direction.

2.3 Second Order Problem

The foregoing leading-order results are relatively well known, although the approaches used by various authors can vary appreciably. The objective here is to investigate the form of the next terms in the expansion in the outer layer and how these potentially interact with the inner-layer solution. Starting with the inner layer, it may be shown upon substituting equations (2.6) into (2.2) that

$$\frac{\partial \tau}{\partial y^+} = -\frac{U_e U_e'}{Re\, u_\tau} + \frac{u_\tau U_e'}{Re\, U_e} (U^+)^2 + ..., \tag{2.16}$$

to leading order (assuming U^+ is independent of x). Upon integration, using equation (2.9), it follows that

$$\tau = u_\tau^2 - \frac{U_e U_e'}{Re\, u_\tau} y^+ + \frac{u_\tau U_e'}{Re\, U_e} \int_0^{y^+} U^{+2} dy^+ + ..., \tag{2.17}$$

and for y^+ large, using (2.7),

$$\tau \sim u_\tau^2 - \frac{U_e U_e'}{Re\, u_\tau} y^+ + \frac{u_\tau U_e'}{Re\, U_e} \left\{ \frac{1}{\kappa^2} y^+ \log^2 y^+ + ... \right\}, \quad \text{as } y^+ \to \infty. \tag{2.18}$$

Rewriting this expression in terms of the outer variable $\eta = y^+/(Re\, u_\tau \Delta_o)$, and utilizing the match condition (2.8), it can be shown that

$$\tau \sim u_\tau^2 \left\{ 1 - \frac{2\beta}{\kappa} \eta \log \eta + ... \right\} + \frac{u_\tau^3}{U_e} \left\{ -\frac{\beta \eta}{\kappa^2} \log^2 \eta + ... \right\} + ..., \tag{2.19}$$

in the limit $Re \to \infty$ with y^+ large and fixed. This suggests that the expansion for the total stress in the outer layer proceeds according to

$$\tau = u_\tau^2 T_1(x, \eta) + \frac{u_\tau^3}{U_e} T_2(x, \eta) + ..., \tag{2.20}$$

where

$$T_1 \sim 1 - \frac{2\beta}{\kappa} \eta \log \eta + ..., \quad T_2 \sim -\frac{\beta}{\kappa^2} \eta \log^2 \eta + ..., \tag{2.21}$$

as $\eta \to 0$. It may now be inferred from equations (2.2) and (2.20) that the corresponding expansion for u is of the form

$$u = U_e + u_\tau \frac{\partial F_1}{\partial \eta} + \frac{u_\tau^2}{U_e} \frac{\partial F_2}{\partial \eta} + ... \tag{2.22}$$

The second and third terms in this expansion must vanish for large η and

$$\frac{\partial F_1}{\partial \eta} \sim \frac{1}{\kappa}\log\eta + C_o, \quad \frac{\partial F_2}{\partial \eta} \sim C_1 \quad \text{as} \quad \eta \to 0. \tag{2.23}$$

The functions C_o and C_1 are functions of x to be found whose values depend on the specific turbulence model adopted, as well as the upstream history of the boundary layer. Note that the second of conditions (2.23) is not logarithmic as $\eta \to 0$, in contrast to the first term. This choice is deliberate in order to avoid complicated higher-order terms in the wall-layer expansion. If $\partial F_2/\partial \eta$ were logarithmic, a term $O(u_\tau^2 / U_e)$ would be required in the wall-layer expansion (2.6). In the present formulation, the first and second terms in expansion (2.22) give rise to the entire logarithmic behavior of the wall-layer profile, that would otherwise require additional terms in the wall-layer expansion. It may be verified that the match condition up to second order now reads

$$\frac{U_e}{u_\tau} = \frac{1}{\kappa}\log(\text{Re}\,u_\tau\Delta_o) + C_i - C_o - \frac{u_\tau}{U_e}C_1 + \dots \tag{2.24}$$

Note that the leading-order expansions presented here are all consistent with past analyses, but there are significant differences in the higher-order terms which deserve consideration. In the conventional approach used by Mellor [6] and Melnik [7], the velocity is expanded in terms of a small parameter ε, which is independent of x and is $O(1/\log \text{Re})$ for $\text{Re} \to \infty$. Substitution of a power series in ε for the velocity ultimately leads to the conclusion that the total shear stress is constant to all orders in ε in the wall layer and, consequently, that the pressure gradient cannot influence the wall layer. However, difficulties exist with this type of expansion procedure in that it: (1) is not capable of including the influence of pressure gradient near the surface, and (2) can require the introduction of considerably more complex expansions in the wall layer. The present approach is believed to be preferable, wherein both the x and Re dependence is contained implicitly in the expansion parameters u_τ and Δ_o.

2.4 Self-Similar Flows

Self-similar motion constitutes an important class of flows which have received considerable experimental and theoretical attention. In such cases $\partial F_1/\partial x = 0$ and α and β in equations (2.13) and (2.15) must be constant to leading order. It is easily shown that for the quantity σ defined in equations (2.13),

$$\frac{d\sigma}{dx} = \alpha - \frac{(\beta - \alpha)}{\kappa}\frac{u_\tau}{U_e} + O\left(\frac{u_\tau^2}{U_e^2}\right), \tag{2.25}$$

where equation (2.14) with $C_o' = 0$ has been used. Equation (2.15) reduces to

$$\frac{dT_1}{d\eta} - (\beta - \alpha)\eta \frac{d^2 F_1}{d\eta^2} + 2\beta \frac{dF_1}{d\eta} = 0, \tag{2.26}$$

and integrating from 0 to η yields

$$T_1 - 1 - (\beta - \alpha)\left\{\eta \frac{dF_1}{d\eta} - F_1\right\} + 2\beta F_1 = 0, \tag{2.27}$$

where condition (2.5) has been used, as well as the conditions $T_1 \to 1$, $F_1 \to 0$ as $\eta \to 0$. Since T_1, $\partial F_1/\partial \eta \to 0$ as $\eta \to \infty$, equation (2.27) defines a relation between α, β and $F_1(\infty)$. A convenient normalization is to pick the constant $F_1(\infty) = -1$, and in equation (2.27) this gives

$$\alpha = 3\beta + 1. \tag{2.28}$$

To leading order, the integral of equation (2.25) gives

$$\sigma = \begin{cases} \alpha(x - x_o) & \text{for } \alpha \neq 0, \\ D_o & \text{for } \alpha = 0, \end{cases} \tag{2.29}$$

where x_o and D_o are arbitrary constants; here x_o may be identified as a virtual origin for the turbulent boundary layer. Since $U_e'/U_e = -\beta/\sigma$, it is easily shown that

$$U_e(x) = \begin{cases} U_1(x - x_o)^{-\beta/(1+3\beta)} & \text{for } \alpha \neq 0 \\ U_1 e^{-x/3D_o} & \text{for } \alpha = 0, \end{cases} \tag{2.30}$$

where U_1 is a constant; these are the only types of external flows which can lead to a self-similar behavior in the boundary layer.

The dependence of skin friction and the outer-layer thickness can be represented in terms of Reynolds number in a variety of ways. First define a Reynolds number based on distance along the surface by $Re_x = Re(x - x_o)U_e$, and then the match condition (2.8) becomes

$$\frac{U_e}{u_\tau} = \frac{1}{\kappa}\log\left\{\alpha\,Re_x\left(\frac{u_\tau}{U_e}\right)^2\right\} + C_i - C_o, \tag{2.31}$$

for $\alpha \neq 0$. Consequently, for large Re_x

$$\frac{u_\tau}{U_e} \sim \frac{\kappa}{\log Re_x}\left\{1 + \frac{\log\log Re_x + E_o}{\log Re_x} + \ldots\right\}, \tag{2.32}$$

where $E_o = -\log(\alpha \kappa^2) + C_o - C_i$, and in addition the outer-layer thickness function has

$$\Delta_o = \alpha(x - x_o)\frac{u_\tau^2}{U_e^2}. \tag{2.33}$$

The logarithmic variation in x is rather slow, and the behavior indicated in equation (2.33) shows the almost linear growth that is characteristic of most empirical correlations for the turbulent boundary-layer thickness.

A Reynolds number can also be defined in terms of the dimensionless displacement thickness by

$$\delta^* = \int_0^\infty \left(1 - \frac{u}{U_e}\right) dy . \tag{2.34}$$

The contribution due to the wall layer is of relatively lower order $O(Re^{-1} u_\tau)$, and substitution of the outer-layer expansion (2.22) leads to

$$\delta^* = \frac{\Delta_o u_\tau}{U_e} \left\{ -F_1(\infty) - \frac{u_\tau}{U_e} F_2(\infty) + ... \right\}, \tag{2.35}$$

and with $F_1(\infty) = -1$ and $F_2(\infty) = 0$, a convenient choice for the outer length scale is

$$\Delta_o = \delta^* U_e / u_\tau . \tag{2.36}$$

Defining a Reynolds number based on displacement thickness by $Re_{\delta^*} = Re \delta^* U_e$, the match condition (2.8) can be written in a simple form

$$\frac{U_e}{u_\tau} = \frac{1}{\kappa} \log Re_{\delta^*} + C_i - C_o, \tag{2.37}$$

and consequently

$$\frac{u_\tau}{U_e} \sim \frac{\kappa}{\log Re_{\delta^*}} \left\{ 1 - \frac{C_i - C_o}{\log Re_{\delta^*}} + ... \right\}. \tag{2.38}$$

It is worthwhile to note that the first term in this expansion gives a reasonably good estimate of u_τ at least for large Re_{δ^*}, as may be readily verified by comparison with experimental data.

2.5 Second-Order Outer Problem

The outer-layer equations for the profile functions $\partial F_2 / \partial \eta$ can be obtained by substituting the expansions (2.20) and (2.22) into equation (2.2) and equating coefficients of (U_τ / U_e). Since the parameter α defined in equations (2.13) involves a gradient of the outer scale Δ_o, it is necessary to expand this as well by writing

$$\alpha = \alpha_o + \frac{u_\tau}{U_e} \alpha_1 + ... \tag{2.39}$$

For self-similar flow, terms involving gradients with respect to x are set equal to zero. To leading order, equation (2.26) is obtained with $\alpha = \alpha_o$, and throughout §2.4, α must be replaced by α_o. The second-order equation follows after some algebra and is

$$\frac{dT_2}{d\eta} + (\alpha_o - \beta)\eta\frac{d^2F_2}{d\eta^2} + 2\beta\frac{dF_2}{d\eta} = \frac{1}{\kappa}(\beta - \alpha_o)\frac{dF_1}{d\eta} - \beta\left(\frac{dF_1}{d\eta}\right)^2$$

(2.40)

$$-\alpha_1\eta\frac{d^2F_1}{d\eta^2} + (\beta - \alpha_o)F_1\frac{d^2F_1}{d\eta^2}.$$

The constant α_1 may be determined in terms of the leading-order solution by integrating equation (2.40) across the boundary layer using equations (2.21) and (2.23) to obtain

$$\alpha_1 = \frac{1}{\kappa}(2\beta + 1) + (\beta + 1)\int_0^\infty \left(\frac{\partial F_1}{\partial \eta}\right)^2 d\eta$$

(2.41)

The second-order problem is to be solved subject to the second of conditions (2.23) and $dF_2/d\eta \to 0$ an $\eta \to \infty$. Numerical solutions [8] for an analogous, but more complicated, set of equations for a three-dimensional boundary layer show that the second-order problem as defined here is self-consistent. However, at least for large values of Re, the corrections to the first-order solutions (for u_τ, for example) are generally relatively small.

2.6 Turbulence Models

The asymptotic analysis outlined in §2.2 has shown that the profile in the wall layer is known as a function of y^+ for a specified turbulence model; for attached turbulent boundary layers, the dependence of U^+ on x is at most parametric. Discussion of specific wall-layer models will be deferred until §3; here the question of appropriate outer-layer models will be addressed. The most common and simplest type of model is an eddy viscosity formulation

$$\tau = \varepsilon\frac{\partial u}{\partial y},$$

(2.42)

for the total stress. Here ε constitutes a total eddy viscosity. Note that the form of equation (2.42) is somewhat different from conventional approaches wherein only the turbulence term $-u'v'$ is modeled using an eddy viscosity coefficient. The two approaches are clearly closely related, but the present method is much more convenient, especially in terms of ensuring a match of total stress within the overlap zone. The stress terms T_1 and T_2 may now be written

$$T_1 = \frac{\varepsilon}{\Delta_o u_\tau}\frac{\partial^2 F_1}{\partial \eta^2}, \qquad T_2 = \frac{\varepsilon}{\Delta_o u_\tau}\frac{\partial^2 F_2}{\partial \eta^2}.$$

(2.43)

A variety of different functions have been used over the years, and here only the simplest will be considered. In a prediction method (see §6), it may be convenient to adopt a different definition of the outer scale than equation (2.36) and, consequently, a specific choice for Δ_o is not made here. An outer eddy viscosity function may be defined by $\varepsilon = U_e \delta^* \hat{\varepsilon}$, where $\hat{\varepsilon}$ is the simple ramp function

$$\hat{\varepsilon} = \begin{cases} \chi & \text{for } \eta > \eta_m, \\ \kappa\eta/\eta_* & \text{for } \eta < \eta_m, \end{cases} \tag{2.44}$$

and

$$\eta_* = U_e\delta^*/(u_\tau\Delta_o), \qquad \eta_m = \chi\eta_*/\kappa . \tag{2.45}$$

Note that for the outer scale Δ_o defined in connection with the similarity solutions by equation (2.36), $\eta_* = 1$. The form of the function (2.44) is dictated by two considerations. The first concerns the outer-region behavior for large η, which must consist of exponential decay to zero; the nature of the profile for large η is determined by the choice of χ. Secondly, as $\eta \to 0$, $T_1 \to 1$, and the requirement that $\partial F_1/\partial\eta$ behave according to equation (2.5) necessitates the linear behavior in equation (2.44) for small η and determines η_* for a specific choice of Δ_o.

The Cebeci-Smith model [9] has been a popular model for turbulent boundary layers, and for this model $\chi = K = 0.0168$, a fixed constant; the model is a variant of one introduced by Mellor and Gibson [10] who used $K = 0.016$. Another similar model is the Baldwin-Lomax [11] model, which was developed because of difficulties that are inherent for certain flow configurations in defining a displacement thickness. This model can also be represented in the form (2.44) with

$$\chi = K\, C_{cp}\, y_{max}\, F_{max}/U_e\delta^* . \tag{2.46}$$

Here C_{cp} is a constant which Baldwin & Lomax [11] suggested should have a value of around 1.6 in order that the method produce the same results as the Cebeci-Smith method for constant pressure transonic boundary layers. Recently, it has been shown [12] that the two models are essentially equivalent for constant pressure boundary layers if

$$C_{cp} = \exp\left(\frac{1}{2}\left(1 + \frac{K}{\kappa^2}\right)\right) \approx 1.74 . \tag{2.47}$$

Here the function F is defined by

$$F = y|\partial u/\partial y|D, \qquad D = 1 - \exp(-y^+/A^+), \tag{2.48}$$

where D is the van Driest damping factor [9], [11] and A^+ is a constant usually assigned a value of 26; in equation (2.46), F_{max} denotes the value of F at the maximum and y_{max} is the corresponding value of y. There may be several local maxima for a given profile, and the relevant one is usually well within the outer part of the boundary layer [12].

The two simplified models given here describe the essence of each model for the outer layer. To deal with the inner layer, a common procedure [9] is to introduce a mixing length formulation close to the wall according to

$$\varepsilon = (\kappa y D)^2 |\partial u/\partial y|. \tag{2.49}$$

An alternative for the inner layer will be considered in §3. Many modifications [9] of the basic model have been made over the years which either have a negligible or an uncertain effect on the computed velocity profile. These include introducing a so-called intermittency factor multiplying the basic ramp function in equation (2.44) and inserting correlations in A^+ to account for various effects such as pressure gradients and transpiration [9], [11]. By and large, these modifications should be regarded as ad hoc correlations (having uncertain but usually small effects on the computed profile for u) and will not be considered further.

2.7 Self-Similar Profiles

For self-similar profiles, a convenient choice for the outer scale is given by equation (2.36) and, using the turbulence model (2.44), equation (2.26) governing the outer profile becomes

$$(\hat{\varepsilon} F_1'')' + (2\beta + 1)\eta F_1'' + 2\beta F_1' = 0, \tag{2.50}$$

subject to the conditions

$$F_1 \to 0, \quad F_1' \sim \frac{1}{\kappa}\log\eta + C_o \quad \text{as} \quad \eta \to 0, \tag{2.51}$$

$$F_1 \to -1, \quad F_1' \to 0 \quad \text{as} \quad \eta \to \infty, \tag{2.52}$$

where C_o is a function of

$$\beta = -\frac{\Delta_o U_e'}{u_\tau} = -\frac{\delta^* U_e U_e'}{u_\tau^2}, \tag{2.53}$$

which is usually referred to as the Clauser pressure gradient parameter [5]. Note that for this choice of Δ_o, $\eta_* = 1$ in equation (2.44).

For constant pressure flow $\beta = 0$ and a simple solution of equation (2.50) may be obtained in closed form with

$$\frac{dF_1}{d\eta} = \begin{cases} -\sqrt{\dfrac{\pi}{2\chi}} e^{-\chi/2\kappa^2} \text{erfc}\left\{\sqrt{\dfrac{1}{2\chi}}\eta\right\}, & \eta > \eta_m, \\[3mm] -\dfrac{1}{\kappa} E_1\left(\dfrac{\eta}{\kappa}\right) + C_o - \dfrac{1}{\kappa}\left(\gamma_o - \log\kappa\right), & \eta \le \eta_m, \end{cases} \tag{2.54}$$

where C_o is the outer log-law constant given by

$$C_o = \frac{1}{\kappa}\left\{\gamma_o - \log\kappa + E_1(\chi/\kappa^2) - \sqrt{\frac{\pi}{2\chi}}e^{-\chi/2\kappa^2}\mathrm{erfc}\left\{\sqrt{\frac{\chi}{2\kappa^2}}\right\}\right\}. \qquad (2.55)$$

Here erfc and E_1 denote the complementary error function and the exponential integral respectively, and $\gamma_o = 0.5772\ldots$ is Euler's constant.

For flows with pressure gradient ($\beta \neq 0$), an analytic solution of equation (2.50) may also be obtained, but a numerical solution is usually more convenient. For small η, a series solution of equation (2.50) is easily found in the form

$$F_1' = \sum_{n=0}^{\infty}b_n\eta^n + \left\{\sum_{n=0}^{\infty}a_n\eta^n\right\}\log\eta. \qquad (2.56)$$

To satisfy conditions (2.51), $a_o = 1/\kappa$ and $b_o = C_o$ and the coefficients satisfy recursion relations [13]

$$a_{n+1} = -\frac{\{2\beta + n(1+2\beta)\}a_n}{\kappa(n+1)^2}, \qquad (2.57)$$

$$b_{n+1} = -\frac{1}{\kappa(n+1)^2}\left\{\frac{a_n}{n+1}(1-2\beta-n(1+2\beta)) + (2\beta+n(1+2\beta)b_n)\right\}. \qquad (2.58)$$

The simplest way to calculate a numerical solution of equation (2.50) is as follows. For a given guess for C_o, values of F_1' and F_1'' may be evaluated at η_m; using the calculated value of $F_1'(\eta_m)$, a numerical solution of equation (2.50) may then be obtained in $\eta_m \leq \eta < \infty$ which satisfies $F_1' \to 0$ as $\eta \to \infty$. In general, the series and numerical solutions will not have a continuous second derivative at η_m. To refine the solution, a second value for C_o is guessed, and a second solution produced in an analogous manner. The two solutions are multiplied each by constants A_1 and A_2 such that $A_2 = 1 - A_1$ (so the second of conditions (2.51) is satisfied) and A_1 is selected so that the composite numerical and series solution has a continuous value of F_1'' at η_m. The solution obtained in this manner is continuous in all derivatives at $\eta = \eta_m$.

2.8 Summary

Models for the inner layer will be described in §3 where an explicit expression for the wall-layer profile will be given. If the outer-layer profile is known, either from equation (2.54) or a numerical solution of equation (2.50), a composite profile across the entire boundary layer may be formed from

$$u = U_e(x) + u_\tau\left\{\frac{\partial F_1}{\partial\eta} + U^+(y^+) - \frac{1}{\kappa}\log y^+ - C_i\right\}. \qquad (2.59)$$

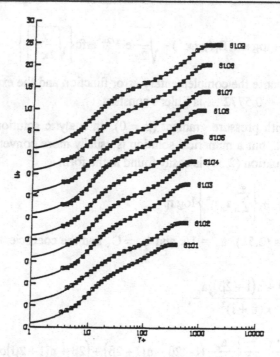

(a) Constant pressure data [15]

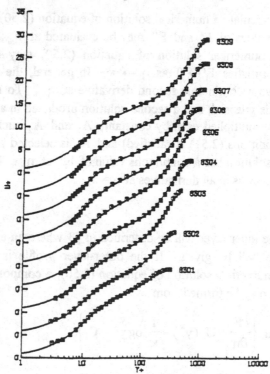

(b) Strong adverse pressure gradient [15]

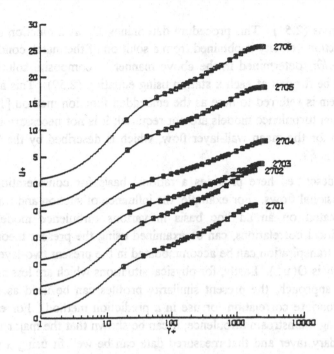

(c) Favorable pressure gradient of Herring and Norbury [16]

Figure 2.1 Comparison of composite theoretical profile with data in near self-similar flows

Some typical examples of how the profiles compare with data are shown in Figures 2.1 for flows developing in a zero, favorable, and adverse pressure gradient. In the computed profiles, values of κ = 0.41, K = 0.0168 and C_i = 5.0 have been used. To compare the results with experimental data, the quoted value of the displacement thickness δ^* was used, and the measured pressure gradients were the only other empirical inputs. The experimental value of δ^* (or some equivalent parameter) is necessary to define the local vertical scale of the boundary layer. It may be observed that the comparison with experiment is quite good and, although the experiments shown in these figures were carried out for self-similar flows, it emerges that reasonable agreement is also obtained with data in situations which are not self-similar. Thus one use of the profiles described here is to represent turbulent boundary-layer profiles at any station where the displacement thickness and pressure gradient are known or can be estimated.

In addition, the similarity profiles can also be used to initiate a full numerical integration of equation (2.15) in the x-direction. Because the approach to prediction in two-dimensions is similar to that in three dimensions, discussion of this aspect is deferred to §6. However, the main idea is that it is only necessary to calculate a numerical solution in the outer region of the boundary layer in a marching procedure which continues in the downstream direction, with the outer profile matched to the asymptotic condition given by the second of equations (2.51). This procedure determines C_o as a function of x, and at any station x, the friction velocity is obtained from a solution of the match condition (2.8). With the profile $\partial F_1/\partial \eta$ determined in the above manner, a composite solution for the velocity profile may be formed at each x station using equation (2.59). This approach to the prediction problem is referred to here as the embedded function method [14] because additional inner-region turbulence models are not required; it is not necessary to compute a numerical solution for the mean wall-layer flow, which is described by the "embedded functions" discussed in §3.

The theory described here provides a rational basis for consideration of other effects in two-dimensional flows. For example, the influence of suction and transpiration, which are often treated on an ad hoc basis in various turbulence models through introduction of empirical correlations, can be examined using the present theory. It may easily be shown that transpiration can be accommodated in the present two-layer structure if the blowing velocity is $O(u_\tau^2)$. Lastly, for physical situations which are less amenable to a rational modeling approach, the present similarity profiles can be used as a basis for determining an appropriate correlation for use in a prediction method. For example, in boundary layers having mainstream turbulence, it can be shown that the major influence is to thicken the boundary layer and that measured data can be well fit using a value of K which is correlated on the mainstream turbulence intensity [13].

References

1.   Schlichting, H. L.: Boundary Layer Theory, McGraw-Hill, 6th Edition (1968).

2.   Hinze, J. O.: Turbulence, McGraw-Hill, 2nd Edition (1975).

3.   White, F. M.: Viscous Flow Theory, McGraw-Hill, 2nd Edition (1996).

4.   Barenblatt, G. I., Chorin, A. J. and Prostokishin, V. M.: Scaling laws for fully-developed turbulent flow in pipes: Discussion of experimental data, Proc. Nat. Acad. Sci. U.S.A., **94** (1997), 773-776.

5.   Fendell, F. E.: Singular perturbation and turbulent shear flow near walls, J. Astro. Sciences **20** (1972), 129-165.

6.   Mellor, G. L.: The large Reynolds number, asymptotic theory of turbulent boundary layers, Int. J. Engng. Sci. **10** (1972), 851-873.

7.   Melnik, R. E.: Turbulent interactions on airfoils at transonic speeds – recent developments, Computation of Viscous-Inviscid Interactions, AGARD-CP-291 (1981), Chapter 10.

8.   Degani, A. T., Smith, F. T. and Walker, J. D. A.: The structure of a three-dimensional turbulent boundary layer, J. Fluid Mech. **250** (1993), 43-68.

9. Cebeci, T. and Smith, A. M. O.: Analysis of Turbulent Boundary Layers, Academic Press (1974).

10. Mellor, G. L. and Gibson, D. M.: Equilibrium turbulent boundary layers, J. Fluid Mech. 24 (1966), 225-253.

11. Baldwin, B. S. and Lomax, H.: Thin layer approximation and algebraic model for turbulent separated flows, AIAA Paper 78-257 (1978).

12. He, J. and Walker, J. D. A.: A note on the Baldwin-Lomax turbulence model, Trans. ASME, J. Fluids Engng. 117 (1995), 528-531.

13. Yuhas, L. J. and Walker, J. D. A.: An optimization technique for the development of two-dimensional steady turbulent boundary layer models, Rep. FM-1, Dept. Mech. Engrg., Lehigh U. (1982); also AFOSR-TR-0417.

14. Walker, J. D. A., Ece, M. C. and Werle, M. J.: An embedded function approach for turbulent flow prediction, AIAA J. 29 (1991), 1810-1818.

15. Anderson, P. S. , Kays, W. M. And Moffatt, R. J.: The turbulent boundary layer on a porous plate: An experimental study of the fluid mechanics for adverse free-stream pressure gradients (1972), Mech. Engng. Rpt. NMT-15, Stanford University.

16. Kline, S. J., Morkovin, M. V., Sovran, G. and Cockrell, P. S.: Proceedings, Computation of turbulent boundary layers - 1968, AFOSR-IFP-Stanford Conference, Dept. Mech. Engng., Stanford University (1969).

3. Wall Layer Models

3.1 Introduction

As shown in §2, the convective terms in the mean wall-layer equations are negligible to leading order, with the primary balance occurring between the Reynolds stress and the viscous term. Therefore, in order to provide closure in the wall layer, it is necessary to model either the velocity profile or the Reynolds stress. Here, a somewhat unconventional model, namely that for the mean profile, will be described. This model applies to attached turbulent shear flows near a wall and is based on the observed dynamics of the time-dependent wall-layer flow. It may be noted that the wall layer has traditionally been represented in prediction methods using a mixing-length model based on an analysis carried out in 1956 by Van Driest [1]. For incompressible flow, this model is of the form

$$\sigma = -\overline{u'v'} = u_\tau^2 \, \ell^2 \, D^2 \left(\frac{\partial U^+}{\partial y^+} \right)^2 \tag{3.1}$$

where $D = 1 - \exp(-y^+/A^+)$ is the damping factor, A^+ is a constant (usually taken to be 26) and the mixing length $\ell = \kappa y^+$. Over the years, a large number of ad hoc corrections have been added to this formulation to account for a variety of effects, such as transpiration and compressibility [2]. However, modern studies of the dynamics of the turbulent boundary layer cast serious doubt on the validity of near-wall mixing length models and also show that the physical picture of the time-dependent flow described by Van Driest [1] is incorrect.

Detailed dynamics of turbulent near-wall flows will be reviewed in §5, where it will be argued that lasting contributions to the Reynolds stress occur in sharp short bursts in a physical process that is extremely complex and formidable to model. Here and instead, the objective is to construct the particular motions which average to produce contributions to the mean quantities, which have orders of magnitude compatible with established asymptotic results and experimental observations. The net result is an analytic expression for the mean profile in the wall layer (as well as the Reynolds stress), which can be utilized in a general boundary-layer prediction method.

3.2 Background

Since the original observations [3] of the cyclic behavior of turbulent near-wall flows, a large number of experimental studies have clearly established the nature of the processes involved (see, for example, [4]). Although some issues regarding cause and effect relationships remain controversial, many aspects of the processes involved are well established. There are two main features which dominate the near-wall flow, namely the wall-layer streaks and the bursting phenomenon. Consider a fixed area of the wall for which the wall layer will be observed to be in a "quiescent state" [3] for a large majority of the time. During the quiescent period, wall-layer streaks will be seen when a visualization medium such as hydrogen bubbles or dye is injected into the wall-layer flow. The streaks are separated by an average spanwise distance λ, and measurements over a range of Reynolds numbers [4] suggest that $\lambda^+ = \text{Re } u_\tau \lambda \cong 100$. The streaks tend to be elongated in the streamwise direction and typically may have a length of 6λ to 10λ. The streamwise velocity in the vicinity of the streaks is generally below the local value of the mean profile, and hence the terminology "low-speed streaks" is commonly used. The streaks are generally understood [4] to be caused by the motion of convected hairpin vortices just outside the viscous wall layer.

The quiescent period locally is observed to initiate with a "sweep event" which is characterized by a penetration of relatively high speed fluid from the outer portion of the boundary layer toward the wall [5]. In the latter stages of the sweep event, the wall-layer streaks appear. At this stage, the wall layer is very thin locally. There then ensues a relatively long quiescent period locally in which the low-speed streaks may be observed, the flow is relatively well-ordered, strong interactions with the outer region do not occur, and the wall layer thickens continuously due to viscous diffusion. In this state, the wall layer may be regarded as a thin viscous region containing a developing unsteady flow, driven by the pressure field at the base of the outer region.

The quiescent period usually terminates locally at isolated spanwise and streamwise locations in an event known as the bursting process. This event invariably initiates near a streak and is characterized by an abrupt, highly localized and violent eruption of wall-layer fluid. Such a process may be regarded as an unsteady viscous-inviscid interaction between the wall layer and the outer layer or, equivalently, as a localized breakdown of the wall-layer flow. During the relatively brief period of breakdown, the two regions interact strongly, and the double-structure of the boundary layer is locally destroyed, as concentrated vorticity from the wall layer is ejected into the outer region. A double structure is quickly restored in the latter stages of the burst as high speed fluid penetrates close to the wall, thereby initiating a new quiescent period.

The cause of the bursting process remains controversial in some quarters. It was originally attributed to some sort of instability phenomenon [3], and this has been a common theme of the modern literature (see, for example, [6]). On the other hand, there is persuasive evidence [4] that the process is provoked not by small disturbances, but by the action of convected hairpin vortices above the wall layer. In any event, if T_q denotes the average duration of the quiescent period for a fixed area of observation of the wall, and T_e denotes the average duration of the bursting process, it is clear from experiments that $T_q \gg T_e$, and hence the average period between bursts $T_b \sim T_q$. Although the time of breakdown T_e is relatively short, the large majority of significant Reynolds stress production (i.e., contributions to $-\overline{u'v'}$) occurs during these events. If direct modeling of $\overline{u'v'}$ is contemplated, it is therefore necessary to analyze a typical eruption of the wall layer, and since this is a strong, three-dimensional, complex unsteady process, the modeling task is formidable; indeed modern progress in this direction has been essentially non-existent. On the other hand, the vast majority of contributions to the mean profile (that would be recorded by a probe, for example) are made during the quiescent period when the time-dependent flow is relatively well-ordered and the modeling problem is more feasible. Here a model for the mean profile will be considered.

3.3 Formulation

Consider a nominally steady two-dimensional turbulent boundary layer and define dimensionless variables with respect to a typical velocity U_o and length L. If $u_\tau(x)$ denotes the local mean dimensionless friction velocity, it is evident from experimental measurements of turbulence intensities that all three velocity components are $O(u_\tau)$ in the instantaneous flow in the wall layer. During a typical quiescent period, the wall layer is well-defined and the characteristic length in the spanwise direction is the mean streak spacing λ. Let (x,y,z) denote Cartesian coordinates in the streamwise, normal and spanwise directions respectively, with corresponding velocity components (u,v,w). Define scaled quantities in the spanwise direction by

$$Z = Re\, u_\tau z/\lambda^+ = z^+/\lambda^+, \quad W = w/(u_\tau\, W_1) = w^+/W_1, \tag{3.2}$$

where $Re = U_o L/\nu$ is the Reynolds number and ν is the kinematic viscosity. Here $W_1 u_\tau$ is to be comparable to the average spanwise velocity at the outer edge of the wall layer during a typical quiescent period; turbulence intensity measurements for w'^2 for large y^+ suggest the constant W_1 is around 2. Scaled normal variables are defined by

$$Y = Re\, u_\tau y(W_1/\lambda^+)^{1/2} = y^+(W_1/\lambda^+)^{1/2}, \quad V = v(\lambda^+/W_1)^{1/2}/u_\tau = v^+(\lambda^+/W_1)^{1/2}, \tag{3.3}$$

which follow from balancing in the continuity equation, as well as the viscous and convective terms in the momentum equation. Scaled streamwise variables are defined by

$$X = x/L_x, \quad U = u/(U_1\, u_\tau) = u^+/U_1, \tag{3.4}$$

where L_x is a characteristic length in the x-direction; U_1 is a constant such that $U_1 u_\tau$ is a typical flow speed at the outer edge of the wall layer. Typical values for U_1 of 13 ~ 16 can be estimated from the logarithmic law for large $y^+ \cong 30 \sim 100$, and it is evident that U_1 is large with respect to W_1. Finally, the time scale follows from balancing the time derivatives with the viscous terms

$$T = Re\, u_\tau^2 W_1 t/\lambda^+ = t^+ W_1/\lambda^+, \tag{3.5}$$

and the pressure is expanded as

$$p = p_\infty(x) + \rho u_\tau^2 U_1^2 p_1 + \rho u_\tau^2 U_1 W_1 p_2/\lambda^+, \tag{3.6}$$

where $p_\infty(x)$ is the local mainstream pressure and ρ is the density. Upon substitution of these scalings into the three-dimensional Navier-Stokes equations, it is found [5] that two Reynolds numbers and a dimensionless parameter γ appear, defined by

$$\text{Re}_x = u_\tau \, \text{Re} \, L_x, \quad \text{Re}_\lambda = W_1 \lambda^+, \quad \gamma = \lambda^+ U_1 / W_1. \tag{3.7}$$

Situations where $\text{Re}_x \to \infty$ are of interest here.

As discussed in [5], at least two cases are relevant to the turbulent boundary layer, namely (i) $\gamma / \text{Re}_x \to 0$ and (ii) $\gamma / \text{Re}_x = O(1)$ in the limit $\text{Re}_x \to \infty$. In the second case, it is easily shown that the governing wall-layer equations are of the boundary-layer type and reduce to the classical three-dimensional unsteady laminar boundary-layer equations in the limit $\text{Re}_\lambda \to \infty$. Such equations describe a developing unsteady viscous flow induced by the pressure field due to some compact disturbance convected above the wall layer. Solutions for this type of problem are complex and can only be found numerically, but an important consequence is that disturbances which can provoke separation and the eruption of a laminar boundary layer will have the identical effect on the turbulent wall layer.

The first relevant limit corresponds to situations where $(U_1 \lambda)/(L_x W_1) \ll 1$ and, consequently, implies motion which is of relatively long extent with respect to the mean streak spacing. When the governing equations are written in the conventional "plus" variables, they are of the form (in the limit $\text{Re}_x \to \infty$)

$$\frac{\partial u^+}{\partial t^+} + v^+ \frac{\partial u^+}{\partial y^+} + w^+ \frac{\partial u^+}{\partial z^+} = -p^+ + \frac{\partial^2 u^+}{\partial y^{+2}} + \frac{\partial^2 u^+}{\partial z^{+2}}, \tag{3.8}$$

$$\frac{\partial v^+}{\partial t^+} + v^+ \frac{\partial v^+}{\partial y^+} + w^+ \frac{\partial v^+}{\partial z^+} = -\frac{\partial p_1}{\partial y^+} + \frac{\partial^2 v^+}{\partial y^{+2}} + \frac{\partial^2 v^+}{\partial z^{+2}}, \tag{3.9}$$

$$\frac{\partial w^+}{\partial t^+} + v^+ \frac{\partial w^+}{\partial y^+} + w^+ \frac{\partial w^+}{\partial z^+} = -\frac{\partial p_1}{\partial z^+} + \frac{\partial^2 w^+}{\partial y^{+2}} + \frac{\partial^2 w^+}{\partial z^{+2}}, \tag{3.10}$$

$$\frac{\partial v^+}{\partial y^+} + \frac{\partial w^+}{\partial z^+} = 0. \tag{3.11}$$

Here $p^+ = -U_e U_e'(x)/\text{Re} u_\tau^3$, where U_e is the mainstream velocity, and p_1 is a pressure function to be determined. The presence of the wall-layer streaks during a typical quiescent period indicates that there are planes at certain spanwise locations across which there is no spanwise flow to leading order. Therefore, to consider the motion between a typical pair of streaks located at $z^+ = 0$ and $z^+ = \lambda^+$, it is possible to write

$$w^+(y^+, z^+, t^+) = -\sum_{n=1}^\infty \frac{\partial f_n}{\partial y^+} \sin\left\{ \frac{2n\pi z^+}{\lambda^+} \right\}, \tag{3.12}$$

and from the continuity equation (3.11), it follows that v^+ has a Fourier representation as a cosine series. The functional coefficients of these series have $f_n = \partial f_n /\partial y^+ = 0$ at $y^+ = 0$ and far from the wall a simple condition consistent with turbulence intensity measurements is

$$\frac{\partial f_1}{\partial y^+} \to W_1, \quad \frac{\partial f_n}{\partial y^+} \to 0, \quad n \neq 1 \quad \text{as} \quad y^+ \to \infty. \tag{3.13}$$

The above equations are sometimes referred to as 2 ½ dimensional equations since the motion in the cross-flow plane (y^+z^+) develops independently from that in the streamwise direction, but the solution for v^+ and w^+ feeds into equation (3.8) for u^+. Similar equations have been considered by other authors (e.g. [6], [7], [8]), in some cases with different conditions for y^+ large.

There are two major influences on the development of the streamwise profile u^+; these are the nature of the boundary condition at large y^+, and the effect of the evolving flow in the cross-flow plane. Solutions for u^+ which are symmetric about $z^+ = 0$ have the general form

$$u^+ = u_o(y^+, t^+) + \sum_{n=1}^{\infty} u_n(y^+, t^+)\cos\left\{\frac{2n\pi z^+}{\lambda^+}\right\}, \tag{3.14}$$

and a set of differential equations for the u_n may be derived in terms of the f_n [9]. Full numerical solutions for the u_n have been obtained [7], [8], revealing interesting topological features in the field for u, as well as evolving inflectional points which may signal the possibility of the onset of instability. However, the motions represented in (3.12), and by the u_n for $n \geq 1$, will make no net contribution to the mean quantities for w and u; this is because following each wall-layer breakdown and concomitant sweep, the low-speed streaks appear at new spanwise locations in a distribution which is random over a large number of cycles. Lasting contributions to the mean profile are made by the component u_o in equation (3.14) and for large λ^+, the equation satisfied by this component is given [9] approximately by

$$\frac{\partial u_o}{\partial t^+} = -p^+ + \frac{\partial^2 u_o}{\partial y^{+2}}. \tag{3.15}$$

Solutions of this equation have been considered by various authors (see, for example [10]) and the resulting turbulence theories are sometimes referred to as "surface renewal" models.

Here the objective is to seek all possible solutions of equation (3.15), which when time-averaged over a representative cycle, will produce an expression for the mean profile U^+ satisfying the appropriate asymptotic condition (2.7) for y^+ large, as well as the correct conditions at $y^+ = 0$. General similarity solutions for u_o, as well as the u_n, have been considered [9]; these solutions constitute a complete set of eigenfunctions which, in principle, can represent any initial condition at the beginning of the quiescent state. There is, however, only one relevant solution that behaves logarithmically for large y^+, and this rapidly dominates all other eigenfunctions as t^+ increases. This solution has the form [9]

$$u_o = \left\{\frac{a_o}{4}\log\tau + A_o\right\}\text{erf}\eta + \frac{2a_o}{\sqrt{\pi}}\Xi(\eta) - p^+\tau\left\{(2\eta^2 + 1)\text{erf}\eta - 2\eta^2 + \frac{2}{\sqrt{\pi}}\eta e^{-\eta^2}\right\}, \quad (3.16)$$

where a_o and A_o are constants; here η and τ are similarity variables defined by

$$\eta = \frac{y^+}{2\sqrt{t^+ + t_o^+}}, \quad \tau = t^+ + t_o^+, \quad (3.17)$$

and t_o^+ is an arbitrary constant, corresponding to an uncertainty in the origin of time. In addition, the function Ξ is defined by

$$\Xi(\eta) = \int_0^\eta e^{-\xi^2}\int_0^\xi e^{t^2}\int_0^t e^{-x^2}dx\,dt\,d\xi, \quad (3.18)$$

and has the following behavior

$$\Xi(\eta) \sim \frac{\sqrt{\pi}}{4}\left\{\log\eta + \frac{\gamma_o}{2} - \frac{1}{2}\sum_{j=1}^\infty \frac{(2j-1)!!}{j2^j\eta^{2j}}\right\} \quad \text{as} \quad \eta \to \infty, \quad (3.19)$$

where $\gamma_o = 0.57211....$ is Euler's constant. For $\eta = 0(1)$, $\Xi(\eta)$ may be evaluated from

$$\Xi(\eta) = \frac{e^{-\eta^2}}{4}\sum_{j=1}^\infty \frac{2^j d_j \eta^{2j+1}}{(2j+1)!!}, \quad (3.20)$$

where $d_j = d_{j-1} + 1/j$ and $d_1 = 1$; the series (3.20) is uniformly convergent for all η, but is particularly useful to evaluate Ξ for small to moderate values of η.

3.4 The Time-Mean Profile

An approximation to the time-mean profile in the wall layer may be found by computing a time-average of u_o by

$$U^+ = \frac{1}{T_B}\int_0^{T_B} u_o dt, \quad (3.21)$$

where T_B denotes the average period between bursts, which corresponds approximately to the average duration of the quiescent state. The result of this calculation [9] is

$$U^+ = \left\{1 + \frac{t_o^+}{T_B^+}\right\}\left\{R(T_B^+)Q(\tilde{y}^+) + Z(\tilde{y}^+) + P(T_B^+)W(\tilde{y}^+)\right\}$$

$$- \frac{t_o^+}{T_B^+}\left\{R(0)Q(\tilde{y}_o^+) + Z(\tilde{y}_o^+) + P(0)W(\tilde{y}_o^+)\right\}, \quad (3.22)$$

where the variables $\tilde{y}^+$ and $\tilde{y}_o^+$ are defined by

$$\tilde{y}^+ = \frac{1}{2}y^+(T_B^+ + t_o^+)^{-1/2}, \quad \tilde{y}_o^+ = \frac{1}{2}y(t_o^+)^{-1/2}, \tag{3.23}$$

and the principal functions in (3.22) are

$$R(\xi) = A_o + (a_o/4)\log(\xi + t_o^+), \tag{3.24}$$

$$Q(y) = (2y^2 + 1)\text{erf}\,y + 2\pi^{-1/2}y\,e^{-y^2}, \tag{3.25}$$

$$Z(y) = 2\pi^{-1/2}a_o\{(2y^2 + 1)\,\Xi(y) + y\,\Xi'(y)$$

$$- (\sqrt{\pi}/8)(6y^2 + 1)\text{erf}\,y - (3/4)ye^{-y^2}\}. \tag{3.26}$$

The functions in the profile (3.22) associated with the mainstream pressure gradient are only significant when the parameter $p^+ = -U_e U_e'/(\text{Re}u_\tau^3)$ is $O(1)$ (i.e., a flow approaching time-mean separation) and are defined by

$$P(\xi) = -(2p^+/3)(\xi + t_o^+), \tag{3.27}$$

$$W(y) = (y^4 + 3y^2 + \tfrac{3}{4})\text{erf}\,y + \pi^{-1/2}y(y^2 + 5/2)\,e^{-y^2} - 3y^2. \tag{3.28}$$

The profile (3.22) contains the parameters t_o^+, T_B^+, a_o and A_o, which can be related to one another through known conditions that the mean profile must satisfy. In order to satisfy the condition (2.7) at large y^+, it is easily shown that

$$a_o = \frac{2}{\kappa}, \quad C_i = A_o + \frac{1}{\kappa}\left(\frac{\gamma_o}{2} - \log 2\right) - \frac{1}{2}p^+\left(T_B^+ + 2t_o^+\right). \tag{3.29}$$

In addition U^+ must satisfy the two compatibility conditions $\partial^2 U^+/\partial y^{+2} = p^+$, $\partial^3 U^+/\partial y^{+3} = 0$ at $y^+ = 0$, which follow from the time-mean streamwise momentum equation and its normal derivative evaluated at the wall. These conditions lead to two further relations

$$\left(T_B^+ + t_o^+\right)^{1/2}\left\{R(T_B^+) - \kappa^{-1} + P(T_B^+)\right\}$$

$$- \left(t_o^+\right)^{-1/2}\left\{R(0) - \kappa^{-1} + P(0)\right\} = \frac{1}{2}\sqrt{\pi}T_B^+, \tag{3.30}$$

$$\left(T_B^+ + t_o^+\right)^{-1/2}\left\{R(T_B^+) + 3P(T_B^+)\right\} - (t_o^+)^{-1/2}\left\{R(0) + 3P(0)\right\} = 0. \tag{3.31}$$

For given values of κ, C_i, values of a_o, A_o, T_B^+ and t_o^+ are readily found by solving the system (3.29) – (3.31) numerically. For example, for $\kappa = 0.41$ and $C_i = 5.0$, it is easily shown that $t_o^+ = 0.00801$, $T_B^+ = 110.2$; this value of T_B^+ is compatible with direct experimental measurements [9] of the burst period. Because T_B^+ is generally large with respect to t_o^+, an expansion of equations (3.30) and (3.31) yields the approximate relation

$$C_i \sim \frac{1}{2}\left(\pi T_B^+\right)^{1/2} + \frac{1}{\kappa}\left\{1 + \frac{\gamma_0}{2} - \frac{1}{2}\log(4T_B^+)\right\} + \frac{2}{3}p^+T_B^+ + ..., \qquad (3.32)$$

which can be used to initiate a refined numerical solution of equations (3.30) and (3.31). Fortran routines for the evaluation of U^+ are given in the Appendix.

3.5 Results

In Figure 3.1, calculated results for a time-mean profile for $\kappa = 0.41$, $C_i = 5$ are shown along with the instantaneous contributions (3.16) at various stages through the cycle. This type of instantaneous relaxing flow is commonly seen in experimental measurements during the quiescent period between bursts. For non-zero values of p^+, the instantaneous profiles are accelerated near the wall-layer edge for favorable pressure gradients and decelerated for adverse pressure gradients [9]. Note that the pressure gradient must be very large or u_τ very small (in an adverse pressure gradient) in order for p^+ to be $O(1)$. For values of $p^+ \cong 0.5$, a process of transitory stall is predicted, wherein flow reversal occurs at the wall part way through the cycle, even though the mean profile still shows attached flow. This phenomenon is typically observed upstream of a point of time-mean separation [11].

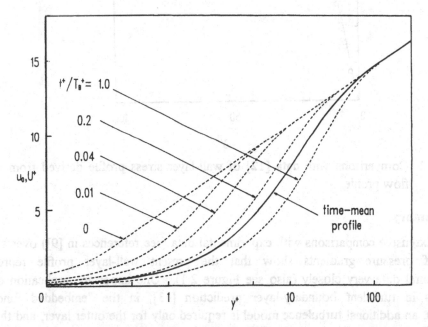

Figure 3.1 Time-mean wall-layer velocity profile and average instantaneous profiles for constant pressure flow.

As indicated previously, the present approach is somewhat unconventional in that the Reynolds stress is the quantity which is usually modeled. However in the wall layer, a knowledge of the mean profile implies a knowledge of Reynolds stress and vice versa. It is of interest here then to compute the Reynolds stress function, and from equation (2.11), it follows for p^+ small that

$$\sigma_1 = -\frac{\overline{u'v'}}{u_\tau^2} = 1 - \frac{\partial U^+}{\partial y^+}, \tag{3.33}$$

which may easily be evaluated using (3.22). Comparisons with some typical wall-layer data [12] are shown in Figure 3.2.

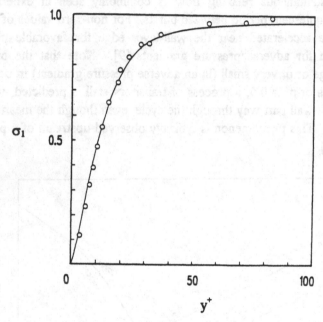

Figure 3.2 Comparisons with data [12] of wall-layer stress profile derived from mean-flow profile.

3.6 Summary

Extensive comparisons with experimental data (see references in [9]) over a wide range of pressure gradients show that the present wall-layer profile represents experimental data very closely (also see Figure 2.1). One important application of the model is in turbulent boundary-layer prediction [13]; in the "embedded function" approach, an additional turbulence model is required only for the outer layer, and there is no need to carry out a numerical solution in the wall layer. As a calculation proceeds downstream in the outer layer (cf. equation 2.15), the outer velocity must be matched to a logarithmic profile at the edge of the wall layer. The approach is very efficient and can result in a total savings of 50% in the number of mesh points used, as compared to a conventional scheme, using a Van Driest inner layer turbulence model, which continues a computation all the way to the wall. This point will be discussed further in §6.

References

1. Van Driest, E. R.: On turbulent flow near a wall, J. Aero. Sci. **23** (1956), 1007-1011, 1036.

2. Cebeci, T. and Smith, A. M. O.: Analysis of turbulent boundary layers, Academic Press, New York (1974).

3. Kline, S. J., Reynolds, W. C., Schraub, F. A. and Rundstadler, P. W.: The structure of turbulent boundary layers, J. Fluid Mech. **30** (1967), 741-773.

4. Smith, C. R., Walker, J. D. A., Haidari, A. H. and Sobrun, U.: On the dynamics of near-wall turbulence, Phil. Trans. R. Soc. Lond. A **336** (1991), 131-175.

5. Walker, J. D. A.: Wall layer eruptions in turbulent flows, Structure of Turbulence and Drag Reduction (A. Gyr, ed.), Springer-Verlag (1990), 109-118.

6. Chapman, D. R. and Kuhn, G. D.: The limiting behavior of turbulence near a wall, J. Fluid Mech. **170** (1986), 265-292.

7. Walker, J. D. A. and Herzog, S.: Eruption mechanism for turbulent flows near walls, Proc. 2nd Int. Symp. on Transport Phenomena in Turbulent Flows (M. Hirata and V. Kasagi, eds.) (1988), 145-156.

8. Brinckman, K. W. and Walker, J. D. A.: Breakup and instability in a high Reynolds number flow, AIAA Paper 96-2156 (1996).

9. Walker, J. D. A., Abbott, D. E., Scharnhorst, R. R. and Weigand, G. G.: Wall-layer model for the velocity profile in turbulent flows, AIAA J. **27** (1989), 140-149.

10. Black, T. J.: An analytical study of the measured pressure field under supersonic turbulent boundary layers, NASA CR-888 (1968).

11. Sandborn, V. A. and Liu, C. Y.: On turbulent boundary layer separation, J. Fluid Mech. **32** (1968), 293-304.

12. Schubauer, G. B.: Turbulent processes as observed in boundary layer and pipe, J. Appl. Physics **25** (1954), 188-196.

13. Walker, J. D. A., Ece, M. C. and Werle, M. J.: An embedded function approach for turbulent flow prediction, AIAA J. **29** (1991), 1810-1818.

APPENDIX

```
      IMPLICIT DOUBLE PRECISION (A-H,O-Z)
C   *** THIS PROGRAM IS A TEST CASE TO ILLUSTRATE THE USE OF THE
C   *** FUNCTIONS CIN AND UP TO CALCULATE THE VELOCITY PROFILE
C   *** IN THE WALL LAYER OF A TURBULENT FLOW
C
      DIMENSION Y(51)
C
C   *** THE CYCLE TIME PARAMETER TB, VON KARMAN CONSTANT XKAP, WALL
C   *** LAYER PRESSURE GRADIENT PARAMETER PPLUS MUST BE SUPPLIED.
C   *** FOR TEMPERATURE PROFILES IN THE WALL LAYER ROOTPR IS THE
C   *** SQUARE ROOT OF THE PRANDTL NUMBER; FOR VELOCITY ROOTPR =1.
C   *** FOR A CONSTANT PRESSURE FLOW: TB=110.2,XKAP=0.41,PPLUS=0.
C   *** AND ROOTPR=1. GIVE A STANDARD PROFILE WITH CIN = 5.0.
C
      TB = 110.2D0
      S=DSQRT(TB)
      XKAP=0.41D0
      PPLUS=0.D0
      ROOTPR=1.D0
C
C   *** DEFINE THE Y LOCATIONS WHERE THE PROFILE IS DESIRED.
C
      N0=11
      Y(1)=0.D0
      DO I=2,N0
      Y(I)=Y(I-1)+10.D0
      END DO
C
C   *** FOR DEFINED VALUES OF TB,XKAP,PPLUS AND ROOTPR A CALL IS MADE
C   *** TO CIN TO DETERMINE THE INNER LOG-LAW CONSTANT CI (AS WELL AS
C   *** THE SIMILARITY PARAMETER TNOTP).
C   *** IF A SPECIFIC VALUE OF CI IS DESIRED FOR WHICH EITHER TB OR
C   *** XKAP ARE CONSIDERED UNKNOWN, THIS MAY BE ESTABLISHED BY
C   *** USING ITERATIVE CALLS TO CIN.
C
      CIN=CINF(S,TNOTP,XKAP,PPLUS,ROOTPR)
C
      PRINT 40,TB,XKAP,PPLUS,TNOTP,ROOTPR,CIN
40    FORMAT(1X,' TB  = ',D15.6,5X,'KAPPA = ',D15.6,/,1X,'PPLUS = ',
     +    D15.6,5X,'TNOTP = ',D15.6,5X,'ROOTPR = ',D15.6,//,1X,
     +    ' CIN  = ',D15.6,///)
C
C   *** SUCCESSIVE CALLS TO FUNCTION UP FOR DEFINED VALUES OF S=SQRT(TB),
C   *** XKAP,TNOTP,ROOTPR AND YP (YPLUS) GIVE THE WALL LAYER VELOCITY
C   *** AT THE DESIRED LOCATIONS
C
      DO I=1,N0
      YP=Y(I)
      U=UP(YP,S,TNOTP,CIN,XKAP,PPLUS)
      PRINT 10,YP,U
      END DO
C
10    FORMAT(1X,2E15.6)
      STOP
      END
      DOUBLE PRECISION FUNCTION CINF(S,TNOTP,XKAP,PPLUS,ROOTPR)
C
C   *** FUNCTION CINF USES THE WALL COMPATABILITY CONDITIONS TO
C   *** CALCULATE THE INNER REGION CONSTANTS CINER AND TNOTP FOR A
C   *** GIVEN VALUE OF TB FOR BOTH THE TEMPERATURE AND THE VELOCITY
C   *** PROFILE IN THE INNER LAYER OF A TURBULENT BOUNDARY LAYER.
C   *** ROOTPR = 1. FOR THE VELOCITY PROFILE CASE.
```

```
C  *** PPLUS = 0. FOR THE TEMPERATURE PROFILE CASE;S=SQRT(TB)
C
   IMPLICIT DOUBLE PRECISION (A-H,O-Z)
   DATA GM,ROOTPI,EPSI,ITMX/-0.404539348109179D0,1.77245385090552D0
  +    ,1.D-10,20/
   JC=0
   X1=0.5D0*S*ROOTPR*ROOTPI*XKAP
   S2=S*S
   X2=2.D0*XKAP*S2*PPLUS/3.D0
   AL=DEXP(-1.D0-X1-X2)

   DO K=1,ITMX
    X3=DLOG(AL)
    X4=AL+1.D0
    AL2=AL*AL
    X5=DSQRT(1.D0-AL2)
    C=AL
    AL=AL-(X1*X5+X4*X3+1.D0-AL+X2*(AL2+4.D0*AL+1.D0)/X4)/(-AL*X1/X5+
  +    X3+1.D0/AL+X2*(AL2+2.D0*AL+3.D0)/(X4*X4))
    IF(DABS(AL/C-1.D0).LE.EPSI)GO TO 20
   END DO

   JC=1
20 AL2=AL*AL
   X4=1.D0-AL2
   TNOTP=AL2*S2/X4
   X5=DSQRT(X4)
   CINF=X1/X5+GM-0.5D0*DLOG(S2+TNOTP)+1.D0/(AL+1.D0)
  +    +(0.25D0*X2*(1.D0-9.D0*AL2)-PPLUS*TNOTP*XKAP*(3.D0*AL2-4.D0
  &    *AL+1.D0)/3.D0)/X4
   CINF=CINF/XKAP
   IF(JC.EQ.0)GO TO 30
   WRITE(6,40)K,TNOTP,CINF,AL,C
   STOP
C
40 FORMAT(1X, 28HNO CONVERGENCE IN CINF AFTER,I4,11H ITERATIONS,/,
  +    1X,7HTNOTP =,D15.5,10X,6HCINF =,D15.5,10X,4HAL =,D15.5,10X
  +    ,5HALP =,D15.5,/)
C
30 RETURN
   END
   DOUBLE PRECISION FUNCTION UP(YP,S,TNOTP,CIN,XKAP,PPLUS)
C
C  *** FUNCTION UP CALCULATES THE TIME-MEAN TEMPERATURE OR VELOCITY
C  *** PROFILE IN THE WALL LAYER OF A TURBULENT BOUNDARY LAYER
C  *** FOR SPECIFIED VALUES OF THE PARAMETERS LISTED BELOW:
C  ***    YP - Y+, SCALED WALL-LAYER COORDINATE
C  ***    S  - S, CYCLE TIME PARAMETER (SQRT OF BURST PERIOD TB)
C  ***    TNOTP - T0+,SIMILARITY PARAMETER
C  ***    CIN - INNER REGION LOG-LAW CONSTANT CI OR BI
C  ***    XKAP - KAPPA (VON KARMAN CONSTANT) OR KAPPA-THETA
C  ***    PPLUS - WALL LAYER PRESSURE GRADIENT PARAMETER.(PPLUS
C  ***         IS ZERO FOR CONSTANT PRESSURE OR TEMPERATURE)
C  *** NOTE: A CALL (OR CALLS) TO UP SHOULD NORMALLY BE PRECEDED BY
C  ***     A CALL TO CINF WHICH COMPUTES CIN AND TNOTP,GIVEN S.
C
   IMPLICIT DOUBLE PRECISION (A-H,O-Z)
   DATA X1,SRPI/-0.404539348109179D0,1.772453850905516D0/
   P(X)=-2.D0*PPLUS*(X+TNOTP)/3.D0
   R(X)=CIN+(0.5D0*DLOG(X+TNOTP)-X1)/XKAP+0.5D0*PPLUS*(S2
  +    +2.0D0*TNOTP)
   Q(X,Y,ZA)=(2.D0*X*X+1.D0)*Y+2.D0*X*ZA/SRPI
   Z(X,Y,ZA)=4.D0*((2.D0*X*X+1.D0)*XI(X)+X*XIP(X))
  +    -0.125D0*SRPI*(6.D0*X*X+1.D0)*Y-0.75D0*X*ZA)/(SRPI*XKAP)
   W(X,Y,ZA)=(X**4+3.D0*X*X+0.75D0)*Y+X*(X*X+2.5D0)*ZA/SRPI
  +    -3.D0*X*X
C
```

```
C    *** PRECIS IS THE VALUE OF X SUCH THAT EXP(-X*X) MAY BE COMPUTED
C    *** WITHOUT INCURRING AN UNDERFLOW; VALUE IS MACHINE DEPENDENT
C
C
     PRECIS=13.45D0
C
     S2=S*S
     TPS2=S2+TNOTP
     H=0.5D0*YP/DSQRT(S2+TNOTP)
     H0=0.5D0*YP/DSQRT(TNOTP)
     ERFH=DERF(H)
     ERFH0=DERF(H0)
     EXPH=0.D0
     EXPH0=0.D0
     IF(H.LT.PRECIS)EXPH=DEXP(-H*H)
     IF(H0.LT.PRECIS)EXPH0=DEXP(-H0*H0)
     UP=TPS2*(R(S2)*Q(H,ERFH,EXPH)+Z(H,ERFH,EXPH))
   +    -TNOTP*(R(0.D0)*Q(H0,ERFH0,EXPH0)+Z(H0,ERFH0,EXPH0))
     IF(PPLUS.EQ.0.D0)GO TO 10
     IF(H.LT.PRECIS)GO TO 20
     UP=UP-0.5D0*(S2*S2+2.D0*S2*TNOTP)*PPLUS
     GO TO 10
  20 IF(H0.LT.PRECIS) THEN
       WH=W(H,ERFH,EXPH)
       WH0=W(H0,ERFH0,EXPH)
     ELSE
       WH=W(H,ERFH,EXPH)
       WH0=H0**4+.75D0
     END IF
     UP=UP+TPS2*P(S2)*WH-TNOTP*P(0.D0)*WH0
  10 UP=UP/S2
     RETURN
     END
     DOUBLE PRECISION FUNCTION XI(X)
     IMPLICIT DOUBLE PRECISION (A-H,O-Z)
     DATA SRPI,GAM0,EPSI/1.77245385090552D0,0.5772156649015329D0,
   +    1.D-10/
C
C    *** FUNCTION XI EVALUATES THE TRIPLE INTEGRAL IN THE UNSTEADY
C    *** WALL LAYER MODEL. THE TOLERANCE EPSI IS THE NUMBER OF
C    *** SIGNIFICANT FIGURES DESIRED FOR XI AND ITS DERIVATIVE; THIS
C    *** VALUE IS NORMALLY MACHINE DEPENDENT; NON-ZERO VALUES ARE RETURNED
C    *** ONLY FOR POSITIVE ARGUMENTS, FOR NEGATIVE ARGUMENTS THE SYMMETRY
C    *** PROPERTIES OF XI AND XIP SHOULD BE USED PRIOR TO THE CALL TO THIS
C    *** ROUTINE
C
     XI=0.D0
     IF=1
     IF(X.LE.0.D0)RETURN
     GO TO 1
     ENTRY XIP
     XI=0.D0
     IF=2
     IF(X.LE.0.D0)RETURN
   1 X2=X*X
     FAC=2.D0*X2
     M=100
     SUM=0.D0
     SUMT=0.D0
     TERM=1.D0
     IF(X.GE.5.3D0)GO TO 110
     IF(IF.EQ.2)GO TO 100
     TERM=X
     ALPHA=1.D0
     DO I=1,M
      TERM=TERM*FAC/DFLOAT(2*I+1)
      SUM=SUM+TERM*ALPHA
```

```
      IF(DABS((SUM-SUMT)/SUM).LT.EPSI)GO TO 3
      ALPHA=ALPHA+1.D0/DFLOAT(I+1)
      SUMT=SUM
      END DO
 3    XI=0.25D0*DEXP(-X2)*SUM
      RETURN
100   DO I=1,M
      TERM=TERM*FAC/DFLOAT(2*I-1)
      SUM=SUM+TERM/DFLOAT(I)
      IF(DABS((SUM-SUMT)/SUM).LT.EPSI)GO TO 5
      SUMT=SUM
      END DO
 5    XI=0.25D0*DEXP(-X2)*SUM
      RETURN
110   IF(IF.EQ.2)GO TO 120
      DO I=1,M
      TERM=TERM*DFLOAT(2*I-1)/FAC
      TERMA=TERM/DFLOAT(I)
      IF(TERMA.LT.EPSI)GO TO 7
      SUM=SUM+TERMA
      END DO
 7    XI=SRPI*(DLOG(X2)+GAM0-SUM)/8.D0
      RETURN
120   DO I=1,M
      TERM=TERM*DFLOAT(2*I-1)/FAC
      IF(TERM.LT.EPSI)GO TO 9
      SUM=SUM+TERM
      END DO
 9    XI=SRPI*(1.D0+SUM)/(4.D0*X)
      RETURN
      END
```

4. Heat Transfer in Incompressible Flow

4.1 Introduction

Many turbulent boundary-layer flows of technological interest involve heat transfer from the fluid to the wall or vice versa. The case of high-speed compressible flow will be addressed in §7. Here, fluid motion which is incompressible, in the sense that density variations are negligible to leading-order, will be addressed. If T_∞ denotes the constant mainstream temperature and T_w is the wall temperature, the situations addressed correspond to forced convection heat transfer in the parameter range

$$M_e^2 \ll \frac{T_\infty - T_w}{T_w} \ll 1, \tag{4.1}$$

where M_e denotes the mainstream Mach number; in this parameter range, it is easily confirmed that the effects of viscous dissipation in the thermal energy equation are not significant to leading order.

The general objective of a heat transfer analysis is to predict the energy flux q_w^* from the fluid to the wall, or equivalently, the local Stanton defined by

$$St = \frac{q_w^*}{\rho c_p U_e^*(x)(T_\infty - T_w)}, \tag{4.2}$$

where ρ and c_p are the density and specific heat at constant pressure respectively, and q_w^* and $U_e^*(x)$ denote the dimensional wall heat flux and mainstream velocity, respectively. If all lengths and velocities are made dimensionless with respect to a representative length L and speed U_o, and a dimensionless temperature difference is defined by $\theta = (T - T_w)/(T_\infty - T_w)$, where T is the mean temperature, the governing equation for θ is

$$u \frac{\partial \theta}{\partial x} + u(1 - \theta) \frac{T_w'}{T_\infty - T_w} + v \frac{\partial \theta}{\partial y} = \frac{\partial q}{\partial y}, \tag{4.3}$$

where q is a total heat flux (toward the wall) given by

$$q = \frac{1}{Re\,Pr} \frac{\partial \theta}{\partial y} + \phi. \tag{4.4}$$

Here, $Re = U_o L/\nu$ is the Reynolds number, Pr is the Prandtl number and ϕ is a dimensionless turbulent heat flux defined by

$$\phi = \frac{-\overline{v'T'}}{U_o(T_\infty - T_w)}, \tag{4.5}$$

in terms of a long-time average of the fluctuating normal velocity v' and temperature T'. The boundary conditions associated with equation (4.3) are

$$\theta = 0 \quad \text{at} \quad y = 0, \quad \theta \to 1 \quad \text{as} \quad y \to \infty, \tag{4.6}$$

and, in addition, ϕ must vanish at $y = 0$ and for y large.

4.2 Background

The turbulent thermal boundary layer is known to be a composite double layer consisting of: (1) an inner wall layer where viscosity and the influence of conduction are important to leading-order, and (2) an outer region where the fluid motion is rotational, but effectively inviscid, and where the turbulent transport of energy is dominant (with respect to conduction). As discussed in §2 and §3, the velocity profile in the overlap zone between the two layers is logarithmic with constant slope (cf. equation 2.7). Most experimental measurements in the thermal turbulent boundary layer have been performed for fluids having $Pr = O(1)$ and, as expected, the thicknesses of the thermal outer and inner layers are comparable to their hydrodynamic counterparts. In addition, it is observed from experiment, as well as arguments based on dimensional analysis, that in the overlap zone between the two thermal layers, the mean temperature should be of the form

$$\theta = St(x, Re, Pr) \ U_e(x) \ \theta^+(y^+, Pr)/u_\tau \ (x, Re) + ..., \tag{4.7}$$

with

$$\theta^+ \sim \frac{1}{\kappa_\theta} \log y^+ + B. \tag{4.8}$$

Like the law of the wall, this relation should be regarded as essentially an empirical result. The quantity κ_θ is analogous to the von Kármán constant κ, but estimates of κ_θ have varied widely from one data set to another and there is no agreement on a universal value. This problem was exacerbated by the substantial difficulties in early experimental work in taking accurate measurements of temperature in a thermal boundary layer (see the discussion in [1]). Modern experimental work [2], [3] suggests that κ_θ depends on the local pressure gradient. Here, it will be shown that κ_θ depends on local flow conditions and, in general, is a function of x to be determined.

Generally, the forced convection problem is more difficult than the flow problem because an additional term, the turbulent heat flux ϕ must be modeled. The task is complicated by the fact that $\overline{v'T'}$ is more difficult to measure accurately than $\overline{u'v'}$ and data for $\overline{v'T'}$ is relatively sparse and sometimes of uncertain reliability, especially near the surface. The most common types of model employed, at least in the outer portion of the boundary layer, were originally suggested by Bousinesq in 1877, and these relate the turbulence quantities to the mean profiles by relationships of the form

$$-\overline{u'v'} = \tilde{\varepsilon}\,\frac{\partial u}{\partial y}, \qquad\qquad -\overline{v'T'} = \tilde{\varepsilon}_T\,\frac{\partial T}{\partial y}, \qquad\qquad (4.9)$$

where $\tilde{\varepsilon}$ and $\tilde{\varepsilon}_T$ are referred to as "eddy viscosity" and "eddy conductivity" coefficients which must be modeled. Traditionally, such models have been modified to have a mixing-length form in the wall layer. In the present development, the wall layer will be modeled using a similar methodology to that described in §3, and simple outer models of the form (4.9) will be considered.

The modeling problem for thermal turbulent boundary layers has usually been approached using some form of the Reynolds hypothesis made by O. Reynolds in 1874; this states that the same complex turbulent transport processes are responsible for the turbulent exchange of momentum and the turbulent exchange of heat. There have been many interpretations of this analogy. Two popular versions of the hypothesis are as follows:

(1) For geometrically similar flows, a simple relation between skin friction and surface heat transfer exists of the form $St = R_o\,u_\tau^2\,/\,U_e^2$; here R_o is an unknown constant referred to as the Reynolds analogy factor;

(2) Since the turbulent exchange of momentum and heat are equivalent processes, the ratio of their exchange coefficients is $O(1)$ and

$$\tilde{\varepsilon}_m = Pr_t\,\tilde{\varepsilon}_t, \qquad\qquad (4.10)$$

where Pr_t is an unknown quantity referred to as the turbulent Prandtl number.

There have been many attempts to model the thermal turbulent layer using various versions of the Reynolds analogy (see, for example, the discussion in [1]). As discussed in [5], computed values of the surface heat transfer appear to be somewhat insensitive to the value of Pr_t in the outer region, and many prediction methods use a uniform constant value of around 0.90 in the outer layer [4], [5]. If indeed Pr_t were a universal constant, the thermal modeling problem would be eliminated according to equation (4.10). However, detailed comparisons with data do not support the concept of a constant Pr_t in the wall layer [2], [6] and in many modern prediction schemes, models for Pr_t, which vary with y across the boundary layer are introduced; while this is certainly one approach to the modeling problem, the entire concept embodied in equation (4.10) must be questioned if it is necessary to introduce correlations for Pr_t which vary across the boundary layer.

In the present section, models will be introduced for the outer region but independent of a Reynolds analogy argument. The constant Pr_t model will be examined subsequently for consistency with the asymptotic expansions that will be developed.

4.3 The Thermal Wall Layer

The situation where Pr is $O(1)$ is relevant to many engineering applications where the associated fluid is either air or water, and here all thermal thicknesses will be comparable to the hydrodynamic ones. The dynamics of the turbulent boundary layer are described in §3 and §5, where it is argued that the turbulent wall layer undergoes a cyclic process of breakdown, followed by a relatively long period of slow viscous growth. Consider, for example, the case of a hot wall. As discussed in §3, the sweep event, which is characterized by a deep penetration of outer-layer fluid toward the wall, will bring relatively cold fluid near the surface; as a result, heat transfer rates will be high due to both conduction and convection effects. As the quiescent period evolves, the surface heat transfer rate will gradually diminish with the growth of the wall layer. In an eruption of the wall layer leading to breakdown, relatively hot fluid will be ejected into the outer part of the layer. The sweep and ejection processes just described account for the fact that a turbulent layer is much more efficient at heat transfer than the laminar boundary layer.

Since the Pr is $O(1)$, the thickness of the thermal wall layer will be $O(\text{Re}^{-1} u_\tau^{-1})$, and because the velocities in the wall layer are small, it follows from equation (4.3) that $\partial q / \partial y = 0$ to leading order in the wall layer; thus $q = q_w$ and this establishes the gauge function for ϕ from equation (4.4). An analysis similar to that described in §3 for the mean velocity profile may be carried out for the mean temperature distribution [1], and this suggests a dependence for θ on the variable

$$y_\theta^+ = \sqrt{\text{Pr}}\, y^+ = \sqrt{\text{Pr}}\, \text{Re}\, u_\tau y. \tag{4.11}$$

The gauge function for θ may be obtained from equation (4.4), with $q = q_w$ evaluated at $y_\theta^+ = 0$ where $\phi = 0$. It follows that ϕ and θ may be written to leading order in the wall layer as

$$\theta = q_* \theta^+(y_\theta^+) + \dots, \quad \phi = q_w \phi_1(y_\theta^+) + \dots, \tag{4.12}$$

where $q_* = q_w / u_\tau$, and the convenient normalization

$$\frac{\partial \theta^+}{\partial y_\theta^+} = \sqrt{\text{Pr}} \quad \text{at} \quad y_\theta^+ = 0, \tag{4.13}$$

has been adopted; note that the dependence on x in equations (4.12) is at most parametric and for convenience is omitted there. The leading-order equation in the wall layer is

$$\frac{1}{\sqrt{\text{Pr}}} \frac{\partial \theta^+}{\partial y_\theta^+} + \phi_1 = 1, \tag{4.14}$$

and once again if the turbulence term ϕ_1 is known, the profile θ^+ is determined and vice versa. These functions are subject to the conditions

$$\theta^+ \sim \frac{1}{\kappa_\theta} \log y_\theta^+ + B_i, \quad \phi_1 \to 1 \text{ as } y^+ \to \infty, \tag{4.15}$$

as well as equation (4.13) and $\phi_1 = 0$ at $y_\theta^+ = 0$. It may be noted that q_w may be expressed in terms of the Stanton number defined by equation (4.2) as follows:

$$q_w = U_e(x)St, \quad q_* = \frac{q_w}{u_\tau} = \frac{St}{u_*}, \quad u_* = \frac{u_\tau}{U_e(x)}. \tag{4.16}$$

4.4 The Outer Layer and Matching

The form of the inner-layer expansions (4.12) suggests the following outer-layer expansions

$$\theta = 1 + q_* \, \Theta_1(x,\eta) + ..., \quad q = q_w \, Q_1(x,\eta) + ..., \tag{4.17}$$

where $\eta = y/\Delta_o$ and Δ_o is the outer-layer velocity scale (cf. §2). Here the scaling for the outer thermal layer is the same as that for the corresponding hydrodynamical layer, since Pr is assumed O(1). Note that the first of equations (4.17) is a defect law and that in order for a match with the inner layer, the profile function Θ_1 must be logarithmic for small η. The functions in equation (4.17) must satisfy

$$\Theta_1 \sim \frac{1}{\kappa_\theta} \log \eta + B_o, \quad Q_1 \to 1 \text{ as } \eta \to 0, \tag{4.18}$$

as well as Θ_1, $Q_1 \to 0$ as $\eta \to \infty$.

Matching of the expressions for the profiles in equations (4.12) and (4.17) leads to the requirement that

$$1 + q_* \left\{ \frac{1}{\kappa_\theta} \log\left(\frac{y}{\Delta_o}\right) + B_o \right\} \sim q_* \left\{ \frac{1}{\kappa_\theta} \log\left(\text{Re} \, u_\tau y \sqrt{Pr} \right) + B_i \right\}. \tag{4.19}$$

But a relation (2.8) is available from the match condition for the velocity profile, and it follows that

$$1 + (B_o - B_i)q_* = \kappa \left\{ \frac{U_e}{u_\tau} - C_i + C_o \right\} \frac{q_*}{\kappa_\theta} + \frac{q_*}{2\kappa_\theta} \log Pr, \tag{4.20}$$

which is the full match condition for the temperature profile. But q_*, $(u_\tau/U_e) \to 0$ as Re $\to \infty$ and, consequently, the dominant terms in equation (4.20) give

$$q_* = \frac{\kappa_\theta}{\kappa} u_*. \tag{4.21}$$

The equation relates the surface heat transfer to the skin friction and the slopes of the velocity and temperature profiles in the logarithmic zone.

Different types of models may be adopted for the thermal problem, but in all situations, the full match condition (4.20) must be satisfied. In the first type of model, namely that developed by Weigand [1], κ_θ is not a universal constant and exhibits a dependence on the pressure gradient. To an extent, this is to be expected since the bursting process, and hence the surface heat transfer rate, does depend on the local pressure gradient. Weigand [1] has shown that κ_θ can be correlated on the Clauser pressure gradient parameter β, but also discusses a second (and generally preferable) alternative. In this approach, the leading order asymptotic result (4.21) is accepted as defining the surface heat transfer and then the balance of the match condition (4.20) defines a relation for κ_θ according to

$$\frac{\kappa_\theta}{\kappa} = \frac{C_i - C_o - \frac{1}{2\kappa}\log Pr}{B_i - B_o}.$$

(4.22)

In the profile models that will be described here, B_o is an implicit function of κ_θ and, consequently, equation (4.21) defines a nonlinear relation for κ_θ. A second type of model, the constant turbulent Prandtl number, will be considered subsequently.

It is evident from equation (4.20) that q_* is $O(u_*)$, and substitution of the expansions (4.17) into equation (4.3) and neglecting terms $O(u_*)$ leads to

$$\frac{\partial Q_1}{\partial \eta} + (\alpha - \beta)\eta \frac{\partial \Theta_1}{\partial \eta} + \sigma \left\{ \frac{T_w'}{T_\infty - T_w} + \frac{\kappa_\theta'}{\kappa_\theta} \right\} \Theta_1 = \sigma \frac{\partial \Theta_1}{\partial x},$$

(4.23)

where α, β and σ are defined in equations (2.13), and the primes denote differentiation with respect to x. In deriving equation, the result that

$$\frac{q_*'}{q_*} = \frac{\kappa_\theta'}{\kappa_\theta} + u_* \left(\frac{\beta - \alpha - C_o'}{\kappa\sigma} \right) + ...,$$

(4.24)

has been used, which follows from equation (4.21) and (2.8).

4.5 Self-Similarity

For a self-similar flow, it follows from §2.4 concerning the velocity profile that α and β are constant with $\alpha = 3\beta + 1$; in addition, $\kappa_\theta' = 0$, and

$$\frac{\sigma T_w'}{T_\infty - T_w} = \mu,$$

(4.25)

where μ is constant. But from equation (2.29), $\sigma = \alpha(x - x_o)$, and equation (4.25) may be integrated to give

$$|T_\infty - T_w| = T_1(x - x_o)^{-\mu/(3\beta+1)},$$

(4.26)

where T_1 is a constant. Consequently, self-similar thermal flows are possible if the wall temperature is constant ($\mu = 0$) or behaves according to a power law ($\mu \neq 0$); in addition, the turbulence model must be such that Θ_1 becomes independent of x, at least under the above circumstances. The outer-layer equation now becomes

$$\frac{dQ_1}{d\eta} + (2\beta + 1)\eta \frac{d\Theta_1}{d\eta} + \mu\Theta_1 = 0. \tag{4.27}$$

4.6 Turbulence Models

Simple eddy viscosity and conductivity models for the total stress and heat flux have the form

$$\tau = \varepsilon \frac{\partial u}{\partial y}, \qquad q = \varepsilon_h \frac{\partial \Theta}{\partial y}. \tag{4.28}$$

The Cebeci-Smith and Baldwin-Lomax models for eddy viscosity were described in §2.6, and here an analogous model for eddy conductivity will be described. In the outer part of the boundary layer, a model having $\varepsilon_T = K_h U_e \delta^*$ was considered by Weigand [1]. In order to produce the logarithmic behavior in equation (4.18), it may be easily verified that it is mandatory that $\varepsilon_h \sim \Delta_o U_\tau \kappa_\theta \Delta_o \eta$ as $\eta \to 0$. A simple ramp model analogous to equation (2.44) is given by $\varepsilon_h = U_e \delta^* \hat{\varepsilon}_h$

$$\hat{\varepsilon}_h = \begin{cases} K_h & \eta > \tilde{\eta}_m \\ \kappa_\theta \eta / \eta_* & \eta < \tilde{\eta}_m, \end{cases} \tag{4.29}$$

where η_* is defined in equation (2.45) for a general outer scale Δ_o and $\tilde{\eta}_m = K_h \eta_* / \kappa_\theta$. Extensive comparisons with subsonic temperature profile data were carried out by Weigand [1], who suggests that K_h is almost a universal constant having a value $K_h = 0.0245$; this value will be used throughout the data comparisons shown here. Note that κ_θ is not a universal constant but depends on local conditions through equations (4.21) and (4.22).

Another popular type of model is the constant turbulent Prandtl number wherein $\varepsilon = Pr_t \varepsilon_h$ and Pr_t is assumed constant across the entire outer layer; it may easily be inferred that this implies

$$\frac{K}{K_h} = \frac{\kappa}{\kappa_\theta} = Pr_t, \tag{4.30}$$

and it is common in prediction methods to assume $Pr_t \cong 0.9$. As shown in ref. [7], this type of model is self-consistent with the present asymptotic theory if $Pr_t = 0.872$.

Various approaches [1] have been used by various authors to model the wall layer, most of which are based on some version of a mixing-length formulation with some type of Pr_t approximation. However, results inferred from experimental data [6] strongly suggest that constant Pr_t models are not satisfactory for the wall layer, and various semi-empirical formulae are often used in prediction algorithms such that Pr_t varies across the wall layer. Here another alternative will be considered. As discussed in §3, the wall layer undergoes a cyclic process of relatively slow viscous growth punctuated by local breakdowns; a model for the mean profile U^+ was obtained through consideration of typical motions during a quiescent period and a subsequent time-average of the results. Evidently while the quiescent period evolves for a heated wall, for example, heat will diffuse into the wall layer in much the same manner as vorticity. A similar analysis to that described in §3 shows that the dominant term in the temperature distribution satisfies a heat conduction equation during the quiescent period. It can be shown [1] that the wall-layer mean profile is also given by equation (3.22) but with y^+ replaced by y_θ^+ in equations (3.22). In addition, T_B^+ on the right side of equation (3.30) is replaced by $T_B^+ \sqrt{Pr}$. The fact that the leading-order equation in the time-dependent wall-layer is of the heat conduction type and contains Pr as a parameter accounts for the factors of $\sqrt{Pr}$ in the model for θ^+ and the coordinate y_θ^+ in equation (4.11).

For self-similar flows, equation (4.27) governing the outer-layer temperature is

$$\frac{d}{d\eta}\left(\hat{\epsilon}_h \frac{d\Theta_1}{d\eta}\right) + (2\beta + 1)\eta \frac{d\Theta_1}{d\eta} + \mu\Theta_1 = 0, \tag{4.31}$$

with boundary conditions given by (4.18) with $\Theta_1 \to 0$ as $\eta \to \infty$. For constant pressure flow $\beta = 0$ and for $\mu = 0$, an analytical solution can be found similar to that described in §2.7. Alternatively, a numerical solution may be used for general values of μ and β as described in §2.7.

4.7 Conclusions

Self-similar profiles can be useful to represent data in a self-similar flow or as an approximate profile to initiate a boundary-layer prediction method. A composite profile across the boundary layer may be written

$$\theta = 1 + q_*\left\{\Theta_1 + \theta^+ - \frac{1}{\kappa_\theta}\log y_\theta^+ - B_i\right\}. \tag{4.32}$$

Extensive comparisons [1] of the self-similar profile with experimental data show excellent agreement, and support the use of the outer model given in equation (4.29). However, experience suggests [7] that the particular value of K_h used in the outer model is much less important to the shape of the outer temperature profile than getting the behavior for $\hat{\epsilon}_h$ correct as $\eta \to 0$. For this reason, similar results may be obtained using the constant turbulent Prandtl number model given by equation (4.30), provided a value of $Pr_t = 0.872$ is used [7]. It is worthwhile to emphasize, however, that even in this case, the formulation used for the thermal wall layer in the present theory is not a constant turbulent Prandtl model.

In a prediction method based on the embedded function approach discussed in §2 and §6, it is necessary to advance the solution of the outer-layer momentum equation (2.15) and equation (4.23) in the downstream direction, while continuing to match the present wall-layer solutions.

In Figure 4.1, some typical results using the present theory and a prediction algorithm similar to that described in §6 are shown and compared to experimental data

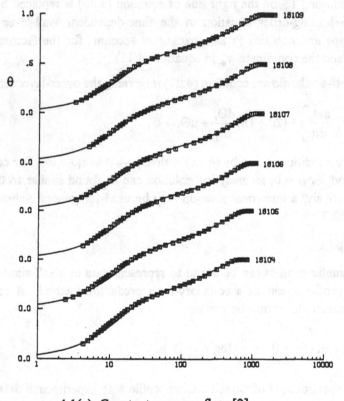

4.1(a) Constant pressure flow [2].

Figure 4.1 Predicted temperature profiles compared to data measured in a low-speed subsonic boundary layer; the first data station is the lowest number. Note the shifted origins on the θ axis.

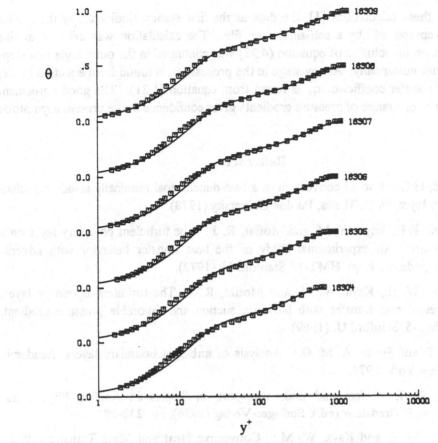

y^+

4.1(b) Strong adverse pressure gradient [2].

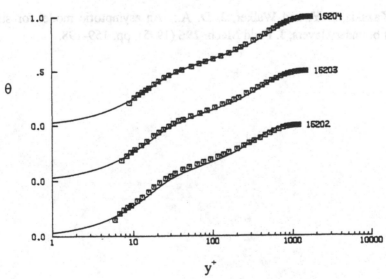

y^+

4.1(c) Favorable pressure gradient [3].

[2], [3]. In these calculations [1], the data at the first station (indicated by the lowest number) is represented by a self-similar profile. The calculation was initiated at this station, and then the solution of equation (4.23) was continued in the outer layer in a step-by-step manner numerically. At any stage in the process, κ_θ is found from equation (4.22) and the heat transfer coefficient q_* is found from equation (4.21). The good agreement with the data over a range of pressure gradients gives confidence in the present asymptotic approach.

References

1. Weigand, G.G.: Forced convection in a two-dimensional nominally steady turbulent boundary layer, Ph.D. Thesis, Purdue University (1978).

2. Blackwell, B.F., Kays, W. M. and Moffat, R. J.: The turbulent boundary layer on a porous plate: An experimental study of the heat transfer behavior with adverse pressure gradients, Rept. HMT-16 Stanford U. (1972).

3. Thielbahr, W. H., Kays, W. M. and Moffat, R. J.: The turbulent boundary layer: Experimental heat transfer with blowing, suction and favorable pressure gradient, Rept. HMT-5, Stanford U. (1969).

4. Cebeci, T. and Smith, A. M. O.: Analysis of turbulent boundary layers, Academic Press, New York, 1974.

5. Launder, B. E.: Heat and mass transport, in Topics in Applied Physics 12, Turbulence, P. Bradshaw (ed.), Springer-Verlag (1976), pp. 232-287.

6. Crawford, M. E. and Kays, W. M.: Convective Heat and Mass Transfer (1980), McGraw-Hill.

7. He, J., Kazakia, J. Y. and Walker, J. D. A.: An asymptotic model for supersonic turbulent boundary layers, J. Fluid Mech. 295 (1995), pp. 159-198.

5. Dynamics of Turbulent Shear Flows

5.1 Introduction

Since the late 1950's, when experiments at Stanford University showed that a considerable degree of structure exists in the near-wall region of turbulent flows, the processes involved have been under intense investigation, especially from an experimental standpoint. There are two main features which appear to dominate the physics of the instantaneous wall-layer flow, namely: (i) the low-speed streaks and (ii) the bursting process. When any visualization medium, such as dye, is introduced into the boundary layer near the wall, it is observed to be swept into relatively long streaks that are essentially aligned in the streamwise direction, where the instantaneous velocity is generally below that of the local mean streamwise velocity; these zones are termed "low-speed" streaks, and in the spanwise direction they are observed to alternate with relatively high speed zones. The relatively long periods, when a portion of the wall layer will be observed containing streaks, are referred to as the quiescent period; the terminology suggests that, at least locally, the thin wall layer is being driven by events transpiring in the outer layer and is for the moment responding passively to such events. By contrast, the bursting process appears as a strongly interactive event that usually initiates near a low-speed streak and culminates in a wall-layer eruption and local breakdown of a hitherto relatively well-ordered near-wall flow. The eruptive process is invariably followed by a sweep event in which relatively high-speed fluid from the outer part of the boundary layer penetrates close to the wall, and a new quiescent period begins locally.

The early observations of such events provoked considerable interest since now there was a clear suggestion that the flow development near the wall was not random (as previously believed) and that the observed cyclic events might be deterministic in some sense. The terminology "coherent structure" evolved to describe the repeatable events and generic structures that were believed to be responsible for the observed phenomena. Unfortunately, it soon became apparent that the processes governing the wall-layer development were very complex, involving convected disturbances whose very nature was controversial and whose effects were open to diverse interpretations [1]. With the advent of direct numerical simulations of turbulent channel and boundary-layer flows, it became possible to carry out an intensive interrogation of these generated data sets to search for generic structure [2], [3]. Unfortunately, this objective was difficult to achieve for a variety of reasons, and issues related to the dynamics of the turbulent wall layer remain controversial today [4]. Experiments at high Reynolds number are very difficult to carry out because of the relative thinness of the shear layers; nevertheless, in an experiment, one is dealing with true physical phenomena, and the greatest challenge is to devise accurate and innovative ways to visualize and/or measure the relevant events. Since such phenomena take place in an unsteady, complex, three-dimensional flow field, the tasks involved are formidable and experimental costs of a definitive study at high Reynolds number can be prohibitive. The alternative of direct numerical simulations (DNS) of turbulence appears very attractive at first glance, and there has been considerable success using this approach over the past fifteen years. However, here one is confronted with two new problems. Such simulations generate considerable quantities of three-dimensional data, and the interpretation of such data sets is difficult. However, a more serious problem is that the representation of the physical phenomena is necessarily inexact, and the data sets contain numerical error, the level of which is difficult to quantity [5] (in some cases this may be substantial). As the Reynolds number increases, the important physical processes occur in progressively thinner layers and sometimes in events which are sharply focused in space; such phenomena require increased numerical resolution locally during certain intervals of time. At present, DNS calculations are possible only for small portions of turbulent flows at relatively low Reynolds numbers, which are usually within the transitional range and well outside the usual regimes encountered in engineering practice.

Despite the aforementioned problems, it has been possible to make real progress over the past two decades, and in this chapter a model of the dynamics of the turbulent boundary layer will be described. This model has been constructed from a number of experimental studies and various recent theoretical results concerning unsteady flows, as well as syntheses [2], [3] of DNS results. It is worthwhile to note that much of the experimental and DNS work has concentrated on attempting to define the kinematics of the near-wall turbulence, by in effect cataloging the sequence of typical events that were believed to be important. The task of delineating the dynamics is more difficult since it is then necessary to identify the cause-and-effect relationships associated with processes in a manner which is still consistent with the kinematical observations.

It should be remarked that although there are competing views [4] on what the most important processes are, it is now reasonably well understood that the turbulent wall layer is not deterministic in the usual sense. Although certain typical events may be observed repeatedly, it is not, in general, possible to calculate accurately a turbulent flow at high Reynolds for an indefinite period of time, and in this sense such flows may be regarded as chaotic. Nevertheless the interrogation of certain typical processes is important because these appear to make the most important contributions to quantities that are relevant in the mean equations. For example, the wall-layer bursting appears to make the dominant and lasting contributions to the Reynolds stress $-u'v'$; unfortunately, the processes involved are so complex that direct modeling of the Reynolds stress appears problematic at present. Indeed, despite the large experimental and DNS efforts over the years, there has been relatively little accrued from this work that can be used in prediction methods for the mean flow. At the same time, there are other valid reasons for developing an understanding of the important dynamics relating, for example, to boundary-layer control. To achieve drag reduction or reduce noise, it is necessary to decrease bursting, while heat transfer rates may be augmented by increasing bursting. In such situations, innovative control methods might well be based on rationally altering the cyclic processes in the turbulent wall layer.

5.2 Hairpin Vortices

Because the motions in a turbulent boundary layer involve complex three-dimensional time-dependent phenomena, progress in resolving the critical aspects of the dynamics of the processes involved has been rather slow. Indeed, there have been major disagreements about even the kinematics of the typical events. It was evident, however, at a fairly early stage that vortices must play a key role [7], [8]. Theodorsen [7], well in advance of the first detailed observations of coherent structure, proposed that vortices of the type shown in Figure 5.1 must be an important flow module in turbulent near-wall flows. The symmetric structure shown in Figure 5.1 is referred to as a hairpin vortex, and similar structures are easily observed today in transitional and fully turbulent flows. The symmetric hairpin vortex is somewhat of an idealization and, because the turbulent boundary layer is a complex environment with many competing influences, the majority of such vortices are expected to exhibit various degrees of asymmetry, as suggested in Figure 5.1(b). Indeed, in a synthesis of DNS results for turbulent boundary-layer and channel flows [3], it was found that the vast majority of the vortex structures were distinctly asymmetric. Robinson [3] actually prefers to classify different parts of the vortex, referring to the head (see Figure 5.1) as a "vortex arch" and the vortex leg as a "quasi-streamwise vortex"; here the latter terminology implies a vortex for which the majority of the associated vorticity is aligned in the streamwise direction.

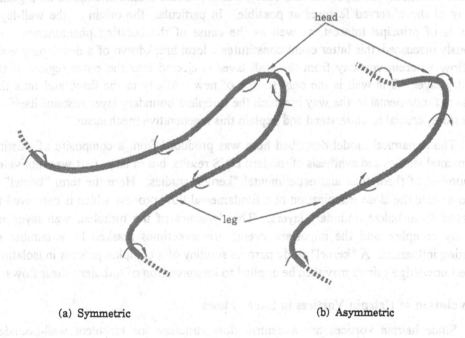

(a) Symmetric   (b) Asymmetric

Figure 5.1 Schematic diagram of typical hairpin vortex configurations, with the sense of vorticity indicated by arrows.

In general, the vortices observed near the wall in a turbulent boundary layer appear to be of relatively small scale nearest the wall. With increasing distance from the wall, the scale of the vortices invariably appears to expand, and this growth to larger scale across the boundary layer is a persistent characteristic. Indeed, in Robinson's [3] work, a multitude of vortex legs relatively closely-spaced were normally observed in the region nearest the wall, in a zone comprising the upper portions of the time-mean wall layer. Further up in the time-mean overlap zone, the legs were interspersed with small vortex heads. Still further up in the outer part of the boundary layer, the predominant structure is the vortex head or "arch" having a significantly increased spanwise scale.

In the following sections, a conceptual model of the dynamics of the turbulent boundary layer will be described briefly, with further details and descriptions being presented elsewhere [6], [9]. It will be argued that the most important events in the near-wall region can be explained in terms of how hairpin vortices interact with the background shear flow, with one another and, most critically, with the viscous flow near the surface. It may be noted that many aspects of turbulent boundary-layer structure are still controversial, and a number of competing viewpoints have appeared in a recent monograph [4]. It is, however, worthwhile to bear in mind that there has been extensive experimental study of the kinematics of such flows, as well as recent study of DNS results (which is admittedly an imperfect data base), and any model should attempt to tie together as many of the observed features as possible. In particular, the origin of the wall-layer streaks is of principal interest, as well as the cause of the bursting phenomenon. As previously discussed, this latter event constitutes a local breakdown of a developing wall-layer flow wherein vorticity from the wall layer is ejected into the outer region of the boundary layer. The wall is the only source of new vorticity in the flow, and thus this process is fundamental to the way in which the turbulent boundary layer sustains itself. It is, therefore, crucial to understand and explain this regenerative mechanism.

The dynamical model described here was produced from a composite of existing experimental studies and synthesis of modern DNS results, but in large part was motivated by a number of theoretical and experimental "kernel" studies. Here the term "kernel" is used to denote the close examination of a fundamental flow process which is perceived to be related to turbulent boundary layers. The dynamics of the turbulent wall layer are extremely complex, and the important events are sometimes masked by a number of competing influences. A "kernel" study permits scrutiny of a complex process in isolation, and the knowledge gained may then be applied to interpretation of turbulent shear flows.

5.3 Evolution of Hairpin Vortices in Shear Flows

Since hairpin vortices are a central flow structure for turbulent wall-bounded flows, it is important to understand where and how such vortices develop. Turbulent flows generally originate in a transition zone upstream. When the external flow is contaminated with significant disturbances, concentrated vortices already exist, which then regenerate themselves in a viscous-inviscid interaction with the viscous flow near the wall;

such phenomena are usually classified as "bypass" transition because the turbulence develops almost directly without the appearance of Tollmein-Schlichting (TS) waves. In controlled transition, the external flow field is kept almost disturbance-free; here TS waves appear at sufficiently large Reynolds number, and a complicated period of amplification and nonlinear effects occurs leading eventually to the production of hairpin vortices. In either case, the latter stages of transition are generally characterized by a large number of hairpin vortices that are then convected into the fully turbulent regime.

In §5.5, the process of the production of new vortices (and turbulence) through a viscous-inviscid interaction with the near-wall flow will be discussed. Here a complementary process will be described which leads to the evolution of new vortex structures; this process, however, may be thought of as a reorganization and concentration of the existing vortex field.

The wall layer of a turbulent boundary layer is a region which experiences relatively high streamwise shear; this occurs on an almost continuous basis because the wall layer is in a quiescent state most of the time, and at any instant the streamwise velocity profile is not dramatically different from the mean profile, especially for large y^+. To understand how a hairpin vortex can develop, Hon and Walker [10] considered the evolution of a symmetric disturbance in an otherwise two-dimensional vortex which was being convected in a uniform shear flow. The evolution of the disturbance was carefully computed [10] using the Biot-Savart law, and a characteristic pattern emerged which is shown schematically in Figure 5.2 and which is essentially independent of the initial

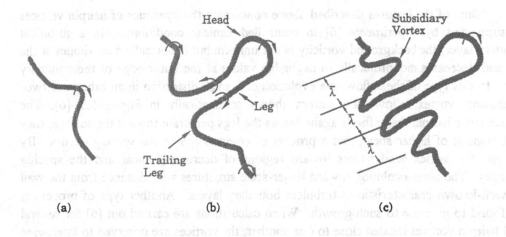

Figure 5.2 Evolution of a symmetric hairpin vortex in a shear flow. (a) Initial distortion; (b) development of vortex legs and head; (c) evolution of subsidiary vortices and penetration of the legs toward the surface

distortion configuration. Here the flow is from left to right, and the speed of the shear flow increases from bottom to top. As indicated in Figure 5.2, a vortex head quickly forms and rises from the surface, subsequently bending back in the shear flow. At the same time, vortex legs form and move progressively down toward the wall. As time increases, the head moves further from the wall and, as the legs reach downward encountering regions of lower streamwise speed, the streamwise extent of the vortex continually increases. The original disturbance also produces similar hairpin structures in the spanwise direction, and these are referred to as subsidiary vortices [6]. The characteristic spacing between vortices, denoted as $\overline{\lambda}$ in Figure 5.2, depends mainly on a parameter ε, which is a ratio of vortex circulation to the strength of the background shear flow; smaller values of ε generally produce smaller spacings $\overline{\lambda}$.

The development shown in Figure 5.2 is a nonlinear, essentially inviscid interaction which occurs for a variety of initial distortions, even those which are asymmetric [6]. This generic behavior can be summarized as follows. When a vortex is convected in a shear flow, the characteristic hairpin shape will evolve from portions of the vortex where the curvature is relatively large, even when that portion is very small; the majority of such vortices will be asymmetric. The legs always move progressively toward regions of decreasing velocity while the heads move toward high speed regions. As the hairpin shape evolves, increasing portions of the vortex are converted from spanwise to streamwise vorticity, and the disturbance spreads in the streamwise direction, as well as the spanwise direction, through the evolution of subsidiary vortices. The relative spacing of the subsidiary vortices is a function of the strength of the background shear, with the spacing decreasing with increasing shear.

Many of the features described above concerning the dynamics of hairpin vortices are supported by experiments [6] in controlled laminar conditions. In a turbulent boundary layer, the background vorticity is not uniform but is normally a maximum at the wall and decreases monotonically to negligible values at the outer edge of the boundary layer. In this type of shear flow, the evolution of a small distortion in an otherwise two-dimensional vortex follows the pattern shown schematically in Figure 5.3 [6]. The characteristic hairpin shape forms again, but as the legs penetrate toward the surface, they reach regions of higher shear, and a process of compression of the spacing occurs. By contrast, the vortex head moves toward regions of decreasing shear and the spacing increases. The latter evolution toward larger-scale structures with distance from the wall is a well-known characteristic of turbulent boundary layers. Another type of process is also found to give rise to such growth. When calculations are carried out [6] for several small hairpin vortices located close to one another, the vortices are observed to intertwine and form a hairpin vortex of larger spanwise scale (which is usually asymmetric).

More detailed descriptions of the above results are given elsewhere [6], [9], [11], as well as detailed comparisons with Robinson's synthesis of the DNS results [3], and here only a brief summary will be given. Near the surface in a turbulent boundary layer, the

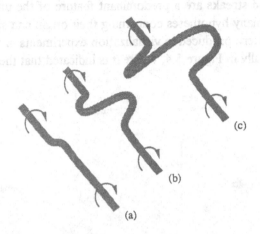

Figure 5.3   Evolution of a hairpin vortex in a turbulent shear flow.   (a) Initial distortion;
(b) development of head and legs; (c) narrowing of leg spacing and expansion
of head.

flow is populated by a large number of hairpin vortices which are of relatively small
spanwise extent (typically on the order of $\Delta z^+ \cong 100$, where $z^+ = Re\ u_\tau\ z$) and which have
varying strengths; these vortices compete with one another for position and upon close
approach can intertwine to produce larger and/or stronger vortices. Each vortex attempts
to send legs downward into the wall layer, giving rise to lateral vortex stretching and local
intensification of the vorticity. The combination of the large number of vortices and the
narrowing of the distance between the legs (in a region of increased shear) gives the
impression of an increased number of streamwise vortices near the surface [3], [6], [11].
The stronger hairpin vortices which successfully penetrate near the surface will act to
provoke a viscous-inviscid interaction, in a regenerative process that produces new hairpin
vortices; this process will be described in §5.5.

The behavior of the upper part of the hairpins is rather different. As the vortex
heads rise to regions of decreased shear, they expand to larger spanwise scale and, at the
same time, the vortex cores relax, leading to a decrease in local vorticity. As the vortices
migrate away from the intense tangle of vorticity near the surface, they also may
intertwine in a process which leads to larger scales. The processes described here all lead
to migration of vorticity away from the surface and a coalescence into weaker vortices of
larger structure. Indeed the large overturning motions that occur in the outer layer are
probably comprised of large agglomerations of vortices that originated in the intensely
active region near the surface. In this sense, the outer part of the boundary layer may be
regarded as a "graveyard" for turbulence, comprised of recirculating clouds of vorticity
that have escaped the active regions near the surface.

5.4 Low-Speed Streaks

The low-speed streaks are a predominant feature of the unsteady wall-layer flow, and there have been many hypotheses concerning their origin and significance [3], [4], [6], [11]. The type of pattern produced in visualization experiments in plan view and end view is depicted schematically in Figure 5.4, where it is indicated that the streaks constitute

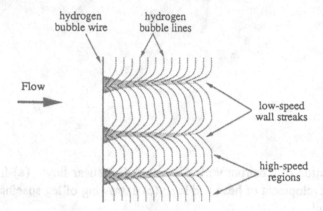

Figure 5.4a Schematic of low-speed streak flow structure that occurs in the turbulent wall layer as seen in plan view.

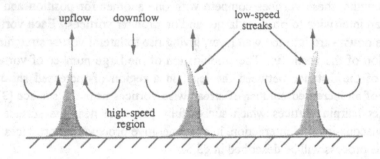

Figure 5.4b Schematic of instantaneous velocity field seen in end view (looking upstream) in the cross-flow plane.

relatively long regions (in the streamwise direction) of induced upwelling. As suggested in Figure 5.5, the most plausible explanation of the wall-layer streaks is that they are the passive trail of a hairpin vortex which instantaneously is moving over the viscous near-wall region [6], [11]. Of course, not every hairpin vortex in the boundary layer will produce a streak, and the observed pattern in the wall layer at any instant indicates the presence of only those vortices closest to the wall. In fact, there is a large number of vortices, but the effects of most of the vortices on the wall-layer flow are shielded by those closest to the surface. A streak will only initiate when either a vortex is convected close enough to the surface or when a vortex leg penetrates through the myriad of vortices in the overlap region and thereby initiates an intimate contact with the near-wall flow. A streak will terminate when the causative vortex departs from the immediate vicinity of the surface. The observed meandering and lateral oscillations observed [6] for streaks is believed to be associated with a weak jostling action associated with vortices which have been convected from further upstream and that overrun existing streaks.

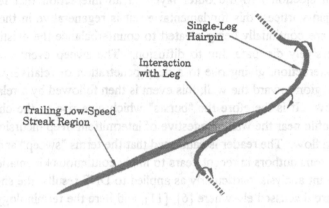

Single-Leg Hairpin

Interaction with Leg

Trailing Low-Speed Streak Region

Figure 5.5   Schematic of the processes whereby an asymmetric hairpin vortex induces a low-speed streak.

For the reasons stated above, it may be concluded that the majority of a wall-layer streak is not dynamically significant and simply marks the trail of a hairpin vortex which had previously passed above. The average spanwise spacing of $\lambda^+ \cong 100$ reflects the population and relative spacing of those hairpin vortices closest to the surface. Because the local streamwise velocity within the overlap zone is much greater than typical spanwise velocities in the turbulence, the streaks are relatively long with respect to the spanwise spacing. As the causative hairpin vortex moves downstream, the leading edge of the streak traces a path emanating from the side of the trailing vortex leg where the flow is away from the surface. If the hairpin vortex remains in contact with the wall layer, it is likely that a viscous-inviscid interaction will be provoked near the trailing leg as discussed in the next section; this interaction will also result in the termination of a streak. Note that the literature contains many arguments as to whether the wall-layer streaks represent "active" or "inactive" motions. In the physical model discussed here, most of the streak constitutes an inactive trail, but the leading-edge portion has the potential to be very active. Indeed, the lifting of a wall-layer streak is a strong indication that an eruptive event is about to occur in the immediate vicinity.

5.5 Bursting and Regeneration

Now consider the process of wall-layer breakdown, which in many ways is the most important of the processes discussed thus far. The wall is the only source of new vorticity, and the physical mechanism of how new vorticity is transmitted into the flow is the key aspect in understanding how a turbulent boundary layer sustains itself. During a typical quiescent state, when the wall-layer flow is relatively well-ordered containing the low-speed streaks, vorticity from the wall diffuses slowly into the wall layer. The local breakdown of the wall layer results in a focusing and concentration of existing vorticity that culminates in an ejection into the outer layer, in an interaction that leads to the creation of new hairpin vortices; this fundamental event is regenerative in the sense that new hairpin vortices are continually being created to counterbalance the existing hairpins above, which progressively dissipate due to diffusion. The sweep event occurs in the latter stages of the interaction, giving rise to a deep penetration of relatively high-speed fluid from the outer region toward the wall; this event is then followed by a relatively long period of relaxing flow. It is therefore the "bursts" which give rise to the characteristic shape of the mean profile near the wall, suggestive of intermittent deep incursions of high-speed, almost uniform flow. The reader is cautioned that the terms "sweep" and "ejection" have been used by various authors in recent years to imply continuous kinematic processes associated with quadrant analysis, particularly as applied to DNS results; the shortcomings of these approaches are discussed elsewhere [6], [11], and here the terminology is used in the classical sense to suggest discrete events of limited duration.

The key to understanding the cause of the intermittent observed wall-layer breakdowns lies in an appreciation of the nature of the viscous response to a vortex in motion above a wall. In order to understand this response, a substantial number of fundamental "kernel" studies have been carried out which are summarized and reviewed elsewhere [6], [11], [12]. Any vortex in motion above a wall induces a region of adverse pressure gradient on the viscous boundary layer on the side of the vortex where the induced flow is away from the wall. When the vortex Reynolds number $Re = \Gamma/\nu$ is large (which occurs for even a relatively low circulation Γ when the kinematic viscosity ν is small), a series of complex events is provoked within the boundary layer. Usually a region of recirculation develops in the surface layer, and this is quickly and abruptly followed by unsteady separation. Here the term "separation" is used in the modern sense to imply an event in which the boundary layer interacts for the first time with the external flow in a significant way. The nature of unsteady separation events has only been understood in relatively recent times [12], [13]. The process initiates rather abruptly, as a sharp streamwise compression and focusing of the boundary-layer vorticity into a band which is very narrow in the streamwise direction. As the boundary layer starts to separate and thereby leave the surface, it does so as a rapidly rising "spike" containing sharply compressed vorticity. The calculation of the subsequent viscous-inviscid interaction that occurs is presently beyond the scope of modern computational methods, but in recent years, it has proved possible to interrogate the nature of the interaction in well-designed experiments [12]. The practical experimental task is difficult because the phenomena involved are most easily visualized in a frame of reference moving with the vortex, and the associated events happen very rapidly. Observations [12] suggest that the most common form of the interaction is a roll-up of the focused plume that emerges from the boundary layer into a new vortex structure.

The response of a viscous flow to a convected hairpin vortex at high Reynolds number is an especially formidable problem since it involves: (1) a deforming, three-dimensional vortex in the mainstream flow, and (2) a complex unsteady interacting three-dimensional boundary layer. At present no reliable solutions exist for such a flow at high Reynolds number. There is, however, a general three-dimensional theory [14] concerning three-dimensional boundary-layer separation and, based on this theory, as well as a considerable amount of experience with relevant two-dimensional flows, it is possible to speculate on the potential separation processes due to a convecting hairpin vortex.

Consider the hairpin vortex shown schematically in Figure 5.6. In the region behind the vortex head, the streamwise pressure gradient induced near the wall is adverse. In addition to the left of the vortex leg (where the induced flow is away from the wall), the spanwise pressure gradient is adverse. In general, the vortex head is furthest from the wall and, as it moves downstream, the head moves further away from the surface; thus the streamwise adverse pressure gradient weakens and is believed to be less significant than the spanwise adverse pressure gradient near the leg. Furthermore, since the leg moves continually downward toward the wall, an intensifying effect will occur. At high Reynolds numbers, the aforementioned adverse pressure gradients will at some stage provoke separation of the viscous layer at the surface. A general theory of three-dimensional separation [14] suggests that this will occur along a U-shaped ridge, in the form of a rapidly rising tongue of fluid containing compressed vorticity. The expected form of the subsequent viscous-inviscid interaction is shown in Figure 5.6 where the erupting tongue rolls over into a new hairpin vortex. This regenerative process has been observed in careful experimental observations of the effects of convected hairpin vortices [6]. A similar process is also observed in the low-Reynolds-number DNS results synthesized by Robinson [2], [3], where new "arches" (vortex heads) were invariably observed to form either behind an existing "arch" or near a "quasi-streamwise" vortex (leg).

In a turbulent boundary layer, newly created vortices may interact with the parent vortex or neighboring vortices; alternately, such vortices may provoke further eruptions downstream to produce additional hairpins. Note that when the turbulent boundary layer is visualized, using various methods [6], [11] in a plane normal to the flow direction and the wall, essentially only narrow-band breakdowns, consisting of thin eruptive spires, are observed; these spires are believed to be an instantaneous slice through the three-dimensional separation structure shown schematically in Figure 5.6.

5.6 Summary

In the present chapter, it has been argued that the main features of a turbulent boundary-layer flow can be explained in terms of how hairpin vortices interact with: (1) the background shear flow, (2) other vortices, and (3) the viscous near-wall flow. Further details may be found in references [6], [9] and [11].

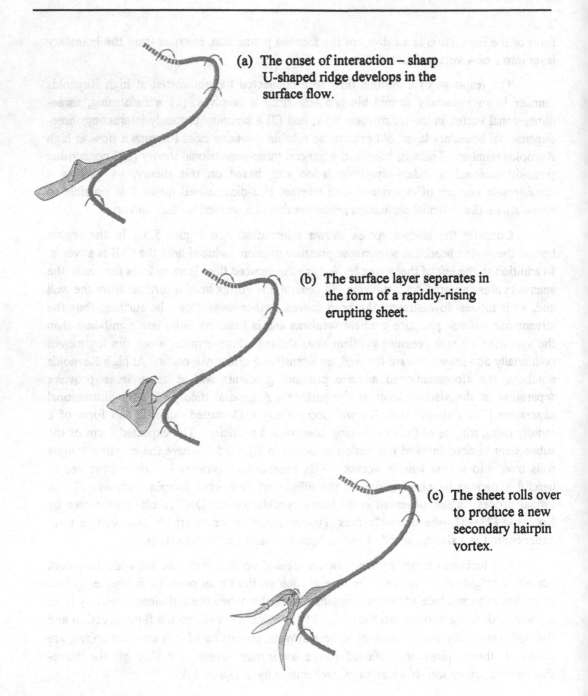

(a) The onset of interaction – sharp U-shaped ridge develops in the surface flow.

(b) The surface layer separates in the form of a rapidly-rising erupting sheet.

(c) The sheet rolls over to produce a new secondary hairpin vortex.

Figure 5.6 The generation of secondary vortices through a viscous-inviscid interaction provoked by the apparent asymmetry hairpin vortex.

References

1. Walker, J. D. A.: Turbulent flow structure near walls, Phil. Trans. Roy. Soc. Lond. A **336** (1991).

2. Robinson, S.K.: Coherent motions in the turbulent boundary layer, Ann. Rev. Fluid Mech. **23** (1991), 601-639.

3. Robinson, S.K.: The kinematics of turbulent boundary layer structure, NASA Tech. Memo 103859 (1991).

4. Panton, R. L.: Self-sustaining mechanisms of turbulent boundary layers, Computational Mechanics Publications, Southampton, U.K. (1997).

5. Zang, T. A.: Numerical simulation of the dynamics of a turbulent boundary layer: perspectives of a transition simulator, Phil. Trans. Roy. Soc. Lond. A **336** (1991), 95-102.

6. Smith, C. R., Walker, J. D. A., Haidari, A. H. and Sobrun, U.: On the dynamics of near-wall turbulence, Phil. Trans. Roy. Soc. Lond. A **336** (1991), 131-173.

7. Theodorsen, T.: Mechanism of turbulence. Proc. 2nd Midwestern Conf. on Fluid Mechanics, Bull. No. 149, Ohio State University (1952).

8. Black, T. J.: An analytical study of the measured wall pressure field under supersonic turbulent boundary layers, NASA-CR 888 (1968).

9. Smith, C. R. and Walker, J. D. A.: Sustaining mechanisms of turbulent boundary layers: The role of vortex development and interactions in self-sustaining mechanism of wall turbulence, R. L. Panton (ed.), Computational Mechanics Publication (1997).

10. Hon, T. L. and Walker, J. D. A.: Evolution of hairpin vortices in a shear flow, Computers and Fluids **20** (1991), 343-358.

11. Smith, C. R. and Walker, J. D. A.: Turbulent wall-layer vortices, in Fluid Vortices, S. I. Green (ed.), Kluwer Acad. Pub. (1995), 235-290.

12. Doligalski, T. L., Smith, C. R. and Walker, J. D. A.: Vortex interactions with walls, Ann. Rev. Fluid Mechs. **26** (1994), 573-616.

13. Cowley, S. J., van Dommelen, L. L. and Lam, L. L.: On the use of Lagrangian variables in descriptions of unsteady boundary-layer separation", Phil. Trans. Roy Soc. Lond. A **333** (1991), 343-378.

14. Van Dommelen, L. L. and Cowley, S. J.: On the Lagrangian description of unsteady boundary-layer separation. Part 2. General Theory, J. Fluid Mech. **210**, 593-626.

6. Three-Dimensional Turbulent Boundary Layers

6.1 Introduction

Three-dimensional turbulent boundary layers are common in engineering applications, and in recent years there have been increasing efforts to develop turbulence models and calculation procedures for such flows. Three-dimensional strongly-interactive flows usually involve some type of turbulent boundary-layer separation and, although a limited number of experiments have been carried out, it is fair to say that relatively little is known concerning the physics and structure of such flows. Consequently, the situations considered here will be confined to the case of attached turbulent boundary layers, where the boundary layer is thin and its development is primarily driven by the external pressure field near the surface.

The essential features of an attached three-dimensional boundary layer are depicted schematically in Figure 6.1. Here streamline coordinates $(x_1 x_2)$ lie in the surface, with x_1 measuring distance in the local mainstream direction and x_2 measuring

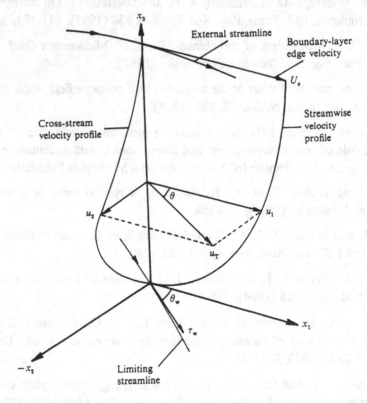

Figure 6.1 Schematic of a three-dimensional boundary-layer velocity profile.

spanwise distance in a direction normal to x_1. The coordinate x_3 measures distance normal to the surface and completes a locally orthogonal coordinate system. In most circumstances, the external streamline at the edge of the boundary layer is curved, and this produces a cross-stream pressure gradient which is positive for the case shown in Figure 6.1. Let the velocity vector within the boundary layer be denoted by $\bar{u}_T$; this vector is tangential to the external streamline at the boundary-layer edge but rotates in the direction of decreasing cross-stream pressure gradient as the distance to the wall decreases. In the streamline coordinate system, therefore, a cross-stream velocity component develops, and this is denoted in Figure 6.1 as u_2. Typically in an attached boundary layer, u_2 increases from zero at the boundary-layer edge to a maximum part way through the layer and then falls to zero at the wall in order to satisfy the no-slip condition. The behavior of the velocity component in the streamwise direction u_1 is generally observed to be similar to that observed in two-dimensional flows such that a double structure is exhibited with an outer layer, a wall layer, and a logarithmic variation in between [1], [2]. To a certain extent, this is expected in light of the turbulent dynamics described in §5; the principal effect of the three-dimensionality is to deflect the trajectory of the hairpin vortices in the turbulence near the wall; however, the central influence of the hairpin vortices on the viscous wall layer, namely that of provoking eruptions, is not significantly altered.

The angle θ between $\bar{u}_T$ and the streamwise direction is called the velocity skew angle, and since u_1 and u_2 both vanish at the wall, the wall skew angle θ_w must be determined from a limiting process using L'hôpital's rule. Consequently, θ_w also describes the angle between the streamwise direction and the wall shear stress; it constitutes the total angle through which the velocity vector rotates across the boundary layer. The total shear stress vector generally has components in each of the x_1 and x_2 directions and also rotates away from a value of θ_w at the wall.

The asymptotic structure of the turbulent three-dimensional boundary layer has been considered recently [3], [4] and in this chapter a summary of the main results will be described, as well as an efficient numerical procedure for the calculation of such flows [5]. It is worthwhile to note that many issues relating to the three-dimensional case have been controversial, partially due to low-Reynolds-number effects in experimental data and the difficulties inherent in taking reliable data in a more complex environment, especially near the surface. Some of these issues are discussed in [4]. First, a large number of "law of the wall" formulae containing a logarithmic variation have been proposed [6] for the cross-flow profile mostly on an *ad hoc* basis, but the nature of the cross-flow profile throughout the boundary layer was not clear until relatively recently [5]. Secondly, the extent of near-wall collateral flow has been debated vigorously; here the term "collateral" implies that the flow is in the same direction as the wall shear stress. Some estimates of collateral flow vary from locations very close to the wall to those well outside the wall layer; one theoretical study [7] suggested that the motion in the wall layer is collateral to all orders. Lastly, the location of the point of maximum cross-stream velocity has been at issue, with some authors [8], suggesting it is close to the wall within the wall layer, while others [7] argue it occurs within the outer layer. These issues will be subsequently discussed here; see also [4], [5].

6.2 Governing Equations

Define dimensionless variables with respect to a representative length L and flow speed U_o and define the Reynolds number as $Re = U_o L / \nu$. In dimensionless variables, the Reynolds-averaged boundary-layer equations written in a streamline coordinate system are

$$\frac{\partial}{\partial x_1}(h_2 u_1) + \frac{\partial}{\partial x_2}(h_1 u_2) + h_1 h_2 \frac{\partial u_3}{\partial x_3} = 0, \tag{6.1}$$

$$\frac{u_1}{h_1}\frac{\partial u_1}{\partial x_1} + \frac{u_2}{h_2}\frac{\partial u_1}{\partial x_2} + u_3 \frac{\partial u_1}{\partial x_3} - K_2 u_1 u_2 + K_1 u_2^2 = \frac{U_e}{h_1}\frac{\partial U_e}{\partial x_1} + \frac{\partial \tau_{13}}{\partial x_3}, \tag{6.2}$$

$$\frac{u_1}{h_1}\frac{\partial u_2}{\partial x_1} + \frac{u_2}{h_2}\frac{\partial u_2}{\partial x_2} + u_3 \frac{\partial u_2}{\partial x_3} - K_1 u_1 u_2 + K_2 u_1^2 = K_2 U_e^2 + \frac{\partial \tau_{23}}{\partial x_3}, \tag{6.3}$$

where the u_i denote velocities in the x_i direction (see Figure 6.1). The metric coefficients in the x_1 and x_2 directions, h_1 and h_2, are independent of x_3 and $h_3 = 1$ [9]; the curvatures are defined by

$$K_1 = -\frac{1}{h_1 h_2}\frac{\partial h_2}{\partial x_1}, \qquad K_2 = -\frac{1}{h_1 h_2}\frac{\partial h_1}{\partial x_2}, \tag{6.4}$$

and τ_{13} and τ_{23} are the total shear stresses in the x_1 and x_2 directions defined by

$$\tau_{i3} = \sigma_{i3} + \frac{1}{Re}\frac{\partial u_i}{\partial x_3}, \quad i = 1, 2, \tag{6.5}$$

where σ_{i3} denotes the dimensionless Reynolds stresses.

In a streamline coordinate system $u_1 \to U_e(x_1, x_2)$, $u_2 \to 0$ at the edge of the boundary layer, and to evaluate the boundary-layer solution, it is also necessary to specify the function $q(x_1, x_2)$ defined by

$$qU_e = \lim_{x_3 \to 0} \frac{\partial U_{3e}}{\partial x_3}, \tag{6.6}$$

where U_{3e} denotes the normal component of velocity in the external flow. In general, it is easily shown that for a streamline coordinate system $h_1 = 1/U_e$, while from the continuity equation, the cross-stream metric satisfies

$$K_1 = -\frac{U_e}{h_1}\frac{\partial h_2}{\partial x_1} = \frac{\partial U_e}{\partial x_1} + q. \tag{6.7}$$

For the special case $q = 0$, it follows that $h_2 = 1/U_e$; in general, however, equation (6.7) defines a differential equation for h_2 which may be integrated along individual streamlines for a given external flow and is therefore considered known.

6.3 The Wall Layer

In view of the discussion of turbulent dynamics in §5, as well as a considerable body of experimental evidence, the three-dimensional attached turbulent boundary layer is expected to exhibit a double structure similar to that observed in two-dimensional boundary layers. The friction velocity $u_\tau = \tau_w^{1/2}$ is defined in the usual manner, and the wall skew angle is given by

$$\tan \theta_w = \tau_{23}/\tau_{13} \quad \text{at} \quad x_3 = 0, \tag{6.8}$$

where, in general, $\theta_w = \theta_w(x_1, x_2, \text{Re})$, and θ_w is to be found. With the wall layer coordinate defined by $y^+ = \text{Re}\, u_\tau x_3$, the following expansions for the total shear stress in the wall layer are suggested:

$$\tau_{13} = u_\tau^2 \cos\theta_w \tau_1(x_1, x_2, y^+) + ..., \quad \tau_1 = 1 \quad \text{at} \quad y^+ = 0, \tag{6.9}$$

$$\tau_{23} = u_\tau^2 \sin\theta_w \tau_2(x_1, x_2, y^+) + ..., \quad \tau_2 = 1 \quad \text{at} \quad y^+ = 0. \tag{6.10}$$

For large y^+, the viscous stresses are expected to decrease, and the turbulent shear stresses are anticipated to be the dominant parts of the total stress; consequently, the expansions for these quantities are also of the form

$$\sigma_{13} = u_\tau^2 \cos\theta_w \sigma_1(x_1, x_2, y^+) + ..., \quad \sigma_1 = 0 \quad \text{at} \quad y^+ = 0, \tag{6.11}$$

$$\sigma_{23} = u_\tau^2 \sin\theta_w \sigma_2(x_1, x_2, y^+) + ..., \quad \sigma_2 = 0 \quad \text{at} \quad y^+ = 0. \tag{6.12}$$

Upon substitution of these expansions in equation (6.5), it follows that the expansions for the leading terms for the streamwise and cross-stream velocities are of the form

$$u_1 = u_\tau \cos\theta_w U^+(x_1, x_2, y^+) + ..., \quad U^+ = 0 \quad \text{at} \quad y^+ = 0, \tag{6.13}$$

$$u_2 = u_\tau \sin\theta_w \Omega^+(x_1, x_2, y^+) + ..., \quad \Omega^+ = 0 \quad \text{at} \quad y^+ = 0, \tag{6.14}$$

where the conditions at $y^+ = 0$ are the no-slip requirements. In addition, because of the definition of y^+, the profile functions U^+ and Ω^+ must satisfy

$$\frac{\partial U^+}{\partial y^+} = \frac{\partial \Omega^+}{\partial y^+} = 1 \quad \text{at} \quad y^+ = 0. \tag{6.15}$$

Upon substitution of the expansions (6.9), (6.10), (6.13) and (6.14) into the momentum equations (6.2) and (6.3), it is readily shown that

$$\frac{\partial \tau_1}{\partial y^+} = O\left(\frac{1}{\text{Re}\, u_\tau^3 \cos\theta_w}\right), \quad \frac{\partial \tau_2}{\partial y^+} = O\left(\frac{1}{\text{Re}\, u_\tau^3 \sin\theta_w}\right). \tag{6.16}$$

It will subsequently be demonstrated that the right sides of both these equations vanish in the limit Re $\to \infty$. It follows that the total shear stress is constant across the wall in direction and magnitude with

$$\tau_1 = \tau_2 = 1 , \tag{6.17}$$

for all y^+. This in turn implies that

$$\sigma_1 + \frac{\partial U^+}{\partial y^+} = 1, \qquad \sigma_2 + \frac{\partial \Omega^+}{\partial y^+} = 1. \tag{6.18}$$

It may be noted that the conditions for σ_1 and σ_2, as well as U^+ and Ω^+ are the same. Furthermore, the second of equations (6.18) is identically satisfied if

$$\sigma_2 = \sigma_1, \qquad \Omega^+ = U^+, \tag{6.19}$$

which implies that the flow in the wall layer is collateral to leading-order; in addition, because the convective terms are negligible to leading-order, any potential dependence of U^+ on (x_1, x_2) is at most parametric. It is evident that the wall-layer flow, in a plane defined by the direction of the wall shear stress and the normal to the wall, is identical to that in a two-dimensional flow to leading order. Consequently, the asymptotic behavior of U^+ is

$$U^+ \sim \frac{1}{\kappa} \log y^+ + C_i \quad \text{as} \quad y^+ \to \infty, \tag{6.20}$$

where κ and C_i are the von Kármán and log-law constants that are generally believed to have universal values of $\kappa = 0.41$ and $C_i = 5.0$. Equations (6.13), (6.14), (6.19) and (6.20) are essentially equivalent to the "three-dimensional law-of-the-wall" proposed by Johnston [8]. It should be noted that models for the cross-flow equation other than (6.19) are mathematically possible but are not believed to be realistic. If, for example, the Reynolds stress σ_2 were arbitrarily specified, it follows that Ω^+ is completely determined from the second of equations (6.18). However, such a model generally implies a flow in which the velocity, velocity gradient, and Reynolds stress vectors all have different skew angles and are not aligned except at the wall itself. However, the convective terms cannot influence the wall-layer motion to leading order, and the main influence on the motion there is the wall shear stress. It therefore appears that such models are not physically realistic in the limit Re $\to \infty$ as the wall layer shrinks to zero thickness.

It should be noted that the issue of the normal extent of collateral flow in the wall layer has been controversial in the literature, and it must be emphasized that higher-order effects (associated with the pressure gradient) are not, in general, negligible for the finite Reynolds numbers occurring in typical experiments; these higher-order effects may well cause the velocity and shear stress profiles to skew across the wall layer. The leading order results chosen in equation (6.19) do not, however, imply that the wall-layer flow is collateral to all orders, as previously suggested by Goldberg and Reshotko [7].

6.4 The Outer Layer

In accordance with experiment, and in analogy with two-dimensional layers, consider a defect law for the streamwise velocity in the outer layer according to

$$u_1 = U_e + u_\tau \cos\theta_w \frac{\partial F_1}{\partial \eta}(x_1, x_2, \eta) + ..., \qquad (6.21)$$

where $\eta = y/\Delta_o$ is the scaled outer variable, and Δ_o is a parameter characteristic of the local boundary-layer thickness. The defect function $\partial F_1/\partial \eta$ must have the following behavior

$$\frac{\partial F_1}{\partial \eta} \sim \frac{1}{\kappa}\log\eta + C_o \quad \text{as} \quad \eta \to 0; \qquad \frac{\partial F_1}{\partial \eta} \to 0 \quad \text{as} \quad \eta \to \infty, \qquad (6.22)$$

and matching of the streamwise velocity using (6.13), (6.20) − (6.22) yields the match condition

$$\frac{U_e}{u_\tau \cos\theta_w} = \frac{1}{\kappa}\log(\text{Re}\, u_\tau \Delta_o) + C_i - C_o. \qquad (6.23)$$

This establishes a relation between u_τ, θ_w and the outer scale Δ_o, and shows that $u_\tau \cos\theta_w / U_e$ is $O(1/\log \text{Re})$ in the limit $\text{Re} \to \infty$.

Now consider the possible form of the expansion for the cross-stream velocity. Taking equations (6.14) and (6.19) and rewriting the asymptotic form of the cross-flow velocity in terms of η using equation (6.23), it is easily shown that

$$u_2 \sim U_e \tan\theta_w + u_\tau \sin\theta_w \left\{ \frac{1}{\kappa}\log\eta + C_o \right\} + ... \quad \text{as} \quad \eta \to 0, \qquad (6.24)$$

in order to ensure matching with the wall layer. However, in the outer layer, the perturbation from the defect law is $O(u_\tau \cos\theta_w)$ and the cross-stream velocity is expected to be of comparable magnitude; balancing with the leading term in (6.24) yields

$$u_\tau \cos\theta_w \sim U_e \tan\theta_w \qquad (6.25)$$

But, $(u_\tau \cos\theta_w)/U_e \to 0$ in the limit $\text{Re} \to \infty$ and consequently θ_w is also small in this limit. A scaled wall skew angle θ_* may be defined by

$$\theta_* = \tan\theta_w \left(\frac{U_e}{u_\tau \cos\theta_w} \right), \qquad (6.26)$$

for which θ_* is $O(1)$ as $\text{Re} \to \infty$. The wall layer expansions (6.11) − (6.14) may be rewritten in terms of θ_* as

$$\sigma_{13} = u_\tau^2 \cos\theta_w \sigma_1 + ..., \qquad \sigma_{23} = \frac{u_\tau^3}{U_e}\theta_* \cos^2\theta_w \sigma_1 + ..., \qquad (6.27)$$

$$u_1 = u_\tau \cos\theta_w U^+ + ..., \qquad u_2 = \frac{u_\tau^2}{U_e}\theta_* \cos^2\theta_w U^+ + ..., \qquad (6.28)$$

from which it is evident that the cross-stream components are smaller than their streamwise counterparts by $O(u_\tau / U_e)$. Note that since θ_w is small, the cosine terms in equations (6.27) and (6.28) may be replaced by unity, but since there is no special advantage to this, the angular dependence will be left explicit. The higher-order terms in the wall-layer expansions have been considered by Degani [10]; those terms associated with the pressure gradient are $O(U_e /(u_\tau \, \mathrm{Re}))$ and $O(U_e^2 /(u_\tau \, \mathrm{Re})^2)$ for shear stress and velocity, respectively, and both are small as $\mathrm{Re} \to \infty$. However, these higher-order terms cause the shear stress and velocity profiles to deviate at *finite* Reynolds numbers from the leading-order collateral flow.

The form of equation (6.24) suggests that the expansion for the cross-flow velocity in the outer layer is

$$u_2 = u_\tau \theta_e \cos \theta_w \left\{ \frac{\partial G_1}{\partial \eta} + \frac{u_\tau}{U_e} \cos \theta_w \frac{\partial G_2}{\partial \eta} + ... \right\}, \tag{6.29}$$

where

$$\frac{\partial G_1}{\partial \eta} \to 1, \qquad \frac{\partial G_2}{\partial \eta} \sim \frac{1}{\kappa} \log \eta + C_o \quad \text{as} \quad \eta \to 0, \tag{6.30}$$

in addition to $\partial G_1 /\partial \eta, \; \partial G_2 /\partial \eta \to 0$ as $\eta \to \infty$. It is evident that the characteristic bulge in the cross-stream profile is a consequence of the asymptotic forms in equation (6.30). In the outer part of the boundary layer, u_2 is dominated by the first term in equation (6.29) which increases as η decreases. However, close to the wall layer the logarithmic variation in the second-order term makes an increasingly negative contribution, and at some location within the logarithmic zone, u_2 reaches a maximum. Below the maximum, the logarithmic term in $\partial G_2 /\partial \eta$ becomes increasingly dominant and acts to reduce the sum of the $O(u_\tau /U_e)$ and $O(u_\tau^2 /U_e^2)$ in the cross-stream velocity to $O(u_\tau^2 /U_e^2)$, thus enabling a match to the wall-layer cross-flow velocity in equation (6.28). Note that this is a similar process to the streamwise outer profile (6.21), where the combination of a $O(1)$ and $O(u_\tau /U_e)$ terms are reduced to $O(u_\tau /U_e)$ in the wall layer. Evidently, it is not necessary to introduce empirical relations for the cross-flow velocity in the wall layer as suggested by other authors (e.g., [6], [7]).

As indicated in equation (6.29), it is necessary to extend the expansions in the outer layer to second order in u_τ /U_e and, for completeness, the extension of the streamwise profile will be considered which is of the form

$$u_1 = U_e(x) + u_\tau \cos \theta_w \frac{\partial F_1}{\partial \eta} + \frac{u_\tau^2}{U_e} \cos^2 \theta_w \frac{\partial F_2}{\partial \eta} + ..., \tag{6.31}$$

where the asymptotic forms of the third term in equation (6.31) are

$$\frac{\partial F_2}{\partial \eta} \sim C_1 \quad \text{as} \quad \eta \to 0, \qquad \frac{\partial F_2}{\partial \eta} \to 0 \quad \text{as} \quad \eta \to \infty. \tag{6.32}$$

The quantities C_o and C_1 in equations (6.22) and (6.32) are functions of (x_1, x_2) which depend on the development in the boundary layer upstream and the particular turbulence model used. The form of the first condition in equation (6.32) is similar to that discussed for the higher-order theory for two-dimensional boundary layers described in §2. Note that in previous analyses (e.g. [7], [11]), a basic expansion parameter denoted by ε_o was taken independent of x and $O(u_\tau / U_e)$ at some representative x station along the plate; a logarithmic asymptotic form for $\partial F_2 / \partial \eta$ was proposed, and this then requires a term $O(\varepsilon_o^2)$ in the wall-layer expansion for the velocity. In the present theory, the expansions are in terms of $u_\tau(x, Re)$, and the entire logarithmic variation in u_1 is captured by the leading terms in both the inner and outer layers. Further comparisons between the two approaches are given by Degani [10]. Upon matching the streamwise velocity u_1, it is easily confirmed that the match condition (6.23) is now refined to

$$\frac{U_e}{u_\tau \cos\theta_w} = \frac{1}{\kappa}\log\{Re\,u_\tau\Delta_o\} + C_i - C_o - \frac{u_\tau \cos\theta_w}{U_e}C_1 + \dots \tag{6.33}$$

It is worthwhile to examine the form of the higher order terms for the total shear stress. It may be shown [4], [10] by substitution in the momentum equations (6.2) and (6.3) and consideration of the (neglected) convective terms that

$$\tau_{13} \sim u_\tau^2 \cos^2\theta_w - \frac{1}{Re\,u_\tau}\frac{U_e}{h_1}\frac{\partial U_e}{\partial x_1}y^+\left\{1 - \frac{u_\tau^2 \cos^2\theta_w}{\kappa^2 U_e^2}\log^2 y^+\right\} + \dots, \tag{6.34}$$

$$\tau_{23} \sim \frac{u_\tau^3\theta_e}{U_e}\cos^2\theta_w - \frac{1}{Re\,u_\tau}K_2 U_e^2 y^+\left\{1 - \frac{u_\tau^2 \cos^2\theta_w}{U_e^2}\frac{\log^2 y^+}{\kappa^2}\right\} + \dots, \tag{6.35}$$

as $y^+ \to \infty$. Note that the higher-order terms in equations (6.34) and (6.35) involve the pressure gradients in the streamwise and cross-stream directions and thereby act to alter the leading-order collateral flow for all flows at finite Reynolds numbers. Using equation (6.34) and writing the inner variable η, it may be shown that the total stress in the streamwise direction in the outer layer must be expanded as

$$\tau_{13} = u_\tau^2 \cos\theta_w\left\{T_1(x_1, x_2, \eta) + \frac{u_\tau \cos\theta_w}{U_e}T_2(x_1, x_2, \eta) + \dots\right\}, \tag{6.36}$$

where, in order to match (6.34)

$$T_1 \sim 1 - \frac{2\beta_s}{\kappa}\eta\log\eta + \dots, \quad T_2 \sim -\frac{\beta_s}{\kappa^2}\eta\log^2\eta + \dots \quad \text{as} \quad \eta \to 0, \tag{6.37}$$

where β_s is a streamwise pressure gradient parameter defined by

$$\beta_s = \frac{\Delta_o}{U_e u_\tau}\frac{1}{h_1}\frac{\partial p_e}{\partial x_1} = -\frac{\Delta_o}{U_e u_\tau}\frac{U_e}{h_1}\frac{\partial U_e}{\partial x}. \tag{6.38}$$

For the cross-stream direction,

$$\tau_{23} = u_\tau^2 \theta_e \cos\theta_w \left\{ \tilde{T}_1(x_1, x_2, \eta) + \frac{u_\tau \cos\theta_w}{U_e} \tilde{T}_2(x_1, x_2, \eta) + ... \right\},$$ (6.39)

where

$$\tilde{T}_1 \sim \frac{2\gamma}{\kappa} \eta \log\eta + ..., \qquad \tilde{T}_2 \sim 1 + \frac{\gamma}{\kappa^2} \eta \log^2\eta + ... \quad \text{as} \quad \eta \to 0,$$ (6.40)

and γ is a cross-stream pressure gradient parameter defined by $\gamma = -\beta_n / \theta_e$ where

$$\beta_n = \frac{\Delta_o}{U_e u_\tau} \frac{1}{h_2} \frac{\partial p_e}{\partial x_2} = -\frac{\Delta_o}{U_e u_\tau} K_2 U_e^2.$$ (6.41)

Since $(U_e / h_2) \partial U_e / \partial x_2 = K_2 U_e^2$. In general, the asymptotic forms (6.37) and (6.40) should be satisfied for any turbulence model adopted for the three-dimensional outer layer flow.

6.5 Turbulence Models

The profile models described in §3 can be used for the three-dimensional wall layer, and here possible models for the outer layer will be considered only. A wide variety of turbulence models exist for three-dimensional flows, many of which represent *ad hoc* extensions of some type of model for two-dimensional boundary layers into three dimensions. The current situation for models for three-dimensional flows is at best unsettled, and here only simple algebraic models will be considered which are consistent with the present asymptotic theory. Define eddy viscosity models of the form

$$\tau_{13} = \varepsilon_1 \frac{\partial u_1}{\partial x_3}, \qquad \tau_{23} = \varepsilon_2 \frac{\partial u_2}{\partial x_3},$$ (6.42)

where ε_1 and ε_2 denote total viscosities (turbulent and kinematic) in the streamwise and cross-stream directions, respectively. It is evident from equation (6.36) that

$$T_1 = \frac{\varepsilon_1}{u_\tau \Delta_o} \frac{\partial^2 F_1}{\partial \eta^2},$$ (6.43)

and it follows from (6.22) that since $T_1 \to 1$ as $\eta \to 0$,

$$\varepsilon_1 \sim u_\tau \Delta_o \kappa\eta + ... \quad \text{as} \quad \eta \to 0.$$ (6.44)

Similarly,

$$\tilde{T}_1 = \frac{\varepsilon_2}{u_\tau \Delta_o} \frac{\partial^2 G_1}{\partial \eta^2}, \qquad \tilde{T}_2 = \frac{\varepsilon_2}{u_\tau \Delta_o} \frac{\partial^2 G_2}{\partial \eta^2},$$ (6.45)

and it follows from equations (6.30) and (6.40) that

$$\varepsilon_2 \sim u_\tau \Delta_o \kappa\eta + ... \quad \text{as} \quad \eta \to 0$$ (6.46)

as well. Turbulence models which are the same in both directions are said to be isotropic, and it is evident from equations (6.44) and (6.45) that the eddy viscosity models described by equations (6.42) must be isotropic, at least for small η. A simple eddy viscosity model has been described in §2 and an extension of this model into three dimensions is

$$\varepsilon_1 = \varepsilon_2 = U_e \delta^* \hat{\varepsilon}, \tag{6.47}$$

where $\hat{\varepsilon}$ is the simple ramp function

$$\hat{\varepsilon} = \begin{cases} \chi & \text{for } \eta > \eta_m \\ \kappa\eta / \eta_* & \text{for } \eta < \eta_m \end{cases}, \tag{6.48}$$

where

$$\eta_* = U_e \delta^* / (u_\tau \Delta_o), \quad \eta_m = \chi\eta_* / \kappa. \tag{6.49}$$

and $\kappa = 0.41$ is the von Kármán constant. For the Cebeci-Smith model, $\chi = K = 0.0168$; a similar expression for the Baldwin-Lomax model is described in §2. Note that, unlike two-dimensional flow, the definition of displacement thickness is ambiguous. However, a length-scale δ^* which is rotationally invariant may be defined by

$$\delta^* = \int_0^\infty \left\{ 1 - \frac{\left(u_1^2 + u_2^2\right)^{1/2}}{U_e} \right\} dy . \tag{6.50}$$

The model described in equation (6.47) is isotropic through the entire outer layer and some data sets suggest that the inferred ratio of eddy viscosities $\varepsilon_2 / \varepsilon_1$ is not unity across the outer layer [4], [12]. Whether or not this is a finite Reynolds number effect is not known but, if necessary, a non-isotropic outer layer model could be introduced in the present formulation in at least two ways, either through introducing non-isotropic, higher-order terms or by adopting a different value of χ in each of the coordinate directions. There does not appear to be a sound basis for making either modification at present, and the model in equations (6.47) and (6.48) should be viewed as the simplest representative (see also [13]) possible turbulence model. Note that corrections for outer-layer intermittency are not included in the model (6.48) since they are found to have negligible influence on the computed profiles.

Self-similar solutions of equations (6.2) and (6.3) have been obtained by Degani et al. [4]. It emerges that such solutions are possible when the mainstream velocity behaves according to a power of distance along a streamline. To calculate such solutions, it is necessary to calculate the numerical solution of four ordinary differential equations for F_1', F_2', G_1', and G_2'; here the prime denotes differentiation with respect to η. It is convenient to define the outer length scale according to $\Delta_o = \delta^* U_e / (u_\tau \cos\theta_w)$ with a Reynolds number based on displacement thickness defined by $Re_{\delta^*} = Re\,U_e \delta^*$, some typical results for $\beta_s = 0.5$ and $\beta_n = -0.2$ are shown in Figure 6.2 for two values of Re_{δ^*}. Here the plotted profiles are the composites

$$\frac{u_1}{U_e} = 1 + \frac{u_\tau \cos\theta_w}{U_e} \left\{ \frac{\partial F_1}{\partial \eta} + U^+(y^+) - \frac{1}{\kappa}\log y^+ - C_i \right\}, \tag{6.51}$$

where U^+ is the wall-layer profile (see §3) and

$$\frac{u_2}{U_e \tan\theta_w} = \frac{\partial G_1}{\partial \eta} + \frac{u_\tau \cos\theta_w}{U_e} \left\{ \frac{\partial G_2}{\partial \eta} + U^+ - \frac{1}{\kappa}\log y^+ - C_i \right\}. \tag{6.52}$$

Note that the second-order velocity term in equation (6.51) is omitted because it is generally found to have a negligible effect on the streamwise velocity profile.

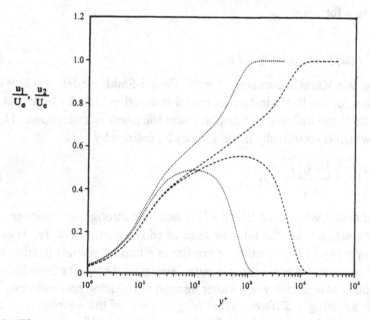

Figure 6.2   The streamwise and cross-stream velocity profiles for $\beta_s = 0.5$ and $\beta_n = -0.2$
            ...., $Re_{\delta^*} = 50{,}000$.

For $Re_{\delta^*} = 5000$, the match condition gives $u_\tau/U_e = 0.0369$, while for $Re_{\delta^*} = 50{,}000$, the computed value is $u_\tau/U_e = 0.0306$. Although both sets of profiles implicitly contain a logarithmic variation, the behavior is not evident in Figure 6.2 at the lower Reynolds number even on a logarithmic scale.   However, at the higher Reynolds number, it is clearly visible for both profiles.   Another noteworthy aspect concerns the apparent extent of collateral flow; for $Re_{\delta^*} = 5000$, it only extends to $y^+ = 5$, whereas the comparable value for $Re_{\delta^*} = 50{,}000$ is $y^+ = 50$.   The results in Figure 6.2 may partially explain why the issue of collateral flow has been controversial in the experimental literature and why the logarithmic behavior has not been observed in the cross-flow profile in data obtained at relatively low Reynolds numbers.

Other aspects of the three-dimensional solutions are discussed by Degani et al. [4] who show, for example, that the maximum in the cross-stream velocity generally occurs in the overlap zone.

6.6 Prediction Methods

In this section, the application of the asymptotic theory for boundary-layer prediction methods will be considered. As previously shown, for attached turbulent boundary-layer flows, the wall layer exhibits a universal similarity form, and the profile $U^+(y^+)$ can be represented by the wall-layer model described in §3; the conventional alternative is a model based on the mixing-length formulation with the Van Driest damping factor (see §3).

A possible selection for the outer scale is $\Delta_o = \delta^* U_e / u_\tau$, which is appropriate for self-similar flows [4]; this scale is analogous to the Levy-Lees transformation for laminar flows in which a "local" self-similar behavior is assumed. The transformation is convenient from a numerical standpoint because generally the boundary layer in a varying pressure gradient will not experience either significant growth or decrease in the η coordinate, and the largest value of η in the numerical mesh can be set at some large fixed number. There is, however, an additional problem for turbulent boundary layers in that the inner and outer layers can grow at slightly different rates. In an embedded function algorithm [14], it is necessary to match a varying outer solution to the self-similar wall layer, and for this purpose, it is convenient to perform the matching at a fixed, but large, value of y^+ as the calculation proceeds.

In general, the outer variable η is related to the inner variable y^+ by $\eta = y^+/(u_\tau \Delta_o \text{Re})$. Let y_m^+ denote the value of y^+ where matching of the outer-layer functions to the wall layer takes place. Typical values for y_m^+ may range from 60 to 120, but the specific value used here has no effect on the calculated results provided y_m^+ is large; however, larger values of y_m^+ normally results in increased savings in the number of mesh points in the outer layer. Suppose that η_2 denotes the fixed value of the smallest value of η used in the numerical mesh for the solution of the outer layer equations. A convenient choice for the outer scale is

$$\Delta_o = (y_m^+ / \eta_2)/(u_\tau/\text{Re}),\tag{6.53}$$

in order to assure that the matching takes place at fixed y_m^+. It can be shown [5] that near the surface at the bottom of the outer layer,

$$\frac{u_1}{U_e} \sim 1 + u_* \left\{ \frac{1}{\kappa} \log \eta + C_o - \frac{2\beta_s}{\kappa^2} \eta \log \eta + ... \right\} \cdot$$

$$+ u_*^2 \left\{ C_1 - \frac{\beta_s}{\kappa^3} \eta \log^2 \eta + ... \right\},\tag{6.54}$$

and

$$\frac{u_2}{U_e} \sim u_*\theta_* \left\{ 1 + \frac{2\gamma}{\kappa^2} \eta \log \eta + ... \right\}$$

$$+ u_*^2\theta_* \left\{ \frac{1}{\kappa} \log \eta + C_o + \frac{\gamma}{\kappa^3} \eta \log^2 \eta + ... \right\}, \tag{6.55}$$

as $\eta \to 0$, where $u_* = u_\tau \cos\theta_w / U_e$, θ_* is defined by equation (6.26), $\gamma = \beta_n / \theta_*$ and β_s and β_n are the streamwise and cross-stream pressure gradient parameters defined by equations (6.38) and (6.41) respectively. The values of C_o and C_1 are to be determined. If η_3 denotes the next mesh point from the wall (after η_2), and η_3 and η_2 are taken sufficiently small so that the asymptotic forms are appropriate, C_o and C_1 may be eliminated to obtain the point-slope conditions:

$$\frac{u_{13} - u_{12}}{U_e} = \frac{u_*}{\kappa} \log\left(\frac{\eta_3}{\eta_2}\right) - \frac{2\beta_s u_*}{\kappa^2} \{ \eta_3 \log \eta_3 - \eta_2 \log \eta_2 \}$$

$$- \frac{\beta_s u_*^2}{\kappa^3} \{ \eta_3 \log^2 \eta_3 - \eta_2 \log^2 \eta_2 \} + ..., \tag{6.56}$$

$$\frac{u_{23} - u_{22}}{U_e} = \frac{u_*^2\theta_*}{\kappa} \log\left(\frac{\eta_3}{\eta_2}\right) + \frac{2\gamma u_*\theta_*}{\kappa^2} \{ \eta_3 \log \eta_3 - \eta_2 \log \eta_2 \}$$

$$+ \frac{\gamma u_*^2\theta_*}{\kappa^3} \{ \eta_3 \log^2 \eta_3 - \eta_2 \log^2 \eta_2 \} + ..., \tag{6.57}$$

where u_{i3}, u_{i2} denote the values of u_i ($i = 1, 2$) at η_3 and η_2 respectively. The conditions can be used in the Thomas algorithm to initiate a back-substitution across the boundary layer using an implicit marching procedure [5].

A full discussion of the numerical procedure is given in reference [5], and here only a brief summary is given. At any station, u_τ and θ_w are determined in a general iterative procedure. First Δ_o and u_* are estimated from the previous station in a streamline coordinate system, and from these the pressure gradient parameters β_s and β_n are calculated. A full numerical solution of the momentum equation using conditions (6.56) and (6.57) produces values of u_1 and u_2 across the entire boundary layer. Using the match condition and rewriting η in terms of y^+, it can be shown [5] that

$$\frac{u_{12}}{U_e} \sim u_* \left\{ \frac{1}{\kappa} \log y_m^+ + C_i \right\} - \frac{u_*\beta_s}{\kappa^2} \eta_2 \log \eta_2 \left\{ 2 + \frac{u_*}{\kappa} \log \eta_2 \right\} + ..., \tag{6.58}$$

$$\frac{u_{22}}{U_e} \sim u_*^2\theta_* \left\{ \frac{1}{\kappa} \log y_m^+ + C_i \right\} - \frac{u_*\beta_n}{\kappa^2} \eta_2 \log \eta_2 \left\{ 2 + \frac{u_*}{\kappa} \log \eta_2 \right\} + ..., \tag{6.59}$$

which permits evaluation of new estimates of u_* and θ_*. The procedure is repeated until convergence; usually only a single iteration is required.

Some typical results for a three-dimensional boundary layer approaching a wedge mounted on the plate are shown in Figure 6.3. Here x_2 denotes individual streamlines originating upstream of the wedge, with $x_2 = 0$ denoting the symmetry plane of the wedge. The results shown as "full calculation" were initiated upstream of the wedge from a laminar flow solution; the calculation proceeds through an assumed transition zone into a fully turbulent regime. The full calculation scheme computes the flow all the way to the wall using a standard mixing-length formulation. The curves labeled "embedded function" are based on the present asymptotic theory and were initiated from the "full calculation" method results once the computation had reached a streamwise station within the fully turbulent zone. Note that there are essentially no differences between the results, even though the embedded function scheme uses 50% or less grid points. Similar results using this approach has been obtained for the other computed flow quantities [5], as well as in compressible flow calculations [14], [15].

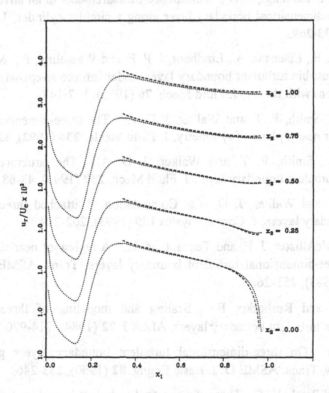

Figure 6.3 Streamwise variation of the friction velocity for the wedge flow for five selected cross-stream locations. ⋯⋯ Full calculation; ──────── embedded-function.

6.7 Summary

Asymptotic theory reveals the structure of the attached three-dimensional turbulent boundary layer and restricts the types of eddy viscosity models that can be used in order to maintain a self-consistent structure. The results can be used to structure an efficient numerical prediction algorithm, which has been termed the "embedded function" method. In this scheme, it is not necessary to use conventional inner region turbulence models or to compute the flow all the way to the wall. Skin friction, heat transfer rates, and wall skew angles can be computed with no degradation in accuracy by simply calculating an outer region numerical solution and matching the results to an analytical wall-layer solution. The algorithm is efficient and accurate and results in substantial savings in mesh points; in principle, the method may easily be adapted in full Navier-Stokes simulations of turbulent flow.

References

1. Fernholz, H. H. and Vagt, J.-D.: Turbulence measurements in an adverse-pressure-gradient three-dimensional boundary layer along a circular cylinder, J. Fluid Mech. **111** (1981), 233-269.

2. Van den Berg, B., Elsenaar, A., Lindhout, J. P. F. and Wesseling, P.: Measurements in an incompressible turbulent boundary layer, under infinite swept-wing conditions, and comparisons with theory, J. Fluid Mech. **70** (1975), 127-147.

3. Degani, A. T., Smith, F. T. and Walker, J. D. A.: The three-dimensional turbulent boundary layer near a plane of symmetry, J. Fluid Mech., **234** (1992), 329-360.

4. Degani, A. T., Smith, F. T. and Walker, J. D. A.: The structure of a three-dimensional turbulent boundary layer, J. Fluid Mech. **250** (1993), 43-68.

5. Degani, A. T. and Walker, J. D. A.: Computation of attached three-dimensional turbulent boundary layers, J. Comp. Physics **109** (1993), 202-214.

6. Pierce, F. J., McAllister, J. E. and Tennant, M. H.: A review of near-wall similarity models in three-dimensional turbulent boundary layers, Trans. ASME I: J. Fluids Engng. **105** (1983), 251-262.

7. Goldburg, U. and Reshotko, E.: Scaling and modeling of three-dimensional pressure-driven turbulent boundary layers, AIAA J. **22** (1984), 914-920.

8. Johnston, J. P.: On three-dimensional turbulent boundary layers generated by secondary flow, Trans. ASME D: J. Basic Engng. **82** (1960), 233-246.

9. Nash, J. F. and Patel, V. C.: Three-dimensional turbulent boundary layers, Atlanta: SBC Technical Book (1972).

10. Degani, A. T.: The three-dimensional turbulent boundary layer – theory and application, Ph.D. Thesis, Lehigh University (1991).

11. Mellor, G. L.: The large Reynolds number asymptotic theory of turbulent boundary layers, Intl. J. Engng. Sci. **10** (1972), 851-873.

12. Anderson, S. D. and Eaton, J. K.: Reynolds stress development in pressure-driven three-dimensional turbulent boundary layers, J. Fluid Mech. **202** (1989), 262-294.

13. Wie, Y. S. and DeJarnette, F. R.: Numerical investigation of three-dimensional separation using the boundary-layer equations" AIAA Paper 88-0617 (1988).

14. Walker, J. D. A., Ece, M. C. and Werle, M. J.: An embedded function approach for turbulent flow prediction, AIAA J. **29** (1991) 1810-1818.

15. Degani, A. T., Walker, J. D. A., Ersoy, E. and Power, G.: On the application of algebraic turbulence models to high Mach number flows, AIAA Paper 91-0616 (1991).

7. Compressible Flow

7.1 Introduction

The prediction of high-speed compressible flows is hampered by many practical problems, some of which are computational in nature; the major difficulties are associated with the fact that the density now varies appreciably across the boundary layer (even for an adiabatic wall), and that there is considerable uncertainty about how well existing turbulence models can represent existing data. For supersonic and hypersonic flows, the data base is not as extensive as for subsonic flows, and the relative thinness of the boundary layer, as well as experimental difficulties associated with taking measurements in a flow with heat transfer, mitigate against obtaining accurate data, especially near the surface. Indeed, there is significant uncertainty concerning the form of the compressible "law of the wall" [1], [2], and inevitably this has led to a large number of different models for compressible flows.

In this section, an emphasis will be on the simplest possible algebraic turbulence models, but many of the asymptotic results discussed are independent of a particular type of turbulence model. Algebraic closures constitute the simplest type of mathematical model capable of representing the influence of turbulence near a wall, but even if higher-order, more complicated closure schemes are contemplated, many of the most critical modeling issues concern the nature of the functional form of the velocity and temperature profiles in the overlap zone between the inner and outer portions of the boundary layer. Examples of modern algebraic models include the Cebeci-Smith [3], Baldwin-Lomax [4] and Johnson-King [5] models. The essence of these models is a simple ramp function for the eddy viscosity function in the outer part of the boundary layer, which behaves linearly near the wall and then abruptly switches to a constant in the outer part of the boundary layer, whose value usually depends on local flow conditions. Eddy viscosity models are conventionally modified to a mixing-length formulation in the wall layer; the mixing length is defined to be linear in distance normal to the wall but is then reduced toward the wall by multiplication by a Van Driest damping factor [3]. Algebraic turbulence models produce good results for attached turbulent flows at low to moderate mainstream speeds [6]; here an extensive base of reliable data exists. Such models are either often used without modification in prediction methods for supersonic turbulent boundary layers or extrapolated in some *ad hoc* manner. Here the emphasis will be on attached turbulent flows, since the uncertainties surrounding the validity of current models for compressible separated flows are even greater than in the incompressible case. The generalization of the "law of the wall" to the compressible case has been controversial, and a number of different forms have been proposed. The most popular form of these is the "effective velocity" approach due to Van Driest [7] which will be reviewed here subsequently. In addition, an alternative approach based on the Howarth-Dorodnitsyn compressibility transformation will be described [2].

7.2 Governing Equations

Let (x^*, y^*) denote Cartesian coordinates, with corresponding mean velocity components (u^*, v^*), measuring distance tangential and normal to the wall, respectively; here and throughout, superscript asterisks denote a dimensional quantity. The mean total enthalpy is defined by $H^* = c_p^* T^* + u^{*2}/2$, where T^* is the static temperature and c_p^*, the specific heat at constant pressure, is assumed constant. Dimensionless variables may be defined in terms of a reference length L_{ref}^*, a velocity U_{ref}^*, a viscosity μ_{ref}^* and a density ρ_{ref}^*. The static temperature and total enthalpy are non-dimensionalized by T_{ref}^* and $c_p^* T_{ref}^*$, respectively, and H is thus given by

$$H = T + \frac{(\gamma - 1)}{2} M_{ref}^2 u^2, \qquad (7.1)$$

where the reference Mach number is defined by $M_{ref}^2 = U_{ref}^{*2} / (\gamma R T_{ref}^*)$, with γ being the constant specific heat ratio and R the gas constant. In dimensionless variables, the governing boundary-layer equations are

$$\frac{\partial}{\partial x}(\rho u) + \frac{\partial}{\partial y}(\rho v) = 0, \qquad (7.2)$$

$$\rho u \frac{\partial u}{\partial x} + \rho v \frac{\partial u}{\partial y} = -\frac{dp_e}{dx} + \frac{\partial \tau}{\partial y}, \qquad (7.3)$$

$$\rho u \frac{\partial H}{\partial x} + \rho v \frac{\partial H}{\partial y} = \frac{\partial q}{\partial y}, \qquad (7.4)$$

where τ and q are the total stress and energy flux, respectively, defined by

$$\tau = \frac{\mu}{Re} \frac{\partial u}{\partial y} + \sigma, \qquad (7.5)$$

$$q = \frac{\mu}{Pr\,Re} \left\{ \frac{\partial H}{\partial y} + \frac{(\gamma - 1)}{2} M_{ref}^2 \left(1 - \frac{1}{Pr} \right) u \frac{\partial u}{\partial y} \right\} + \phi. \qquad (7.6)$$

Here σ and ϕ denote the Reynolds stress and a turbulent heat flux according to

$$\sigma = -\rho \overline{u'v'}, \qquad \phi = -\rho \overline{v'H'}. \qquad (7.7)$$

The turbulence quantities $\overline{u'v'}$ and $\overline{v'H'}$ have been made dimensionless with respect to U_{ref}^{*2} and $c_p^* U_{ref}^* T_{ref}^*$, respectively, and the Reynolds number Re and Prandtl number Pr are given by

$$\text{Re} = \frac{\rho_{ref}^* L_{ref}^* U_{ref}^*}{\mu_{ref}^*}, \quad \text{Pr} = \frac{\mu^* c_p^*}{k^*}, \qquad (7.8)$$

where μ^* and k^* are the dimensional absolute viscosity and thermal conductivity of the gas. The Prandtl number is taken to be $O(1)$ and constant while Re is assumed to be large. Note that a complication not present in the incompressible cases discussed in §2 – §6 is the appearance of the dimensionless density in equations (7.2) – (7.6).

The ideal equation of state is used to relate the mainstream pressure to the density and static temperature variation across the boundary layer giving

$$p_e = \rho T / (\gamma M_{ref}^2). \qquad (7.9)$$

For a steady mainstream flow, the total mainstream enthalpy H_e is constant and is related to the mainstream speed U_e by $U_e^2 = \alpha H_e / \{(\gamma - 1)M_{ref}^2\}$, where α is related to the mainstream Mach number M_e by

$$\alpha = (\gamma - 1)M_e^2 \left\{ 1 + \frac{\gamma - 1}{2} M_e^2 \right\}^{-1}. \qquad (7.10)$$

Since the pressure is constant across the boundary-layer at any streamwise location $\rho T = \rho_e T_e$, and it is easily shown that in the boundary layer

$$\frac{T}{T_e} = \left\{ 1 + \frac{1}{2}(\gamma - 1)M_e^2 \right\} \left\{ \frac{H}{H_e} - \frac{1}{2}\alpha \frac{u^2}{U_e^2} \right\}. \qquad (7.11)$$

Thus the solution of equations (7.3) and (7.4) for u and H will also produce the static temperature from (7.11). The boundary conditions for equations (7.2) – (7.4) are

$$u = v = 0 \quad \text{at} \quad y = 0; \quad u \to U_e \quad \text{as} \quad y \to \infty, \qquad (7.12)$$

$$H = H_w \quad \text{at} \quad y = 0; \quad H \to H_e \quad \text{as} \quad y \to \infty. \qquad (7.13)$$

Finally, the viscosity is taken to be a known function of T alone, such as the Sutherland formula or other empirical correlations [2].

7.3 Conventional Methods

In a conventional prediction scheme [3], algebraic models are used to represent the turbulence quantities by

$$-\overline{\rho u'v'} = \rho\varepsilon\frac{\partial u}{\partial y}, \qquad -\overline{\rho v'H'} = \rho\varepsilon_H\frac{\partial H}{\partial y}, \qquad (7.14)$$

where ε and ε_H are the eddy viscosity and conductivity functions respectively. Conventional models for eddy viscosity use a formula in two tiers. Letting ε_o denote the eddy viscosity in the outer tier, the simplest model is due to Cebeci-Smith [3] with $\varepsilon_0 = K U_e \delta^*$, where $K = 0.0168$ and δ^* is the (incompressible) displacement thickness defined by

$$\delta^* = \int_0^\infty \left(1 - \frac{u}{U_e}\right) dy . \qquad (7.15)$$

In the inner region (comprising part of the outer layer and the wall layer), a mixing-length formulation is adopted with

$$\varepsilon_i = \ell^2 D^2 |\partial u/\partial y|, \qquad (7.16)$$

where $\ell = \kappa y$ is the mixing length and $\kappa = 0.41$ is the von Kármán constant; in addition, D is the damping factor modified for compressibility [3] by

$$D = 1 - \exp\left\{-\frac{\rho_w\mu_w}{\rho\mu}\left(\frac{\rho}{\rho_w}\right)^{3/2}\frac{y^+}{A^+}\right\}. \qquad (7.17)$$

Here A^+ is a constant, usually taken to be equal to 26, and the inner coordinate $y^+ = \rho_w u_\tau \text{Re}/\mu_w$, where u_τ is the usual friction velocity defined in terms of the wall shear stress by $u_\tau = (\tau_w\rho_w)^{1/2}$; here and throughout, a subscript w indicates that a quantity is evaluated at the wall.

A similar model (the Baldwin-Lomax model [4]) was mentioned in §2 and differs from the above model only in the choice of ε_o for the outer tier. It may be noted that equations (7.16) and (7.17) represent an *ad hoc* extrapolation of a turbulence model originally fine-tuned for incompressible flow. A variety of embellishments have been added to the basic model over the years [3] to account for various effects; most of these have negligible impact and will not be considered here. In practice, the edge of the outer and inner tier is determined at any streamwise location by the requirement that both ε_i and ε_o give the same value. For the energy equation, it is common to assume that $\varepsilon_H = \varepsilon/\text{Pr}_t$, where Pr_t is the turbulent Prandtl, normally assumed to be constant having a value of about 0.92. This last assumption is questionable, particularly in the wall layer where measurements [8] suggest a substantial variation in Pr_t near the wall, even in a low speed flow.

A conventional treatment for the inner layer is normally based on the Van Driest transformation which is obtained as follows. In the wall layer, u is expected to be $O(u_\tau)$ (i.e. $u = u_\tau U^+(y^+) + ...$), and upon substitution of the usual definition of y^+ and equation

(7.16) into equation (7.3), the familiar result that $\tau = \rho_w \, u_\tau^2$ is constant across the wall layer is obtained and

$$\frac{\rho}{\rho_w} \kappa^2 y^{+2} D^2 \left(\frac{\partial U^+}{\partial y^+} \right)^2 + \frac{\mu}{\mu_w} \frac{\partial U^+}{\partial y^+} = 1. \tag{7.18}$$

For low values of M_e, the ratios ρ/ρ_w and μ/μ_w are effectively unity to leading order and equation (7.18) may be integrated immediately to give U^+ throughout the wall layer. To account for the density and viscosity ratios in equation (7.18), first introduce the Chapman-Rubesin law $\rho\mu = \rho_w \, \mu_w$, which can be shown to be valid in the wall layer [2]. In addition, near the wall $H \cong H_w + \dots$ to leading order, and it can be shown using equation (7.11)

$$\frac{\rho_w}{\rho} \cong 1 - a^2 U^{+2} + \dots, \qquad a^2 = \frac{(\gamma - 1) \, M_{ref}^2 u_\tau^2}{2 \, H_w}. \tag{7.19}$$

Substitution into equation (7.18) yields

$$\frac{\kappa^2 y^{+2} D^2}{1 - a^2 U^{+2}} \left(\frac{dU^+}{dy^+} \right)^2 + \left(1 - a^2 U^{+2} \right) \frac{dU^+}{dy^+} = 1, \tag{7.20}$$

which may be integrated to find an expression for $U^+(y^+)$ with $U^+ = 0$ at $y^+ = 0$.

It may be readily verified that an incompressible law of the wall

$$U^+ \sim \frac{1}{\kappa} \log y^+ + C_i \quad \text{as} \quad y^+ \to \infty \tag{7.21}$$

will not satisfy equation (7.20) and the "compressible law" according to this formulation is more complex. Setting $D = 1$ for large y^+, it may be easily shown that

$$\frac{1}{a} \sin^{-1}(aU^+) = \frac{1}{\kappa} \log y^+ + C, \tag{7.22}$$

where C is a constant. Alternatively,

$$\frac{1}{A} \sin^{-1} \left(A \frac{u}{U_e} \right) = \frac{u_\tau}{U_e} \left\{ \frac{1}{\kappa} \log y^+ + C \right\}, \tag{7.23}$$

where $A = \sqrt{\alpha / 2}$ and α is given by equation (7.10). The value of C may be found by numerical integration of equation (7.20) and, in models which contain correlations for surface and pressure effects in D [3], the value of C will vary in the streamwise direction.

It may be noted that the behavior in equation (7.23) is similar to the incompressible law of the wall, but at the same time considerably different involving an inverse sine function. The relation (7.23) has sometimes been used to define an "effective velocity" by $\tilde{u} = (1/A)\sin^{-1}(Au/U_e)$ across the entire boundary layer, and Maise and McDonald [9] showed that, when this transformation was applied to compressible data for adiabatic walls, a large amount of data could be compressed to a single curve consisting of an incompressible profile for $\tilde{u}$. The Van Driest transformation in one form or another permeates much of the compressible flow literature. In the next section, an alternative approach will be considered.

7.4 The Howarth-Dorodnitsyn Transformation

There are several aspects of the Van Driest near-wall formulation which may be regarded as disadvantageous. First the transformation is based on a specific mixing-length formulation in a presumed overlap region between the inner and outer layers; the physical basis for this model is not compelling. In contrast to the incompressible case, where both the outer eddy viscosity and the inner mixing-length formulations produce the "law of the wall" in the region where the mixing length is essentially linear, other models (such as an eddy viscosity model) produce a different functional form in the overlap zone [10]. This situation is at best disquieting, and the problem lies in the uncertainty surrounding the form of the appropriate "compressible law of the wall"; unfortunately, the data base for compressible flow is less copious, as well as less definitive on this issue than for incompressible flow. A second problem is that it is difficult to assess the circumstances under which the Van Driest transformation becomes invalid with increasing surface heat flux and/or increasing mainstream Mach number. Lastly, the inverse sine behavior contained in the formulation can be awkward in a prediction algorithm.

An alternative formulation [2] utilizes the Howarth-Dorodnisyn variables which are known to be appropriate variables for the description of compressible laminar boundary layers [11]. Define

$$Y = \int_0^y \frac{\rho}{\rho_o} dy \qquad \tilde{v} = \frac{\overline{\rho v}}{\rho_o} + u \frac{\partial Y}{\partial s}, \qquad (7.24)$$

where ρ_o is a reference density (to be defined). In these variables, equations (7.2) – (7.4) become [2]

$$\frac{\partial u}{\partial x} + \frac{\partial \tilde{v}}{\partial Y} = 0, \qquad (7.25)$$

$$u \frac{\partial u}{\partial x} + \tilde{v} \frac{\partial u}{\partial Y} = \frac{1}{M_e} \frac{dM_e}{dx} \left\{ \frac{H}{H_e} U_e^2 - \frac{1}{2} \alpha u^2 \right\} + \frac{1}{\rho_o} \frac{\partial \tau}{\partial Y}, \qquad (7.26)$$

$$u \frac{\partial H}{\partial x} + \tilde{v} \frac{\partial H}{\partial Y} = \frac{1}{\rho_o} \frac{\partial q}{\partial Y}, \qquad (7.27)$$

where τ and q are defined by equations (7.5) and (7.6) with $\partial/\partial y$ replaced by $\rho/\rho_o\partial/\partial Y$. The boundary conditions are also given by equations (7.12) and (7.13) with $\tilde{v} = 0$ at $Y = 0$.

First consider the wall layer and define the dimensionless friction velocity by

$$u_\tau^2 = \frac{\mu_w}{\rho_w\,Re}\frac{\partial u}{\partial y}\bigg|_{y=0}. \tag{7.28}$$

For compressible flow, the velocity in the wall layer is taken [2] to be a function of ρ_w, ρ_o, μ_o and Y, where ρ_o and μ_o are the density and viscosity evaluated at the reference temperature T_o; this suggests a functional relation of the form

$$u = u_{\tau_o}f(\rho_o u_{\tau_o}Y/\mu_o) + ..., \quad u_{\tau_o} = \left(\frac{\rho_w}{\rho_o}\right)^{1/2}u_\tau. \tag{7.29}$$

Here u_{τ_o} is a general velocity scale which reduces to u_τ in the case of incompressible flow or if the reference condition is defined at the wall (as is the case for laminar boundary layers). The central hypothesis in the alternative analysis [2] is that the generalization to the compressible wall layer is of the form

$$u = u_{\tau_o} U^+(Y^+) + ..., \quad Y^+ = Y/\Delta_i, \tag{7.30}$$

where Y^+ is the scaled inner variable and, as in incompressible flow, U^+ satisfies

$$U^+ = 0, \quad \frac{\partial U^+}{\partial Y^+} = 1 \quad \text{at} \quad Y^+ = 0, \tag{7.31}$$

$$U^+ \sim \frac{1}{\kappa}\log Y^+ + C_i, \quad \text{as} \quad Y^+ \to \infty. \tag{7.32}$$

Here $\kappa = 0.41$ is the von Kármán constant and C_i is the inner-region logarithmic law constant taken to have a universal value of 5.0. The main premise [2] is that the streamwise profile conforms to the universal logarithmic law in an overlap zone, but in terms of equation (7.30) and (7.32), and the scaled Howarth-Dorodnitsyn variable Y. It may be readily verified that the inner scale is given by $\Delta_i = \mu_w\rho_w/(\rho_o^2\,Re\,u_{\tau_o})$ which reduces to the conventional scale $\mu_w/(\rho_w\,Re\,u_\tau)$ for incompressible flow. For compressible flow, Y^+ is proportional to the conventional scaled inner-region coordinate y^+ close to the wall; elsewhere in the wall layer Y^+ contains an implicit density variation and represents, in effect, a density-weighted scaled distance from the wall.

Substitution in equation (7.26) shows that the total stress is constant across the wall layer to leading order with $\tau = \rho_w u_\tau^2 = \rho_o u_{\tau_o}^2$. This suggests the following expansion for the Reynolds stress in the wall layer

$$\sigma = \rho_o \, u_{\tau_o}^2 \sigma_1(Y^+) + ..., \tag{7.33}$$

where $\sigma_1 = 0$ at $Y^+ = 0$ and $\sigma_1 \to 1$ as $Y^+ \to \infty$. As shown in reference [2], the Chapman-Rubesin viscosity law [11] $\rho\mu = \rho_w\mu_w$ may be used in the wall layer, and it is easily demonstrated that the wall-layer equation is

$$\frac{dU^+}{dY^+} + \sigma_1 = 1. \tag{7.34}$$

This is the same result as for incompressible flow (cf. §2), and again the implication is that if the profile U^+ is known, the Reynolds stress is determined and vice-versa.

For the energy equation, it follows from equation (7.27) that $\partial q / \partial Y^+ = 0$ to leading order, and hence $q = q_w$ across the wall layer, where q_w is a dimensionless heat flux (from fluid to wall) at the wall, defined by

$$q_w = \frac{\rho_o u_{\tau_o}}{Pr} \frac{\partial H}{\partial Y^+}\bigg|_{Y^+=0}. \tag{7.35}$$

It can then be inferred [2] that the expansions in the wall layer are of the form

$$H = H_w + \frac{q_w}{\rho_o u_{\tau_o}} \theta^+ + ..., \quad \phi = q_w \, \phi_1 + ..., \tag{7.36}$$

where ϕ_1 and θ^+ are functions of Y^+ and Pr. A similar analysis [2], [12] to that discussed in §3 shows that the profile function θ^+ should exhibit a dependence on $\sqrt{Pr}$ according to

$$\theta^+ = \theta^+\left(Y_\theta^+\right), \quad Y_\theta^+ = Y^+ \sqrt{Pr}, \tag{7.37}$$

and similar results, as obtained in §4 for incompressible flow, follow:

$$\frac{1}{Pr^{1/2}} \frac{\partial \theta^+}{\partial Y_\theta^+} + \phi_1 = 1, \tag{7.38}$$

where θ^+ and Y_θ^+ satisfy

$$\theta^+ = 0, \quad \frac{d\theta^+}{dY_\theta^+} = \sqrt{Pr}, \quad \phi_1 = 0 \quad \text{at} \quad Y_\theta^+ = 0, \tag{7.39}$$

$$\theta^+ \sim \frac{1}{\kappa_\theta} \log Y_\theta^+ + B_i, \quad \phi_1 \to 1 \quad \text{as} \quad Y_\theta^+ \to \infty. \tag{7.40}$$

Note that as in §4, κ_θ plays the role of the von Kármán constant but is not a constant and depends on local flow conditions.

To be consistent with the wall-layer solutions, it is easily inferred [2] that the outer-layer expansions are of the form

$$u = U_e(x) + u_{\tau_e} \frac{\partial F_1}{\partial \eta} + \ldots, \quad H = H_e \{1 + q_* \, \Theta_1 + \ldots\}, \tag{7.41}$$

where H_e denotes the mainstream enthalpy, $q_* = q_w /(\rho_o \, u_{\tau_e} \, H_e)$ and $\eta = Y/\Delta_o$ is the scaled outer variable. The total stress and heat flux are written as

$$\tau = \rho_o \, u_{\tau_e}^2 \, T_1(x,\eta) + \ldots, \quad q = q_w \, Q_1(x,\eta) + \ldots \tag{7.42}$$

These functions must satisfy

$$\frac{\partial F_1}{\partial \eta} \sim \frac{1}{\kappa} \log \eta + C_o, \quad \Theta_1 \sim \frac{1}{\kappa_\theta} \log \eta + B_o; \quad T_1, \, Q_1 \to 1 \tag{7.43}$$

as $\eta \to 0$ and

$$\frac{\partial F_1}{\partial \eta}, \quad \Theta_1, \, T_1, \, Q_1 \to 0 \quad \text{as} \quad \eta \to \infty. \tag{7.44}$$

Matching of the velocity and total enthalpy profiles to the wall-layer solutions leads to the two match conditions

$$\frac{U_e}{u_{\tau_e}} = \frac{1}{\kappa} \log \left\{ \frac{\rho_o^2}{\mu_w \rho_w} \mathrm{Re} \, u_{\tau_e} \Delta_o \right\} + C_i - C_o, \tag{7.45}$$

$$\frac{1 - I_w}{q_*} = \frac{1}{\kappa_\theta} \log \left\{ \frac{\rho_o^2}{\mu_w \rho_w} \mathrm{Re} \, u_{\tau_e} \Delta_o \, \mathrm{Pr}^{1/2} \right\} + B_i - B_o, \tag{7.46}$$

where $I_w = H_w / H_e$. These conditions relate the wall shear and heat flux to the outer-length scale Δ_o and the logarithmic-law "constants". It follows from equation (7.45) and (7.46) that $\left(u_{\tau_e} / U_e \right)$, $q_* \to 0$ as $\mathrm{Re} \to \infty$. In a manner similar to that described for incompressible flow in §4, it can be shown, using equations (7.45) and (7.46), that to leading order

$$\frac{\kappa_\theta}{\kappa} = \frac{q_* U_e}{(1 - I_w) u_{\tau_e}} \quad \text{as} \quad \mathrm{Re} \to \infty. \tag{7.47}$$

This should be compared to the incompressible result (4.21). The asymptotic result (7.47) is used here to define a relation for κ_θ, and it may then be shown that the full match conditions (7.45) and (7.46) reduce to

$$\frac{\kappa_\theta}{\kappa} = \frac{C_i - C_o - (1/2\kappa) \log \mathrm{Pr}}{B_i - B_o}, \tag{7.48}$$

which is the same as equation (4.22) for incompressible flow.

7.5 Turbulence Models

The results described in §7.4 are independent of any particular turbulence model; here, however, specific turbulence models will now be described for both parts of the boundary layer. For the outer layer, simple algebraic models will again be considered, but modifications of the conventional formulae are required to account for compressibility and to ensure that both the velocity and total enthalpy are logarithmic in the scaled Howarth-Dorodnitsyn variable. Simple algebraic models for *total* stress and heat flux are of the form

$$\tau = \rho \varepsilon \frac{\partial u}{\partial y}, \quad q = \rho \varepsilon_H \frac{\partial H}{\partial y}. \tag{7.49}$$

Note that here, as in §2, §4 and §6, it is the total stress and heat flux which are modeled in this manner. This is different from the conventional approach in equations (7.14) where only the turbulence quantities are represented in this manner. The two approaches are clearly closely related, but the present method is much more convenient, especially with regard to the matching of total stress and heat flux in the overlap zone. To ensure a logarithmic behavior as indicated in equations (7.43), it can be shown that ε, for example, must behave as $\rho_0^2 u_{\tau_0} \kappa \Delta_0 \eta / \rho^2$ as $\eta \to 0$; note the dependence on density in this latter expression. A number of compressible extensions of common eddy viscosity and conductivity functions have been considered [2], subject to three requirements: (i) the near-wall behavior must produce the logarithmic behavior in equations (7.43), (ii) the models must reduce to established incompressible models in the limit $M_e \to 0$, and (iii) there should be wide applicability of the models without introducing additional empiricism. There are a number of different functional forms which will satisfy these requirements, but the particular model described here [2] is believed to be very effective over a wide range of Mach numbers. Eddy viscosity and conductivity functions are defined by

$$\varepsilon = \frac{\rho_0 \rho_e}{\rho^2} U_e \delta^* \hat{\varepsilon}(\eta), \quad \varepsilon_H = \frac{\rho_0 \rho_e}{\rho^2} U_e \delta^* \hat{\varepsilon}_H(\eta), \tag{7.50}$$

where δ^* is the kinetic (or incompressible) boundary-layer thickness defined by equation (7.15), and $\hat{\varepsilon}$ and $\hat{\varepsilon}_H$ are simple ramp functions defined by equations (2.44) and (4.29) respectively, but now with $\eta_* = \rho_e U_e \delta^* / (\rho_0 u_{\tau_0} \Delta_0)$. It can be shown by substitution in equations (7.26) and (7.27) that the outer-layer defect profiles satisfy

$$\frac{\partial}{\partial \eta} \left(\hat{\varepsilon} \frac{\partial^2 F_1}{\partial \eta^2} \right) + a(x) \eta \frac{\partial^2 F_1}{\partial \eta^2} + b(x) \left[\left(1 - I_w \right) \frac{\kappa_\Theta}{\kappa} \Theta_1 - 2 \frac{\partial F_1}{\partial \eta} \right] = c(x) \frac{\partial^2 F_1}{\partial x \partial \eta}, \tag{7.51}$$

$$\frac{\partial}{\partial \eta} \left(\hat{\varepsilon}_H \frac{\partial \Theta_1}{\partial \eta} \right) + a(x) \eta \frac{\partial \Theta_1}{\partial \eta} - d(s) \Theta_1 = c(x) \frac{\partial \Theta_1}{\partial x}, \tag{7.52}$$

where the coefficients in these equations are given by

$$a = \frac{(\Delta_0 U_e)'}{\eta_* u_{\tau_0}}, \quad b = \frac{c(x)}{M_e} \frac{dM_e}{dx}, \quad c = \frac{\Delta_0 U_e}{\eta_* u_{\tau_0}}, \quad d = \frac{q_*'}{q_*} c(x), \tag{7.53}$$

and the prime denotes differentiation with respect to x. For a specific choice of Δ_o (see, for example, §6) and reference temperature T_o, these equations may be integrated in the streamwise direction to define a prediction method. As the calculation proceeds downstream, the numerical results must be matched to the wall-layer profiles, and the skin friction and wall heat transfer are evaluated from the match conditions for velocity and total enthalpy. Suitable profiles for the wall layer for U^+ and θ^+ are given by equation (3.22) with y^+ replaced by Y^+ and Y_θ^+ respectively (see also [2]).

Self-similar solutions for constant pressure flow have been examined in [2] and, with the outer scale defined by $\Delta_o = \rho_e U_e \delta^*/(\rho_o u_{\tau_o})$, such solutions are available in an analytical form similar to equation (2.54); at present compressible self-similar solutions for flows with pressure gradient have not been determined. The reference density ρ_o appears throughout this formulation, and a suitable choice for ρ_o and the reference temperature T_o is important to accurately determine the surface properties, as well as the physical extent of the outer layer. Although the wall density ρ_w could be used, this is usually not satisfactory because the density can vary appreciably across the wall layer even in an adiabatic flow. The reference temperature can be defined as an average value across the wall layer or, alternatively, as a mean value. He et al. [2] suggest that defining $T_o = T_o (Y_o^+)$ with $Y_o^+ = 23$ gives good overall correspondence with data over a range of Mach numbers and surface heat transfer conditions.

Some typical comparisons with constant pressure data are shown in Figure 7.1. The labeling scheme corresponds to that of ref. [13]. The theoretical curves are obtained

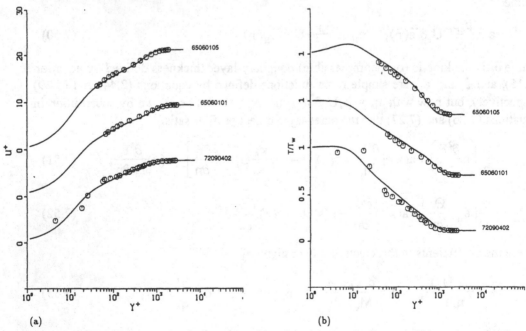

(a)   (b)

Figure 7.1 Comparison of calculated profiles for supersonic flow with heat transfer for:
(a) velocity and (b) static temperature with experimental data (6506 - Young;
7209 - Stone and Gary). Note the shifted origins.

by specifying the wall and mainstream properties, as well as δ^*; the Mach number range for this data is from $M_e = 4.8$ to 7.8, and these particular cases have surface heat transfer. It may be seen that the agreement of the data is quite reasonable, and other data comparisons confirm that the present formulation is a viable alternative to conventional methods [2].

The variation of self-similar profiles for constant pressure flow with Mach number and a fixed value of $Re_{\delta^*} = 10,000$ is shown in Figure 7.2 for an adiabatic wall. The relative growth in the thickness of the wall layer should be noted as M_e increases; as shown in Figure 7.2(a), a smaller percentage of the mainstream speed is achieved at fixed y/δ with increasing M_e. The variation in static temperature becomes more severe in the wall layer as shown in Figure 7.2(b).

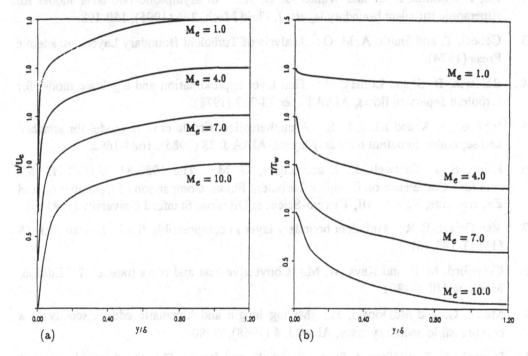

Figure 7.2 Profile behavior for $Re_{\delta^*} = 10,000$: variation of (a) velocity and (b) static temperature with M_e for an adiabatic wall.

7.6 Summary

Extensive data comparisons [2] suggest that the formulation described in Sections 7.4 and 7.5 provides an effective description of high-speed compressible flows through the supersonic regime and into the hypersonic range for flows with and without heat transfer. The results indicate that the hypothesis of a universal law of the wall for velocity is valid in terms of the scaled Howarth-Dorodnitsyn variable. The theory described in [2] may be used as an effective way to represent turbulent mean profiles in high-speed boundary layers, even in situations where the boundary-layer motion is not self-similar. The outer-layer turbulent models described in §7.5 can be used in prediction schemes for compressible flows, and the wall layer can be effectively treated using the embedded function methodology discussed in §6 [10], [14]. Note that the case of the adiabatic wall involves some subtle points that require additional attention, as well as different turbulence models; these issues are addressed in [15].

References

1. White, F. M.: Viscous Flow Theory, McGraw-Hill (1992).

2. He, J. Kazakia, J. Y. and Walker, J. D. A.: An asymptotic two-layer model for supersonic turbulent boundary layers, J. Fluid Mech. **295** (1995), 159-198.

3. Cebeci, T. and Smith, A. M. O.: Analysis of Turbulent Boundary Layers, Academic Press (1974).

4. Baldwin, B. S. and Lomax, H.: Thin layer approximation and algebraic model for turbulent separated flows, AIAA Paper 78-257 (1978).

5. Johnson, J. A. and King, L. S.: A mathematical simple closure model for attached and separated turbulent boundary layers, AIAA J. **23** (1985), 1684-1692.

6. Kline, S. J., Cantwell, B. J. and Lilley, G. M.: The 1980-81 AFOSR-HTTM-Stanford Conference on Complex Turbulent Flows: Comparison of Computation and Experiments, Vols. I - III, Thermo-Sciences Division, Stanford University (1981).

7. Van Driest, E. R.: Turbulent boundary layer in compressible fluids, J. Aero. Sci. **18** (1951), 145.161.

8. Crawford, M. E. and Kays, W. M.: Convective heat and mass transfer, 2nd Edition, McGraw-Hill (1980).

9. Maise, G. and McDonald, H.: Mixing length and kinematic eddy viscosity in a compressible boundary layer, AIAA J. **6** (1968), 73-80.

10. Degani, A. T., Walker, J. D. A., Ersoy, E. and Power, G.: On the application of algebraic turbulence models to high Mach number flows, AIAA Paper 91-0616 (1991).

11. Stewartson, K.: The theory of laminar boundary layers in compressible fluids, Oxford (1964).

12. Weigand, G. G.: Forced convection in a two-dimensional nominally steady turbulent boundary layer (1978), Ph.D. Thesis, Purdue University.

13. Fernholz, H. H. and Finley, P. J.: A critical commentary on mean flow data for two-dimensional compressible turbulent boundary layers (1977), AGARD-AG-223.

14. Walker, J. D. A., Werle, M. J. and Ece, M. C.: An embedded function approach for turbulent flow prediction, AIAA J. **29** (1987), 1810-1818.

15. He, J., Kazakia, J. Y., Ruban, A. I. and Walker, J. D. A.: A model for adiabatic supersonic turbulent boundary layers, Theoretical and Computational Fluid Dynamics **8** (1996), 349-364.

INTERACTING LAMINAR AND TURBULENT BOUNDARY LAYERS

A. Kluwick
Technical University of Vienna, Vienna, Austria

Abstract

This chapter deals with the properties of viscous wall layers in the high Reynolds number limit which are subjected to rapid changes of the boundary conditions. Classical, e.g. hierarchical boundary layer theory in which the driving pressure is imposed by the external inviscid flow then typically leads to difficulties which often can be overcome by an interaction strategy. Examples include flows past bodies of finite length and shock boundary layer interactions. The interaction concept is formulated first for laminar flows but extended later also to the case of turbulent boundary layers.

1   Introduction

Boundary layer theory represents one of the corner stones of modern fluid mechanics. According to the original concept going back to the seminal paper by Prandtl (1904) "Über Flüssigkeitsbewegung bei sehr kleiner Reibung" the calculation of viscous flows in the limit of large Reynolds numbers can be carried out in successive steps dealing with essentially inviscid (external) and viscous dominated (boundary layer) flow regions. This hierarchical structure in which the pressure distribution inside the boundary layer is imposed by the external inviscid flow and thus known in advance at each level of approximation has proved very powerful as outlined in the preceeding sections by H. Herwig, K. Gersten and J.D. Walker, see also Gersten (1989). Nevertheless, however, one must concede that it is possible to obtain a complete solution for a given problem in very rare cases only such as the laminar flow past an aligned semi-infinite plate. Difficulties arise already if one considers a plate of finite length since classical, e.g. hierarchical, theory does not account properly for the transition of the wall boundary layer into the wake. Similar difficulties occur in problems in which the boundary layer is subjected to rapidly changing boundary conditions. As pointed out in the pioneering studies by Neiland (1969), Stewartson (1969), Stewartson and Williams (1969), Messiter (1970) the associated breakdown of boundary layer theory is avoided for a wide class of flows if the pressure disturbances caused by the boundary layer are allowed to affect its properties at the same level of approximation rather than in higher order. The resulting theory of interacting boundary layers has provided significant insight into basic theoretical problems which are untractable by means of the classical concept, most important laminar boundary layer separation.

In the following discussion of laminar and turbulent interacting boundary layers completeness is, of course, not attempted. Rather the role of interaction processes will be demonstrated for a number of specific cases dealing with external forced convection flows. Readers interested in further details including also internal flows and natural convection flows.as well are unsteady effects are referred to the review papers by Stewartson (1974), Kluwick (1979), Adamson and Messiter (1980), Neiland (1981), Smith (1982), Messiter (1983), Sychev, Ruban, Sychev and Korolev (1987).

2   Interacting laminar boundary layers

2.1   Flow near the trailing edge of an aligned flat plate

The fluid motion in the trailing edge region of a flat plate placed into an incompressible fluid with constant kinematic viscosity $\tilde{\nu}$ has been investigated by Goldstein (1930) using classical boundary layer theory. Specifically he showed that the Blasius surface boundary layer can be continued into the wake despite the fact that the wake solution develops a singularity as the trailing edge is approached:

$$v_e \sim \text{prop}\, Re^{-1/2} x^{-2/3}\,, \quad u_{cl} \sim \text{prop}\, x^{1/3}\,, \quad x \to 0^+\,. \tag{2.1}$$

Here x, v_e, u_{cl} and Re denote the distance from the trailing edge nondimensional with the length $\tilde{L}$ of the plate, the normal velocity component at the outer edge of the wake, the wake centreline velocity nondimensional with the freestream velocity $\tilde{u}_\infty$ and the Reynolds number $\tilde{u}_\infty \tilde{L}/\tilde{\nu}$, respectively.

Since $v_e \to -\infty$ as $x \to 0^+$ one expects that large pressure disturbances will be induced in the vicinity of the trailing edge by the associated displacement of the inviscid flow outside the wake. These feedback pressure disturbances can be calculated from second order boundary layer theory and as shown by Messiter (1970) they exhibit the same x-dependence as v_e. Formally they are of $O(Re^{-1/2})$ and thus do not enter the flow description in leading order. It is easily seen, however, that the induced pressure force and the leading order inertia term estimated from the result for u_{cl} become comparable at distances $|x| = O(Re^{-3/8})$ from the trailing edge. At these distances pressure disturbances caused by the interaction of the oncoming surface boundary layer and the wake with the external inviscid flow are large enough to represent a first order rather than a higher order effect.

The properties of the flow inside the trailing edge local interaction region have been studied in detail by Stewartson (1969), Messiter (1970). As shown in Fig. 1 a three tiered structure, now commonly termed triple deck structure, has to be considered. Since the length of the interaction region is so small, the disturbances generated by the trailing edge are essentially inviscid not only in the region outside the boundary layer which is affected by the interaction process (upper deck) but also over most of the boundary layer (main deck). Viscous effects are confined to a thin layer adjacent to the plate surface and the wake centreline (lower deck). As pointed out earlier the length of the interaction region is $O(\varepsilon^3 \tilde{L})$ where

$$\varepsilon = Re^{-1/8} \ll 1\,. \tag{2.2}$$

It is thus small compared to the plate length but large compared to the boundary layer thickness $O(\varepsilon^4 \tilde{L})$. Furthermore, simple order of magnitude estimates yield that the lateral extent of the upper deck is $O(\varepsilon^3 \tilde{L})$ while the thickness of the lower deck is $O(\varepsilon^5 \tilde{L})$.

Turning now to a more detailed description of the flow properties in the local interaction region it is useful to retain the effects of compressibility which are of importance in many applications. Appropriate expansions of the field quantities inside the main deck

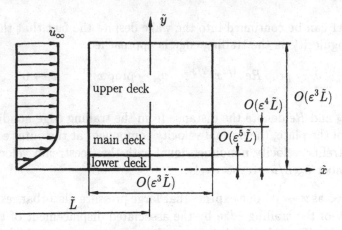

Figure 1: Asymptotic flow field structure near trailing edge.

then are

$$\frac{\tilde{u}}{\tilde{u}_\infty} = u = U_0(y_m) + \varepsilon u_1(x_1, y_m) + \dots ,$$

$$\frac{\tilde{v}}{\tilde{u}_\infty} = v = \varepsilon^2 v_1(x_1, y_m) + \dots ,$$

$$\frac{\tilde{\rho}}{\tilde{\rho}_\infty} = \rho = R_0(y_m) + \varepsilon \rho_1(x_1, y_m) + \dots , \tag{2.3}$$

$$\frac{\tilde{p} - \tilde{p}_\infty}{\tilde{\rho}_\infty \tilde{u}_\infty^2} = p = \varepsilon^2 p_1(x_1, y_m) + \dots$$

with

$$x_1 = \frac{\tilde{x}}{\tilde{L}\varepsilon^3} , \qquad y_m = \frac{\tilde{y}}{\tilde{L}\varepsilon^4} . \tag{2.4}$$

Here $\tilde{\rho}$ and $\tilde{p}$ denote the density and the pressure and the subscript ∞ refers to freestream conditions, while U_0 and R_0 represent the velocity and density profiles in the undisturbed boundary layer at $x = 0$. Substitution of (2.3) into the full Navier Stokes equations shows that the leading order perturbations of the velocity components and the density are unaffected by pressure and viscous forces and can be expressed in terms of a (yet unknown) displacement function $A_1(x_1)$:

$$u_1 = A_1(x_1)U_0'(y_m) , \qquad v_1 = -A_1'(x_1)U_0(y_m) , \qquad \rho_1 = A_1(x_1)R_0'(y_m) . \tag{2.5}$$

Furthermore, since the thickness of the main deck is much smaller than its length the pressure disturbance p_1 does not depend on the lateral distance y_m

$$p_1 = p_1(x_1) . \tag{2.6}$$

According to equation (2.5)

$$u_1 = \rho_1 = 0, \quad v_1 = -A_1'(x_1), \quad y_m \to \infty \qquad (2.7)$$

which indicates that the perturbations of the field quantities in the upper deck region are $O(\varepsilon^2)$. Consequently, the flow there is governed by the linearized gasdynamic equation in a first approximation. Matching of the solution for the pressure and the transverse velocity component v with the main deck results (2.5), (2.6) then yields a relationship between the pressure disturbances $p_1(x_1)$ inside the main deck and the displacement function $A_1(x_1)$. The specific form of this relationship depends on the value of the freestream Machnumber M_∞ and resembles the wellknown Prandtl-Glauert or Ackeret relationship for slender airfoils in subsonic or supersonic parallel flow:

$$M_\infty < 1: \quad p_1(x_1) = \frac{1}{\pi(1 - M_\infty^2)^{1/2}} \oint_{-\infty}^{\infty} \frac{A_1'(\xi)}{x_1 - \xi} \, d\xi, \qquad (2.8)$$

$$M_\infty > 1: \quad p_1(x_1) = -\frac{A_1'(x_1)}{(M_\infty^2 - 1)^{1/2}}. \qquad (2.9)$$

Equation (2.3) together with the estimate for thickness of the lower deck region $y = O(\varepsilon^5)$ suggests the following expansion holding in this viscous sublayer

$$
\begin{aligned}
u &= \varepsilon \bar{u}_1(x_1, y_l) + \dots, \\
v &= \varepsilon^3 \bar{v}_1(x_1, y_l) + \dots, \\
\rho &= R_0(0) + \varepsilon \bar{\rho}_1(x_1, y_l) + \dots, \\
p &= \varepsilon^2 p_1(x_1) + \dots
\end{aligned}
\qquad (2.10)
$$

where

$$y_l = \frac{\bar{y}}{\bar{L}\varepsilon^5}. \qquad (2.11)$$

Owing to the fact that the lower deck is so thin the pressure disturbances there agree with those in the main deck. Furthermore, since the velocities there are low, compressibility effects are negligible small in a first approximation which is thus governed by the classical boundary layer equations of an incompressible fluid. However, these equations are subjected to nonclassical boundary conditions following from the match of the lower deck solution with the main deck and the oncoming unperturbed boundary layer. It is convenient to define the transformed quantities

$$X = C^{-3/8}\lambda^{5/4}|M_\infty^2 - 1|^{3/8}(\tilde{T}_W/\tilde{T}_\infty)^{-3/2}x_1 \,,$$

$$Y = C^{-5/8}\lambda^{3/4}|M_\infty^2 - 1|^{1/8}(\tilde{T}_W/\tilde{T}_\infty)^{-3/2}y_l \,,$$

$$P = C^{-1/4}\lambda^{-1/2}|M_\infty^2 - 1|^{1/4}p_1 \,,$$

$$U = C^{-1/8}\lambda^{-1/4}|M_\infty^2 - 1|^{1/8}(\tilde{T}_W/\tilde{T}_\infty)^{-1/2}\tilde{u}_1 \,, \tag{2.12}$$

$$V = C^{-3/8}\lambda^{-3/4}|M_\infty^2 - 1|^{-1/8}(\tilde{T}_W/\tilde{T}_\infty)^{-1/2}\tilde{v}_1 \,,$$

$$A = C^{5/8}\lambda^{3/4}|M_\infty^2 - 1|^{1/8}(\tilde{T}_W/\tilde{T}_\infty)^{2/3}A_1$$

where $\lambda, C, \tilde{T}_W$ and $\tilde{T}_\infty$ characterize, respectively, the scaled wall shear $Re^{-1/2}\partial u/\partial y|_{y=0}$ $= U_0'(0)$ in the unperturbed boundary layer at $x = 0$, the Chapman Rubesin parameter entering the linear viscosity law $\tilde{\mu}/\tilde{\mu}_\infty = C\tilde{T}/\tilde{T}_\infty$ and the wall- and freestream temperature. The lower deck problem can then be formulated in the following parameter free form

$$U\frac{\partial U}{\partial X} + V\frac{\partial U}{\partial Y} = -\frac{dP}{dX} + \frac{\partial^2 U}{\partial Y^2} \,, \quad \frac{\partial U}{\partial X} + \frac{\partial V}{\partial Y} = 0 \,, \tag{2.13}$$

$$U = V = 0 \,, \ Y = 0 \,, \ X \le 0; \quad \frac{\partial U}{\partial Y} = V = 0 \,, \ Y = 0 \,, \ X > 0 \,, \tag{2.14}$$

$$U = Y \,, \ X \to -\infty \,, \quad U = Y + A(X) \,, \ Y \to \infty \,, \tag{2.15}$$

$$P = \frac{1}{\pi}\oint_{-\infty}^{\infty}\frac{A'(\xi)}{X - \xi}d\xi \quad \text{subsonic flow} \,, \tag{2.16}$$

$$P = -A'(X) \quad \text{supersonic flow} \,. \tag{2.17}$$

Before turning to a discussion of numerical solutions of this problem it is useful to summarize briefly the underlying physical concept. Since the velocities in the lower deck are small compared to the free-stream value, the perturbations of the streamtube cross-section area caused by the induced pressure gradient are much larger there than in the outer part of the boundary layer. To leading order, therefore, the variation of the displacement thickness can be calculated solely from the velocity distribution in the lower deck. The role of the main deck then is a passive one, to transfer the displacement effect exerted by the lower deck unchanged to the upper deck by shifting the unperturbed boundary layer profile away from the wall at the distance $-A(X)$. At the outer edge $Y \to \infty$ of the lower deck, where the velocity varies linearly with Y, this shift leads to a reduction of the velocity by the amount $A(X)$. Outside the boundary layer it causes pressure disturbances of $O(\varepsilon^2)$ and the flow there is governed by the linearized equations of inviscid theory. If the external flow is subsonic A and P are thus related by the Hilbert integral (2.16) while they satisfy the Ackeret relationship (2.17) if the external flow is supersonic.

For the case of subsonic flow numerical solutions to equations (2.13), (2.14), (2.15) and (2.16) have been obtained by Jobe and Burggraf (1974), Veldmann and van de Vooren (1975), Melnik and Chow (1975). The trailing edge problem with external supersonic flow has been solved by Daniels (1974). The distributions of the pressure and the wall shear stress for $M_\infty < 1$ are depicted in Fig. 2. Owing to the interaction process the pressure drop upstream of the trailing edge is smaller than that predicted by second order boundary layer theory and P remains finite as $X \to 0$. The pressure drop at the plate leads to an increase of the wall shear stress $\tau_w = \partial U/\partial Y|_{Y=0}$ and at the trailing edge τ_w is about 35% higher than in the unperturbed boundary layer having $\tau_w = 1$.

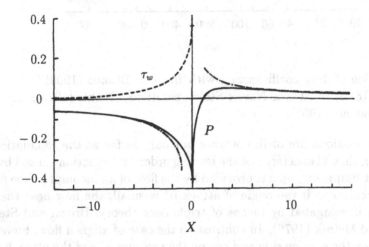

Figure 2: Laminar trailing edge interaction. - - - wall shear, —— pressure, — · — pressure distribution according to second order boundary layer theory.

It is interesting to calculate the drag correction associated with the local interaction process. It is caused by shear stresses which are of the same order of magnitude as those in the unperturbed boundary layer but which are confined to the interaction region of streamwise extent $O(\varepsilon^3)$. This leads to a drag increase of $O(\varepsilon^4\varepsilon^3)$ which is larger than the contribution of $O(\varepsilon^8)$ following from second order boundary layer theory. Consequently, the expansion of the drag coefficient assumes the form

$$c_D = 1.328 Re^{-1/2} + d_2 Re^{-7/8} + O(Re^{-1}) \tag{2.18}$$

where d_2 has to be determined numerically. As shown by Melnik and Chow (1975) this two term asymptotic result is in excellent agreement with the experimental data and numerical predictions based on the full Navier Stokes equations. More detailed comparisons between triple-deck and Navier Stokes solutions have been carried out by Chen and Patel (1987), McLachlan (1991).

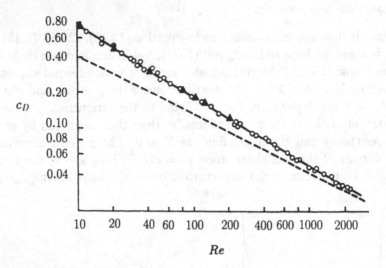

Figure 3: Variation of drag coefficient c_D with Re. - - - Blasius (1908), —— equation (2.18) $d_2 = 2.66$, ▲ Navier Stokes solutions, Dennis (1973); o experiments, Janour (1951).

Trailing edge interactions are of importance not only as far as the calculation of the drag is concerned, they also determine the leading order lift reduction caused by viscous effects. The most simple example is provided by the flow of an incompressible fluid past a flat plate at incidence. If the angle of attack α^* is small, the flow near the trailing edge can still be investigated by means of triple deck theory, Brown and Stewartson (1970), Chow and Melnik (1976). In contrast to the case of aligned flow, however, two triple decks, one on the suction side and one on the pressure side of the plate, have to be considered separately. The requirement that the pressure is continuous at the trailing edge, which serves as a viscous Kutta condition, then determines the circulation Γ of the flow outside the boundary layer and thus the lift coefficient. Comparison of the prediction $p \sim \pm\alpha^*(-x)^{1/2}$, $x \to 0^-$ following from inviscid theory with the triple deck scalings $x = O(\varepsilon^3)$, $p = O(\varepsilon^2)$ shows that triple deck theory applies if $\alpha^* = O(\varepsilon^{1/2})$. As a consequence, the relative change of the circulation Γ due to viscosity is $O(\varepsilon^3)$ and the lift coefficient can be expressed in the form

$$c_L/c_{L\,\text{inv}} \sim 1 - 2\varepsilon^3 a_1(\alpha). \tag{2.19}$$

Here α denotes the stretched angle of attack $\alpha = \varepsilon^{-1/2}\lambda^{-9/8}\alpha^*$ and $c_{L\,\text{inv}} = 2\pi\alpha^*$. The

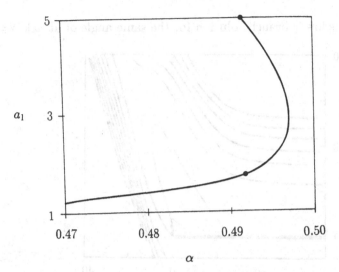

Figure 4: Flat plate at incidence: non-uniqueness of viscous correction to lift coefficient (Korolev (1989)).

function $a_1(\alpha)$ has been determined numerically by Chow and Melnik (1976). If α is small the flow remains attached as in the case of aligned flow. Incipient separation was estimated to occur for $\alpha \approx 0.47$ but numerical difficulties prevented the calculation of solutions up to this value. It has been conjectured (Brown and Stewartson (1970), Chow and Melnik (1976)) that the rapid increase of a_1 near the point of the incipient separation which is associated with a rapid decrease of c_L heralds the onset of catastrophic stall. This point of view was contested by Smith (1983) who provided strong analytical and numerical arguments that in the case of one-sided separating flow boundary layer attachment will occur immediately upstream of the trailing edge. This has been confirmed recently by Korolev (1989) who was able to extend the numerical solutions of the triple deck problem into the separated flow regime. According to these results the wall shear on the upper plate surface at $x = 0^-$ reaches a minimum value $\tau_w(0^-) = 0$ for $\alpha = 0.47$ in agreement with the estimate for incipient separation given by Chow and Melnik (1976). For larger values of α the boundary layer on the upper plate surfaces separates but reattaches upstream of the trailing edge so that $\tau_w(0^-)$ now increases with increasing angle of attack. Furthermore, the calculations indicate that solutions to the local interaction problem do not exist if α exceeds the critical value $\alpha_c \approx 0.497$ and that the relationship between $a_1(\alpha)$ and α is non unique in a certain neighbourhood of $\alpha = \alpha_c$, Fig. 4. Fig. 5 displays the flow pattern corresponding to the upper branch solution with $\alpha = 0.492$. It is seen to exhibit a pronounced tongue just downstream of the separated flow region bounded by the streamline $\psi = 0$ in which fluid is drawn back towards the trailing edge before being turned anticlockwise forward to yield reattachment completely in line with the arguments put forward and the results obtained in Smith (1983). In addition Fig. 5 includes the separation zone corresponding to the lower branch solution for the same angle of attack $\alpha = 0.492$.

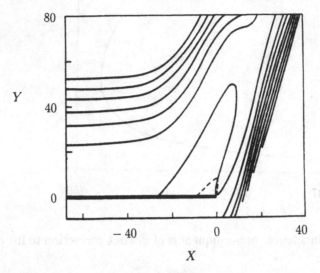

Figure 5: Flat plate at incidence: flow pattern for $\alpha = 0.492$, Fig. 4. —— upper branch solution, - - - separation zone of lower branch solution (Korolev (1989)).

2.2 Separation from a curved wall

The problems discussed so far (as well as related problems of flows past sufficiently slender bodies) have the common property that the fluid motion outside the boundary layer in the limit $Re = \infty$ is described by a fully attached inviscid flow. In contrast, flows past bluff bodies are in general characterized by the presence of pronounced wakes indicating that this is not a valid proposition anymore and that the limiting inviscid solution has to be sought in the class of flows with free streamlines. According to the theory of such flows (see e.g. Gurevich (1966)) the pressure gradient on the body surface in a small neighbourhood of the separation point $x = 0$ satisfies the relationship

$$\frac{dp}{dx} \sim k_0(-x)^{-1/2} + k_1 , \quad x \to 0^- ,$$

$$\frac{dp}{dx} = 0 , \quad x \to 0^+ . \tag{2.20}$$

Here x and p denote the distance from the separation point and the pressure disturbances nondimensional with the body length and $\tilde{\rho}\tilde{U}_0^2$ where $\tilde{U}_0$ is the velocity on the free streamline of the limiting irrotational flow. Furthermore, let κ_0 denote the curvature of the body contour at $x = 0$. The curvature κ of the free streamline can then be written as

$$\kappa \sim k_0 x^{-1/2} + \kappa_0 , \quad x \to 0^+ . \tag{2.21}$$

Equation (2.20) expresses the interesting result that the pressure distribution upstream of the separation point on a smooth surface is of the same form as the pressure distribution upstream of the trailing of a flat plate at incidence locally. In contrast to the latter problem the constant k_0 is not fixed by geometric conditions, however, but depends on the position of the separation point which cannot be determined from inviscid theory alone.

Inspection of equation (2.21) shows that the free streamline intersects with the body surface if $k_0 < 0$ which, therefore, can immediately be ruled out. In contrast, all solutions with $k_0 \geq 0$ are fully consistent with the governing equations and the wall geometry indicating that viscous effects have to be taken into account to determine the relevant value which in turn fixes the location of the separation point. However, when the pressure distribution (2.20) is combined with classical boundary layer theory one faces a severe difficulty. If $k_0 = 0$, which in the case of the flow past a circular cylinder corresponds to the Kirchhoff (1869) solution, the pressure gradient is favourable up to $x = 0$ and this is not consistent with the occurrence of boundary layer separation. On the other hand, if $k_0 > 0$ the adverse pressure gradient at $x = 0$ is singular and, as a consequence, the boundary layer will separate already upstream of $x = 0$ thus yielding a contradiction also. As pointed out first by Sychev (1972) this difficulty can be overcome by taking k_0 positive but small so that the associated singularity of the pressure gradient predicted by equation (2.20) is small too and can be eliminated by taking into account viscous-inviscid interaction. Similar to the trailing edge problem this requires

$$k_0 = \varepsilon^{1/2} \lambda^{9/8} \alpha , \quad \alpha = O(1) . \tag{2.22}$$

The resulting triple deck problem then leads to the boundary layer equations (2.13) subject to the noslip condition at the wall, equation (2.15), the coupling relationship (2.16) and the requirement

$$P \sim -\alpha(-X)^{1/2} , \quad X \to -\infty ,$$
$$P \sim 0 \quad , \quad X \to \infty \tag{2.23}$$

to ensure the proper match with the pressure distribution upstream and downstream of the local interaction zone.

The interaction problem formulated by Sychev (1972) has been investigated numerically first by Smith (1977). He showed that a unique solution exists, Fig. 6, and that the required value of α is approximately

$$\alpha = 0.44 . \tag{2.24}$$

More recent calculations by Korolev (1980) are in complete agreement with this conclusion but predict a slightly smaller value $\alpha = 0.42$. These results indicate that the combination of free streamline theory and triple deck theory as proposed by Sychev (1972) yields a selfconsistent description of laminar boundary layer separation from a curved surface. It thus provides an important element needed to construct global solutions for grossly separated flows. The discussion of such flows is beyond the scope of the present review of basic local interaction processes and the interested is referred to the original contributions by Smith (1977), (1979), (1985), (1987), Herwig (1982), Chernyshenko (1988), Chernyshenko and Castro (1993), Taganov (1968), (1970).

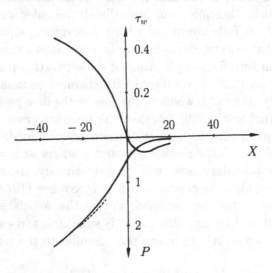

Figure 6: Separation from a smooth surface: pressure- and wall shear distribution near the separation point. - - - equation (2.23).

2.3 Supersonic flow past a compression ramp

Following this brief discussion of examples of local interaction processes occurring in incompressible and subsonic flows let us consider the case that the flow outside the boundary layer is purely supersonic. Specifically, the discussion will concentrate on effects associated with the presence of shocks which impinge on the boundary layer or are generated through viscous-inviscid interaction. Flows of the first kind have been investigated first by Ackeret, Feldmann and Rott (1946), Hakkinen, Greber, Trilling and Arbabanel (1959). Their results show that viscosity leads to a modification of the inviscid pressure distribution not only downstream but also upstream of the shock reflexion point even if the flow remains attached. Clearly, the observed phenomenon of upstream influence is incompatible with classical boundary layer theory. Therefore, the question arises if and how it is included in the triple deck formalism. According to the results summarized in section 2.1 the set of equations which govern the flow upstream of an arbitrary disturbance acting on the boundary layer on a flat plate consists of the boundary layer equations (2.13) subjected to the boundary conditions (Neiland (1969), Stewartson and Williams (1969))

$$U = V = 0 \quad \text{on} \quad Y = 0, \qquad U = Y \quad \text{as} \quad X \to -\infty,$$

$$U = Y + A(X) \quad \text{as} \quad Y \to \infty, \qquad P = -A'(X) \,. \tag{2.25}$$

One solution of the homogeneous set of equations (2.13) and (2.25) is, of course, the trivial solution $U = Y$, $V = A = P = 0$ which describes the unperturbed boundary layer. However, as shown by Neiland (1969), Stewartson and Williams (1969) there exist two additional solutions having $P \sim \delta \exp(KX)$, $X \to -\infty$ where $K = 0.8272$ in agreement with the results obtained by Lighthill (1953) using a different approach. If δ is positive/negative the pressure disturbances increase/decrease with X and the solution, respectively, is termed a compressive/expansive free interaction solution. A physical explanation for the occurrence of the free interaction solution was given even earlier by Oswatitsch and Wieghardt (1941) who noticed that a local pressure increase/decrease and boundary layer thickening/thinning can reinforce each other.

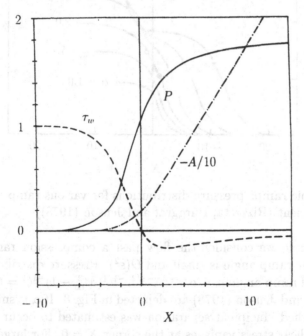

Figure 7: Compressive free interaction solution. —— $P(X)$, - - - $\tau_w(X)$, —·— $-A(X)/10$.

Equations (2.13) and (2.25) have been solved numerically by marching downstream from an initial position by Stewartson and Williams (1969). Some properties of the compression free interaction solution are summarized in Fig. 7. Initially P increases progressively while τ_w decreases and eventually vanishes in a completely regular fashion.

To stabilize the downstream marching procedure in the reversed flow region the Flare approximation ($U\partial U/\partial X = 0$ if $U < 0$) was used. Downstream of the separation point P approaches a constant value $P = P_0 \approx 1.8$ and τ_w vanishes in the limit $X \to \infty$. Since P is almost constant for X large, it follows immediately from the Ackeret relationship (2.17) that $A \sim -P_0 X$ as $X \to \infty$.

The results discussed so far show that the behaviour of a supersonic laminar boundary layer far upstream of any arbitrary disturbance is characterized either by a compressive or an expansive free interaction solution and thus does not depend on the details of the mechanism which provokes the interaction process. For any specific problem, on the other hand, the free interaction solution is but one part of the flow field.

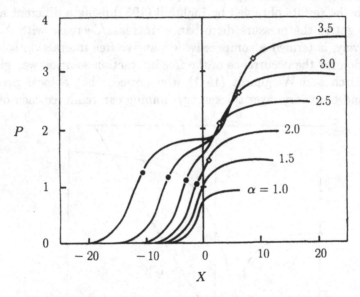

Figure 8: Supersonic ramp: pressure distributions for various ramp angles α. • separation, ◇ reattachment (Rizzetta, Burggraf and Jenson (1978)).

As a specific example we consider the flow past a compression ramp. Triple deck theory applies if the ramp angle is small and $O(\varepsilon^2)$. Pressure distributions for various values of the scaled ramp angle $\alpha = \alpha^* \varepsilon^{-2} \lambda^{-1/2} c^{-1/4} (M_\infty^2 - 1)^{-1/4} = O(1)$ obtained by Rizzetta, Burggraf and Jenson (1978) are depicted in Fig. 8. If α is small the boundary layer remains attached. Incipient separation was estimated to occur for $\alpha \approx 1.57$. In this case the wall shear stress vanishes at the corner $X = 0$. For larger values of α the boundary layer separates upstream of the corner and reattaches on the inclined wall. It should be noted that the shapes of the pressure distributions for $X < 0$ closely resemble each other, in fact the pressure distributions there are generated by a single curve - the compressive free interaction solution - which is shifted upstream as α increases (owing to the fact that information can propagate upstream in the reversed flow region this result is valid in a strict sense only ahead of the separation point). Downstream of the corner the various pressure distributions differ significantly since the ultimate pressure level is imposed by the ramp angle: $P \sim \alpha$, $X \to \infty$.

The ramp problem has been studied independently by Ruban (1978) who obtained numerical solutions up to values $\alpha = 6$ but numerical difficulties prevented the calculations for significantly larger ramp angles. Similar difficulties have been encountered in related triple deck calculations in which a transition from small to large scale separation was attempted. As noticed by Smith (1988) the results of such studies typically exhibit a rapidly increasing maximum of the adverse pressure gradient and minimum of the wall shear once the separated flow region reaches a substantial size indicating the formation of a reversed flow singularity as the controlling parameter of the problem under consideration approaches a critical value. If X_0 denotes the position of this singularity the analysis of Smith (1988) predicts

$$\max{(dP/dX)}|_{X=X_0} \sim (\alpha_c - \alpha)^{-2}, \quad \min{(\tau_w)}|_{X=X_0} \sim -(\alpha_c - \alpha)^{-1}. \qquad (2.26)$$

Calculations performed by Smith and Khorrami (1991) compare favourably with the proposed structure and lead to the estimate that α_c is "somewhat below 9". This implies the previously unexpected result that triple deck theory fails already for moderate values of α and thus cannot be used to elucidate the properties of large scale separated flows by taking the limit $\alpha \to \infty$. It has been argued by Smith (1988) that the inclusion of normal pressure gradient effects in a small neighbourhood of X_0 will be necessary to eliminate the singularity and to proceed to larger ramp angles but this step has yet to be taken.

As a typical example of the solutions obtained by Smith and Khorrami (1991) distributions of the pressure and the wall shear stress for $\alpha = 6.6$ are depicted in Fig. 9. The calculations show that the initially pronounced growth of the separated flow region practically comes to a halt for $\alpha \gtrsim 3.5$ (indeed, comparison with Fig. 8 reveals that the upstream shift is overpredicted by the earlier calculations of Rizzetta et al (1978) if $\alpha \gtrsim 3$ and a similar comment applies to the calculations of Ruban (1978) for slightly larger values of α). Furthermore, the plot of the wall shear stress clearly displays a short zone of positive τ_w inside the separated flow region pointing to the formation of a secondary vortex before triple deck theory breaks down.

The supersonic ramp problem has been reconsidered recently by Cassel, Ruban and Walker (1995). Similar to the studies by Rizzetta et al (1978), Ruban (1978) the unsteady triple deck equations were integrated through to the steady state. Good agreement with the earlier computations was observed for relatively coarse mesh sizes. Calculations with progressively refined mesh sizes, however, showed the occurrence of an instability if $\alpha \gtrsim 3.9$. This instability, manifested in the form of a wave pocket that develops and remains stationary near the corner most likely is associated with the development of an inflection point in the velocity profiles within the reversed flow region.

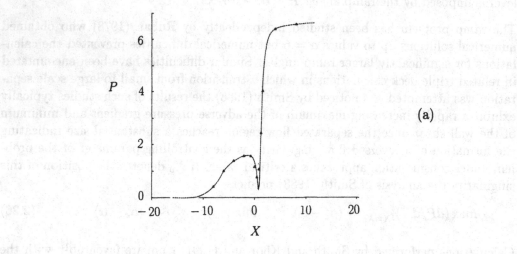

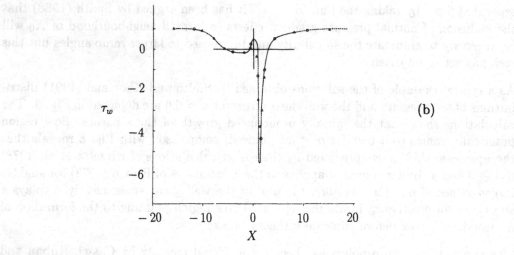

Figure 9: Supersonic ramp, $\alpha = 6.6$: (a) pressure distribution, (b) wall shear distribution (Smith and Khorrami (1991)).

2.4   Transonic effects

In all problems considered so far the flow outside the boundary layer was incompressible, purely subsonic or purely supersonic and governed by linear equations. The linearization of the gasdynamic equation ceases to be valid, however, if the difference between the freestream Machnumber M_∞ and its critical value 1 is so small that the changes of the local Machnumber M caused by the induced pressure disturbances are of the same order as $|1 - M_\infty^2|$. It has been shown by Messiter, Feo and Melnik (1971) that this is the case if the transonic interaction parameter

$$K_0 = \lambda^{-2/5} C^{-1/5} (M_\infty^2 - 1) \varepsilon^{-8/5} \tag{2.27}$$

is of order one rather than large as in purely subsonic or supersonic flows. As in these flows the fluid motion in the lower deck is described by the incompressible form (2.13) of the boundary layer equations if the field quantities are appropriately rescaled:

$$X = C^{-3/10} \lambda^{7/5} K_0^{3/8} (\tilde{T}_w/\tilde{T}_\infty)^{-3/2} \varepsilon^{-12/5} \frac{\tilde{x}}{\tilde{L}},$$

$$Y = C^{-3/5} \lambda^{4/5} K_0^{1/8} (\tilde{T}_w/\tilde{T}_\infty)^{-3/2} \varepsilon^{-24/5} \frac{\tilde{y}}{\tilde{L}},$$

$$U = C^{-1/10} \lambda^{-1/5} K_0^{1/8} (\tilde{T}_w/\tilde{T}_\infty)^{-1/2} \varepsilon^{-4/5} \frac{\tilde{u}}{\tilde{u}_\infty}, \tag{2.28}$$

$$V = C^{-2/5} \lambda^{-4/5} K_0^{-1/8} (\tilde{T}_w/\tilde{T}_\infty)^{-1/2} \varepsilon^{-16/5} \frac{\tilde{v}}{\tilde{u}_\infty},$$

$$P = C^{-1/5} \lambda^{-2/5} K_0^{1/4} \varepsilon^{-8/5} \frac{\tilde{p} - \tilde{p}_\infty}{\tilde{\rho}_\infty \tilde{u}_\infty^2}.$$

These scalings account for the higher sensitivity of transonic external flows against displacement effects exerted by the boundary layer which leads, among others, to a larger streamwise extent of the interaction zone and an increase of the pressure disturbance level. In contrast, however, the flow in the upper deck region is no longer governed by linear relationships but by the nonlinear transonic small disturbance equation. Introducing the scaled upper deck coordinate

$$Z = C^{-1/5} \lambda^{8/5} (\tilde{T}_w/\tilde{T}_\infty) \varepsilon^{-8/5} \frac{\tilde{y}}{\tilde{L}} \tag{2.29}$$

and using the pressure disturbance $P^T(X, Z)$ (scaled in exactly the same way as in the lower deck region) as the dependent variable this equation assumes the form

$$\frac{\partial^2 P^T}{\partial Z^2} + \left[\frac{P^T}{|B_0|} - \text{sign}(B_0)\right]\frac{\partial^2 P^T}{\partial X^2} + \frac{1}{|B_0|}\left(\frac{\partial P^T}{\partial X}\right)^2 = 0\,,$$

$$B_0 = \text{sign}(K_0)\frac{|K_0|^{5/4}}{\gamma + 1}\,. \tag{2.30}$$

Here γ is the ratio of the specific heats. Equation (2.30) has to be solved subject to appropriate boundary conditions for $X^2 + Z^2 \to \infty$ which depend on the problem under consideration and

$$P^T(X, Z = 0) = P(X)\,, \qquad \frac{\partial P^T}{\partial Z} = \frac{d^2 A}{dX^2} \quad \text{on} \quad Z = 0\,. \tag{2.31}$$

Inspection of the equation for P^T shows that it changes type from that of an elliptic to a hyperbolic equation depending on whether the local value of the coefficient of the $\partial^2 P^T/\partial X^2$ term which can be expressed in terms of the local Mach number

$$\frac{P^T(X, Z)}{|B_0|} - \text{sign}(B_0) = \frac{1 - M^2}{1 - M_\infty^2} \tag{2.32}$$

is positive or negative, respectively. This typechange severely complicates the numerical treatment of transonic interactive flows. Significant simplifications are possible, however, if $M > 1$ in the whole interaction region and if there are no incoming disturbances impinging on the boundary layer. It then follows from the simple wave solution of (2.30) that the displacement function $A(X)$ is related to the pressure disturbance $P(X)$ inside the boundary layer by

$$A(X) = -\frac{2}{3}B_0 \int\limits_{-\infty}^{X} \left[1 - \left(1 - \frac{P(\xi)}{B_0}\right)^{2/3}\right] d\xi \tag{2.33}$$

which reduces to the linear Ackeret formula in the limit $B_0 \to \infty$.

On the basis of equation (2.33) transonic freely interacting boundary layers, trailing edge flows and flows past convex corners have been investigated by Bodonyi and Kluwick (1977), (1982), Bodonyi (1979). The simple wave approach can easily be generalized to the case of an impinging shock wave provided that the flow downstream of the reflected shock front is still supersonic, Brilliant and Adamson (1973), Bodonyi and Smith (1986).

Results including mixed transonic flow have been obtained so far for the trailing edge problem of an aligned thin plate only, Korolev (1983), Bodonyi and Kluwick (1997).

Similar to the cases of linear subsonic and supersonic flow the fluid inside the boundary layer accelerates as the trailing edge is approached owing to the absence of the noslip condition in the wake region. This in turn leads to the occurrence of a local Mach number maximum at the inner edge of the upper deck which increases with increasing B_0. As a consequence there must exist a lower critical value B_0^{crit} for which the external flow is purely subsonic. According to the numerical computations by Bodonyi and Kluwick (1997), $B_0^{crit} \approx -0.325$ which in combination with equations (2.27) and (2.30) yields a relationship for the lower critical Mach number

$$M_\infty^{crit} = 1 - (0.4069\ldots)\lambda^{2/5}C^{1/5}(\gamma+1)^{4/5}\varepsilon^{8/5} \qquad (2.34)$$

in the classical gasdynamic sense.

For $B_0 > B_0^{crit}$ a local supersonic pocket develops near the trailing edge which grows in size as B_0 increases further. As a representative result the distribution of the iso-Machlines in the upper deck is displayed in Fig. 10 for $B_0 = -0.28$. The concentration of the iso-Machlines near the downstream boundary of the supersonic pocket strongly points to the existence of a shock front which terminates this region. This shock is expected to increase in length as the freestream Mach number approaches one and finally extend to infinity in the limit $B_0 = 0$, merging smoothly with the results obtained for supersonic upstream conditions.

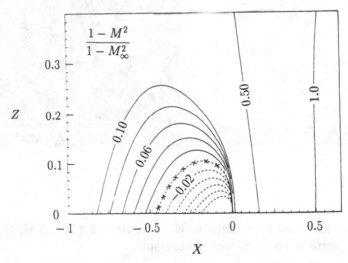

Figure 10: Transonic trailing edge flow: Iso-Machlines for $B_0 = -0.28$, ×–× sonic line.

2.5 Further applications of triple deck theory

The theory of laminar interacting boundary layers outlined briefly so far can be extended and generalized considerably. A few such generalizations will be discussed in this section. For a more complete account of triple deck and related studies the reader is referred to the review articles by Stewartson (1974), Smith (1982), Messiter (1983), Sychev (1987). A comprehensive discussion of unsteady interactive flows with special emphasis on contributions to the theory of laminar boundary layer stability is given in a separate section of the present volume.

With the exception of the transonic and hypersonic flow regimes the upper deck equations are linear. As a consequence, triple deck solutions are not uniformly valid, in general, at large distances from the interaction region if the flow outside the boundary layer is supersonic. Linear theory predicts that the disturbances generated by the interaction process propagate along the unperturbed Mach lines. However, as is well known, even small disturbances may lead to significant distortions of Mach lines at large distances from their origin. To obtain uniformly valid results it is thus necessary to account for nonlinear far field effects. For this purpose the analytical method of characteristics is especially useful. As an example Fig. 11 compares the calculated and observed wave pattern generated by a weak shock which impinges on a flat plate, Kluwick (1985).

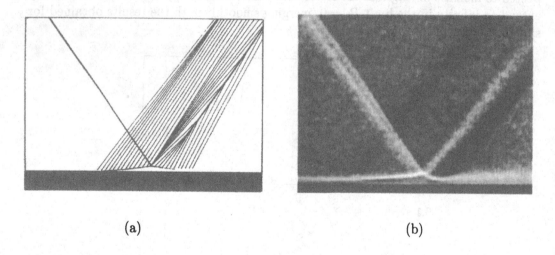

(a)   (b)

Figure 11: Shock boundary layer interaction: $M_\infty = 2$, $Re = 3 \times 10^5$, $\Delta M/M_\infty = 0.12$, (a) calculated wave pattern, (b) Schlieren photograph.

In all studies dealing with laminar interacting boundary layers discussed so far the flow medium was taken to be an incompressible fluid or a perfect gas. In recent years, however, there has been a rapidly growing interest in gases consisting of relatively complex molecules and effects which are associated with the magnitude and sign of the so-called fundamental derivative $\Gamma = (1/c) \cdot (\partial(\rho c)/\partial\rho)|_s$. Here c denotes the thermodynamic sound speed and s denotes the specific entropy. The fundamental derivative of dilute (perfect) gases can be expressed in terms of the ratio γ of the specific heats

$\Gamma = (\gamma + 1)/2$ and, therefore, is seen to satisfy the condition $\Gamma \geq 1$. In contrast, dense gases may have $\Gamma < 1$ and even $\Gamma < 0$ if the specific heats are sufficiently high as, for example, in the case of high molecular hydrocarbons and fluorocarbons. So far studies of such gases have concentrated on situations where viscosity plays a minor role and can be neglected and have revealed a number of interesting new and often anti-intuitive properties. For a summary, the reader is referred to Kluwick (1991). The results obtained indicate that fluids with high specific heats may prove beneficial in a number of practical applications. Before a final conclusion can be drawn it will be necessary, however, to take into account viscous effects and to investigate how inviscid flows are affected by and interact with boundary layers. The first steps in this direction have been taken by Whitlock (1992), Zieher (1993), Park (1994) who obtained numerical solutions of the boundary layer and the full Navier Stokes equations. In addition the scaling laws for viscous inviscid interactions in the dense gas regime have been derived by Kluwick (1994) and verified numerically by Cramer, Park and Watson (1997).

Of course, one of the most important generalizations of the triple deck concept outlined in the proceeding sections concerns the treatment of three-dimensional effects (Smith, Sykes and Brighton (1977), Sykes (1980a)). So far, however, solutions to the linearized version of the lower deck equations have been obtained in almost all the studies dealing with three-dimension flows which precludes the occurrence of separation. More recently, a spectral method has been applied by Duck and Burggraf (1986) to solve the nonlinear sublayer equations. Nonlinear solutions can be calculated more easily if it is possible to exploit symmetry properties as for example in the case of axissymmetric flows which will be considered in a separate section. Significant simplifications are possible also if the variation of the flow quantities vanishes identically in the lateral direction as in the case of the flow past swept wings, Vatsa and Werle (1977), Gittler and Kluwick (1989).

As an example of this latter type of flow, the wall streamline pattern on a swept compression ramp is shown in Fig. 12. Here the corner of the ramp coincides with the Z-axis of the coordinate system and the ramp angle is large enough to cause separation. Far upstream the wall streamlines are parallel to the free stream direction. With increasing values of X, however, they are deflected and eventually they approach the separation line. It can be shown, that the wall streamlines do not merge with the separation line at finite values of Z. As a consequence, the separation line is not an envelope of wall streamlines as has been suggested by a number of authors. Rather, it coincides with a special kind of wall streamlines along which the X-component of the wall shear stress vanishes identically. An analogous result holds for the reattachment line. In contrast to the case of plane flow the streamlines inside the separation bubble are not closed but still there is no mass exchange between the flow outside and inside the recirculation region.

Fig. 13 which shows wall streamlines on a swept indentation the depth of which is increasing with Z (Gittler and Kluwick (1989)) indicates the changes of the wall stream-

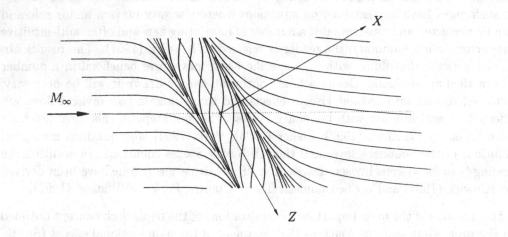

Figure 12: Wall streamline pattern on a swept compression ramp.

line pattern which take place if the wall geometry varies slowly in the spanwise direction. In the upper part of this plot the indentation is so shallow that the flow remains attached while it is deep enough to cause separation if $Z > 0.12\ldots$. Since the depth of the indentation increases with Z the wall streamlines are deflected to the right. This effect is small where the indentation is shallow but the deflexion of the streamlines is seen to increase rapidly as the indentation becomes deeper. Eventually, wall streamlines turn back to enter the separated flow region. Thus a massflux from a region of attached to a region of separated flow is set up close to the wall which is characteristic for so-called open separation. As in the case of the flow past a swept compression ramp, Fig. 12, it can be shown that envelopes of wall streamlines do not exist. In contrast to this case, however, it is no longer possible to find streamlines which have the meaning of separation or reattachment lines. This indicates that the concept of separation- and reattachment lines is of less importance for the understanding of three-dimensional flows than previously assumed.

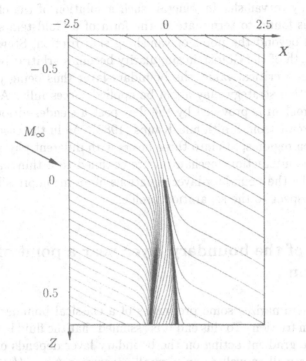

Figure 13: Wall streamline pattern on a swept, slowly varying indentation.

3   Marginal Separation

3.1   Routes towards separation

Asymptotic analysis of high Reynolds number flow has shown that there exist at least two different routes leading to the separation of a laminar boundary layer.

Firstly, a firmly attached laminar boundary layer may be forced to separate due to the presence of a large adverse pressure gradient acting over a short distance. In contrast to classical boundary layer theory, these pressure disturbances are induced by a local interaction process rather than imposed by the external inviscid flow. The interaction region exhibits a triple-deck structure and viscous effects are found to be of importance inside a thin layer adjacent to the wall (lower deck) only. Here the flow is governed by the (non-linear) boundary layer equations of an incompressible fluid. Interaction processes of this kind have been reviewed in section 2 and typical examples include flows over surface mounted obstacles, flows near the trailing edges of bodies of finite length, flows past bluff bodies, etc.

Secondly, the presence of an imposed adverse pressure gradient acting over a distance of order one on the typical boundary-layer length scale may cause the wall shear stress to decrease and finally to vanish. In general, such a solution of the classical boundary layer equations is found to terminate in the form of a Goldstein singularity and cannot be continued beyond the point of vanishing skin friction, Stewartson (1970). However, if the strength of the Goldstein singularity becomes arbitrarily small as some parameter approaches a critical value, the boundary-layer thus being just marginally separated, an interaction strategy may again be applied successfully. A typical example of this kind of problem is provided by the flow past a slender airfoil at incidence, Ruban (1981 a,b), Stewartson, Smith and Kaups (1982). As in the case of triple deck theory, the interaction region splits into three layers with different physical properties. Viscous effects on the interaction mechanism are confined to a thin wall layer where the flow is governed by the boundary-layer equations of an incompressible fluid which are linearized with respect to the separation profile.

3.2   Structure of the boundary layer near a point of vanishing skin friction

This section briefly summarizes some properties of a classical boundary layer near a point of vanishing skin friction. To this end it is assumed that the fluid is incompressible and that the pressure gradient acting on the boundary layer depends on the distance $\tilde{x}$ measured along the wall as well as on a small parameter $\Delta k = (k - k_0) \to 0$. As mentioned earlier, a specific example of such a case is provided by the flow past a thin airfoil where k characterizes the angle of incidence and k_0 denotes the critical value of k at incipient separation, Fig. 14a.

Nondimensional variables are defined as follows

$$x = \frac{\tilde{x}}{\tilde{L}}, \quad y = \frac{\tilde{y}}{\tilde{L}}, \quad u = \frac{\tilde{u}}{\tilde{u}_\infty}, \quad v = \frac{\tilde{v}}{\tilde{u}_\infty}, \quad p = \frac{\tilde{p} - \tilde{p}_\infty}{\tilde{\rho}_\infty \tilde{u}_\infty^2}. \tag{3.1}$$

Here $\tilde{x}$, $\tilde{y}$, $\tilde{u}$, $\tilde{v}$, $\tilde{p}$, $\tilde{\rho}$, $\tilde{\nu}$ and $\tilde{L}$ denote coordinates parallel and normal to the wall, the corresponding velocity components, the pressure, the density, the kinematic viscosity and a characteristic length, respectively. Furthermore, the subscript ∞ indicates the reference state of the various field quantities. Introducing the scaled variables

$$y_m = y Re^{1/2}, \quad v_m = v Re^{1/2}, \quad Re = \frac{\tilde{u}_\infty \tilde{L}}{\tilde{\nu}_\infty} \tag{3.2}$$

and the stream function ψ via $u = \partial\psi/\partial y_m$, $v_m = -\partial\psi/\partial x$ leads to the boundary layer equation in the form

$$\frac{\partial\psi}{\partial y_m}\frac{\partial^2\psi}{\partial x \partial y_m} - \frac{\partial\psi}{\partial x}\frac{\partial^2\psi}{\partial y_m^2} = -\frac{dp}{dx} + \frac{\partial^3\psi}{\partial y_m^3}. \tag{3.3}$$

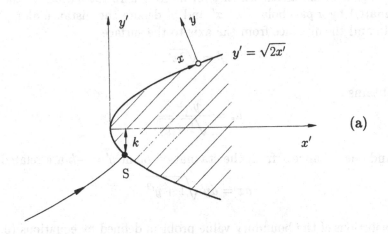

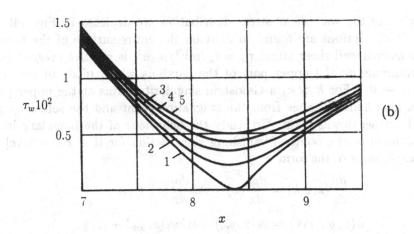

Figure 14: Flow in the nose region of a slender airfoil. (a) notation, (b) **wall shear** distribution for various values of k. 1: $k_0 = 1.1556$, 2-5: $\Delta k = -0.0003$ (Ruban (1981a)).

Equation (3.3) is subjected to the boundary conditions

$$\psi(x,0) = \frac{\partial\psi}{\partial y_m}(x,0) = 0 \ , \qquad \frac{\partial\psi}{\partial y_m}(x,\infty) = u_e(x,k) \ . \tag{3.4}$$

where $u_e(x,k)$ is the slip velocity at the airfoil surface predicted by inviscid theory and

$$\frac{dp}{dx} = -u_e\frac{du_e}{dx} \ . \tag{3.5}$$

In the nose region of the airfoil which is of interest here the surface of the airfoil can be approximated by a parabola. Let x' and y' denote the distance along the axis of the parabola and the distance from the axis to the surface:

$$y' = \sqrt{2x'} \, . \tag{3.6}$$

One then obtains

$$u_e = \frac{y' + k}{\sqrt{1 + y'^2}} \, . \tag{3.7}$$

Finally y' and the distance x from the stagnation point $y' = -k$ are related by

$$dx = dy'\sqrt{1 + y'^2} \, . \tag{3.8}$$

Numerical solutions of the boundary value problem defined by equations (3.1) to (3.8) have been obtained by Ruban (1981 b).

Typical results for the wall shear stress distribution are depicted in Fig. 14b. For $k < k_0 \approx 1.1556$ solutions are found to exist on the entire surface of the parabola. The nondimensional wall shear stress $\tau_w = \tilde{\tau}_w Re^{1/2}/\tilde{\rho}_\infty \tilde{u}_\infty^2$ is positive everywhere but exhibits a minimum on the upper part of the parabola which tends to zero in the limit $(k - k_0) \to 0^-$. For $k > k_0$ a Goldstein singularity forms at the upper part of the parabola at a finite distance from the stagnation point and the solution cannot be continued further downstream. To study the behaviour of the boundary layer in the neighbourhood of the point of vanishing skin friction for $|k - k_0| = |\Delta k| \ll 1$ asymptotic expansions of the form

$$\frac{dp}{dx}(x; \Delta k) = \frac{dp_0}{dx}(x) + \Delta k \frac{dp_1}{dx}(x) + \dots \, ,$$

$$\psi(x, y_m; \Delta k) = \psi_0(x, y_m) + \Delta k \psi_1(x, y_m) + \dots \tag{3.9}$$

are assumed to hold for the pressure gradient and the stream function. Furthermore, since we are interested in the flow properties near $x = x_0$, the point of vanishing wall shear for $\Delta k = 0$, it is useful to write

$$\frac{dp_0}{dx}(x) = p_{00} + (x - x_0)p_{01} + \dots \, ,$$

$$\frac{dp_1}{dx}(x) = p_{10} + (x - x_0)p_{11} + \dots \, . \tag{3.10}$$

As pointed out by Goldstein (1948), the classical boundary layer develops a two tiered structure in the limit $(x_0 - x) \to 0^+$ considered here, Fig. 15. Inside the wall layer (1b) appropriate independent variables are $\xi = (x_0 - x)^{1/4}$ and $\eta = y_m/\xi$ leading to

$$\psi_0 \sim \xi^3 \left[f_{00}(\eta) + \xi^\beta f_{01}(\eta) + \dots + \xi^{2\beta} f_{02}(\eta) + \dots \right], \quad \beta > 0 \, . \tag{3.11}$$

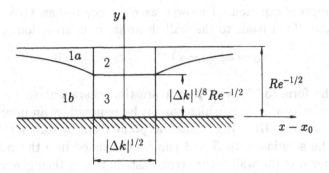

Figure 15: Structure of classical boundary layer near a point of vanishing skin friction.

Substitution of equations (3.10), (3.11) into the boundary layer equations (3.3), (3.4) gives

$$f_{00}''' - \frac{3}{4} f_{00} f_{00}'' + \frac{1}{2} f_{00}'^2 = p_{00} \,,$$

$$f_{00}(0) = f_{00}'(0) = 0 \,,$$

(3.12)

$$f_{01}''' - \frac{1}{8} p_{00} \eta^3 f_{01}'' + \frac{1}{2}(1+\beta) p_{00} \eta^2 f_{01}' - \frac{1}{4}(3+4\beta) p_{00} \eta f_{01} = 0 \,,$$

$$f_{01}(0) = f_{01}'(0) = 0 \,.$$

(3.13)

The solutions of equation (3.12)

$$f_{00} = \frac{1}{6} p_{00} \eta^3$$

(3.14)

describes the leading order term of the separation profile. The solution for f_{01} is independent of the value of β and contains one positive but otherwise arbitrary constant a_0,

$$f_{01} = \frac{1}{2} a_0 \eta^2 \,.$$

(3.15)

Inspection of the equation for f_{02}

$$f_{02}''' - \frac{1}{8} p_{00} \eta^3 f_{02}'' + \frac{1}{2}(1+2\beta) p_{00} \eta^2 f_{02}' - \frac{1}{4}(3+8\beta) p_{00} \eta f_{02} = -\frac{1}{8}(1+4\beta) a_0 \eta^2 \,,$$

$$f_{02}(0) = f_{02}'(0) = 0$$

(3.16)

shows that β must satisfy the relationship

$$\beta = 2m+1 \,, \quad m = 0, 1, 2, 3 \ldots$$

(3.17)

in-order to suppress exponential growth as $\eta \to \infty$, Ruban (1981a). Adopting the lowest eigenvalue $\beta = 1$ leads to the wall shear-stress distribution

$$\tau_w \sim a_0(x_0 - x)^{1/2}, \quad (x_0 - x) \to 0^+. \tag{3.18}$$

Solutions of the form (3.11) with $\beta = 1$ are thus characterized by the presence of a Goldstein singularity at $x = x_0$ which cannot be removed by an interaction strategy as shown by Stewartson (1970). In contrast, as pointed out by Ruban (1981b), Stewartson et al (1982), the solution with $\beta = 3$ can be continued into the region $x > x_0$. The leading order term of the wall shear stress distribution is then given by

$$\tau_w \sim a_0(x_0 - x), \quad (x_0 - x) \to 0^+ \tag{3.19}$$

in agreement with the numerical results obtained for the leading edge separation problem mentioned earlier.

Evaluation of the first order correction ψ_1 caused by small changes Δk of the parameter k from the critical value k_0 leads to the following representation of the stream function inside the near wall region (1b):

$$\psi \sim \xi^3 \left[\frac{1}{6}p_{00}\eta^3 + \xi^3 \frac{1}{2}a_0\eta^2 + \xi^6 \left(\frac{1}{2}b_0\eta^2 - \frac{a_0^2}{120}\eta^5 + \frac{p_{00}a_0^2}{40320}\eta^9\right) + \cdots\right]$$
$$+\Delta k \left[\xi^{-2}\frac{1}{2}a_1\eta^2 + \xi\frac{1}{2}b_1\eta^2 + \cdots\right] \tag{3.20}$$

which is seen to contain three additional constants b_0, a_1 and b_1 which can only be determined by a global analysis.

For completeness, the solution for ψ holding in the outer predominately inviscid region (1a) the boundary layer is briefly noted

$$\psi \sim \psi_{00}(y_m) + (x_0 - x)\psi_{01}(y_m) + \cdots$$
$$+\Delta k \left[(x_0 - x)^{-1}\psi_{10}(y_m) + (x_0 - x)^{-1/4}\psi_{11}(y_m) + \cdots\right]. \tag{3.21}$$

Here $\psi_{00}(y_m)$ characterizes the separation profile for $k = k_0$ which uniquely determines the functions $\psi_{01}(y_m)$, $\psi_{10}(y_m)$ and $\psi_{11}(y_m)$.

Expansion (3.11) for the stream function breaks down if $(x_0 - x) = O(|\Delta k|^{1/2})$ thus indicating that two additional regions (2) and (3), Fig. 15, have to be considered. Viscous effects are of importance in (3) only, which is the continuation of (1b).

Inspection of equation (3.20) suggests the following form of the expansion of ψ

$$\psi \sim \varepsilon^{3/8}\frac{1}{6}p_{00}\overline{y}^3 + \varepsilon^{3/4}\overline{\psi}_1(\overline{x}, \overline{y}) + \varepsilon^{9/8}\overline{\psi}_2(\overline{x}, \overline{y}) + \cdots \tag{3.22}$$

holding in (3) where

$$\bar{x} = (x_0 - x)\varepsilon^{-1/2}, \quad \bar{y} = y_m\varepsilon^{-1/8}, \quad \varepsilon = |\Delta k|. \quad (3.23)$$

Evaluation of the wall shear-stress distribution yields

$$\tau_w \sim a_0 \left[(x_0 - x)^2 + 2\Delta k \frac{a_1}{a_0} \right]^{1/2} \quad (3.24)$$

which matches with the result (3.19) in the limit $\bar{x} \to \infty$.

The investigation of the flow properties for $x > x_0$ closely follows the analysis outlined so far and thus is not repeated here. Most important, it is found that equations (3.19) and (3.24) hold unchanged and thus determine the wall shear distribution for $|x_0 - x| \to 0$. Depending on the value of $a_1\Delta k$ three different cases have to be distinguished.

(a) $a_1\Delta k > 0$: the wall shear-stress distribution obtained for $(x_0 - x) \to 0^+$ can be continued uniquely downstream of x_0.

(b) $a_1\Delta k = 0$: this case can be interpreted as the limit of (a) and the wall shear stress distribution can thus be continued beyond the point of vanishing skin friction although a singularity forms at $x = x_0$.

(c) $a_1\Delta k < 0$: the wall shear-stress distribution terminates with a Goldstein singularity at $x_s = x_0 - (-2a_1\Delta k/a_0)^{1/2}$ upstream of x_0.

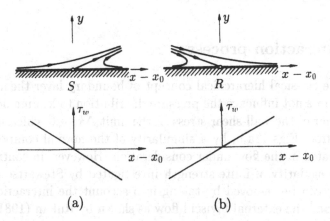

Figure 16: Streamline pattern and wall shear distribution near a separation point (a), reattachment point (b).

In passing it should be noted that the result for case (b) with $\Delta k = 0$ can be motivated in a different way using local solutions of the Navier Stokes equation in the vicinity of a separation point. The wall shear-stress distributions and the form of streamlines associated with regular two-dimensional separation and reattachment derived by Oswatitsch 1958 are sketched in Fig. 16. These distributions represent local solutions of the full Navier Strokes equations but as pointed out by Kluwick, Gittler and Bodonyi (1984) also local solutions of the boundary layer equations. In contrast to the Navier Strokes equations, however, the boundary layer equations do not contain second order derivatives with respect to x. Therefore, if Figs 16a,b are cut along the line $x = 0$ and if the outer and inner parts are recombined as shown in Figs 17a,b the new distributions will again satisfy the boundary layer equations. Fig. 17a corresponds to case (b) of marginal separation discussed earlier while Fig. 17b describes a case of marginal reattachment.

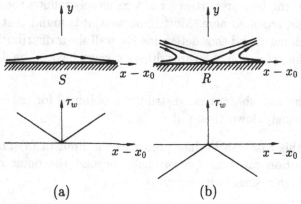

(a)   (b)

Figure 17: Streamline pattern and wall shear distribution corresponding to marginal separation (a), marginal reattachment (b).

3.3   The interaction process

According to the classical hierarchical concept of boundary layer theory the displacement thickness does not influence the pressure distribution to leading oder. Due to the singular behaviour of the wall-shear stress in the limit $\Delta k \to 0$ which is accompanied, as can be seen from Figs 17a,b, by a singularity of the normal component v this assumption is violated in the flow under consideration. However, in contrast to the case of a Goldstein singularity of finite strength investigated by Stewartson (1970) such a weak singularity can be removed by taking into account the interaction between the boundary layer and the external inviscid flow as shown by Ruban (1981b), Stewartson et al (1982).

The fact that the induced pressure disturbances $p_i = O(Re^{-1/2})$ act over a region with streamwise extent of $O(\varepsilon^{1/2})$ gives

$$\frac{dp_i}{dx} = O(Re^{-1/2}\varepsilon^{-1/2})\,. \tag{3.25}$$

Furthermore, since the singularity occurring in the classical boundary-layer solution is determined by the second order term of the expansion for the stream function ψ, the interaction pressure must enter the flow description at this level of approximation. Using equation (3.22) leads to the order of magnitude estimate

$$u\frac{\partial u}{\partial x} \sim \varepsilon^{3/8}\frac{1}{2}p_{00}\overline{y}^2\frac{\partial^2\overline{\psi}_1}{\partial\overline{x}\partial\overline{y}} + \varepsilon^{3/4}\left(\frac{1}{2}p_{00}\overline{y}^2\frac{\partial^2\overline{\psi}_2}{\partial\overline{x}\partial\overline{y}} + \frac{\partial\overline{\psi}_1}{\partial\overline{y}}\frac{\partial^2\overline{\psi}_1}{\partial\overline{x}\partial\overline{y}}\right) + \dots . \qquad (3.26)$$

Comparison of equations (3.25) and (3.26) yields the appropriate scaling of the controlling parameter Δk in terms of the Reynolds number

$$\Delta k = k_1 Re^{-2/5}, \quad k_1 = O(1) . \qquad (3.27)$$

As a consequence, the length of the interaction region is $O(Re^{-1/5})$. As sketched in Fig. 18 the interaction region exhibits a three layer structure. Inside the "lower deck" of thickness $O(Re^{-11/20})$, the acceleration of the fluid is balanced by pressure and viscous forces. Viscous effects are negligible small inside the "main deck", which comprises most of the boundary layer. The role of the main deck is to transmit displacement effects caused by the thin layer of low speed fluid adjacent to the wall thus leading to an inviscid pressure response in the "upper deck" outside the boundary layer which is strong enough to eliminate the singularity at the point of vanishing skin friction.

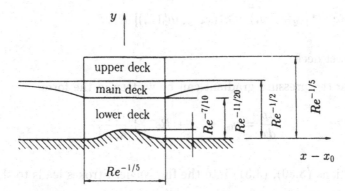

Figure 18: Structure of the interaction region.

While triple deck problems are typically triggered by rapid changes of the boundary conditions to which the viscous wall layer is subjected it is the imposed pressure gradient which plays the key role as far as marginally separated flows are concerned. In contrast to triple deck theory, therefore, disturbances due to surface mounted obstacles, impinging pressure waves etc. do not represent indispensable ingredients of marginal separation. Nevertheless, it is interesting and important to include such effects into the interaction theory, Hackmüller and Kluwick (1989). As an example we consider the effect of a surface mounted obstacle of nondimensional height H generating pressure disturbances of $O(HRe^{1/5})$. If the formation of a separation singularity is to be avoided, these pressure disturbances must be of the same order of magnitude as the pressure perturbation due to interaction:

$$H = O(Re^{-7/10}).$$ (3.28)

To formulate the modifications of the flow properties inside the interaction region necessary to account for a weak deformation of surface it is convenient to introduce the stretched coordinates

$$x_* = Re^{1/5}(x - x_0), \quad y_* = Re^{11/20}y - Re^{-3/20}h_0(x_*)$$ (3.29)

where y_* and $h_0(x_*)$ characterize, respectively, the distance from the surface and the shape of the surface mounted obstacle. Application of the Prandtl transposition theorem then gives the expansions for the velocity components

$$u \sim Re^{-1/10}\frac{1}{2}p_{00}y_*^2 + Re^{-1/4}u_1^*(x_*, y_*) + Re^{-2/5}u_2^*(x_*, y_*) + \ldots,$$

$$v \sim Re^{-3/5}\left[v_1^*(x_*, y_*) + \frac{1}{2}p_{00}y_*^2h_0'(x_*)\right] + \ldots$$ (3.30)

$$+ Re^{-3/4}\left[v_2^*(x_*, y_*) + u_1^*(x_*, y_*)h_0'(x_*)\right] + \ldots$$

holding in the lower deck.

The expansion for the pressure gradient can be written in the form

$$\frac{dp}{dx} = p_{00} + \ldots + Re^{-3/10}\frac{dp_i}{dx_*}.$$ (3.31)

Insertion of equations (3.30), (3.31) into the full Navier Strokes leads to the first order solutions

$$u_1^* = A_1(x_*)y_*, \quad v_1^* = -\frac{1}{2}A_1'(x_*)y_*^2$$ (3.32)

where $A_1(x_*)$ remains undetermined at this level of approximation. From the analysis of the flow properties outside the interaction region it is known, however, that $A_1(x_*)$ has to satisfy the matching condition

$$A_1(x_*) \sim a_0|x_*| + \frac{a_1k_1}{|x_*|}, \quad |x_*| \to \infty.$$ (3.33)

For the sake of brevity, details of the analysis of the main deck are omitted. As far as the upper deck is concerned it suffices to note that the induced pressure disturbances p_i, A_1 and h_0 satisfy the relationship

$$\frac{dp_i}{dx_*} = \frac{U_{00}^2}{\pi p_{00}} \oint_{-\infty}^{\infty} \frac{A_1''(\xi) - p_{00}h_0''(\xi)}{\xi - x_*} d\xi \tag{3.34}$$

where U_{00} is the prescribed velocity at the outer edge of the boundary layer at $x = x_0$. In order to determine the function $A_1(x_*)$ it is necessary to investigate the second order lower deck equations. To this end it is convenient to introduce the transformed quantities

$$\overline{X} = U_{00}^{-4/5} a_0^{2/5} p_{00}^{1/5} x_* , \qquad \overline{Y} = U_{00}^{-1/5} a_0^{1/10} p_{00}^{3/10} y_* ,$$

$$\overline{u} = U_{00}^{-8/5} a_0^{-6/5} p_{00}^{7/5} u_2^* , \qquad \overline{v} = U_{00}^{-1} a_0^{-3/2} p_{00}^{3/2} v_2^* ,$$

$$\overline{A} = U_{00}^{-4/5} a_0^{-3/5} p_{00}^{1/5} A_1 , \qquad \overline{h} = U_{00}^{-4/5} a_0^{-3/5} p_{00}^{6/5} h_0 , \tag{3.35}$$

$$\frac{d\overline{p}}{d\overline{X}} = U_{00}^{-6/5} a_0^{-7/5} p_{00}^{4/5} \frac{dp_i}{dx_*} .$$

It can then be shown that the second order lower deck equations admit solutions which can be matched with the results holding in the main deck only if $\overline{A}$ satisfies the solvability condition

$$\overline{A}^2(\overline{X}) - \overline{X}^2 - 2a = \lambda \int_{\overline{X}}^{\infty} \frac{A''(t) - \overline{h}''(t)}{\sqrt{t - \overline{X}}} dt , \tag{3.36}$$

$$a = k_1 a_1 p_{00}^{2/5} a_0^{-1/5} , \qquad \lambda = (-1/4)!/\sqrt{2}(1/4)! .$$

By applying the transformation

$$X = \lambda^{-2/5}\overline{X} , \quad A = \lambda^{-2/5}\overline{A} , \quad h = \lambda^{-2/5}\overline{h} , \quad \Gamma = -2a\lambda^{-4/5} \tag{3.37}$$

the nonlinear integro-differential equation (3.36) for the unknown first order wall shear stress distribution and the matching conditions (3.33) can be cast into the form

$$A^2(X) - X^2 + \Gamma = \int_{X}^{\infty} \frac{A''(t) - h''(t)}{\sqrt{t - X}} dt , \tag{3.38}$$

$$X \to \mp\infty : A \to \mp X .$$

If the interaction term on the right hand side is neglected one recovers equation(3.24) in transformed form, e.g. $A = \sqrt{X^2 - \Gamma}$. Numerical solutions of the interaction problem with $h \equiv 0$ have been obtained by Ruban (1981b), Stewartson et al (1982), Brown and Stewartson (1983). Typical results are depicted in Fig. 19a. In all cases Γ is positive which means that k is larger than the critical value k_0 following from the investigation of the classical boundary layer equations. We therefore conclude that separation is delayed by the interaction process. If the flow remains attached or if the separated flow region is very short the wall shear stress distribution is almost symmetrical. With increasing length of the recirculation zone, however, the point of minimum wall shear stress is shifted towards the reattachment point. A similar effect occurs in triple deck solutions but also in solutions of the full Navier Stokes equations, Katzer (1989).

In Fig. 19b the wall shear at $x = 0$ is plotted as a function of Γ. It is seen that the interaction problem does not possess a unique solution for all values of Γ. For all values $\Gamma > 0$ for which solutions can be found there exist at least two different solutions (As shown by Kluwick (1989), equation (3.38) can be used to study marginal separation on axisymmetric compression ramps in supersonic flow if Γ is interpreted as a scaled ramp angle. The nonuniqueness of this equation then can be interpreted as continuation of the nonuniques found in the related triple deck problem, Gittler and Kluwick (1987a,b)).

Furthermore, it is interesting to note that equation (3.38) does admit solutions for $\Gamma < \Gamma_c \approx 2.75$ only indicating that a substantial change of the flow pattern must take place if Γ exceeds the critical value Γ_c. How this takes place is still unclear, however.

To demonstrate the effect of surface mounted obstacles we take $h(X)$ to be of the form

$$h(X) = \begin{cases} \alpha(X^2 - 1)^3 & |X| \leq 1, \\ 0 & |X| > 1. \end{cases} \tag{3.39}$$

In Fig. 20a, the the value $A(0)$ of the scaled wall shear stress at the origin is plotted as a function of α for $\Gamma = 1$. Numerical solutions were found by Hackmüller and Kluwick (1989) for $-5.19 \leq \alpha \leq 2.45$. However, if the height of the protrusion/indentation exceeds the critical values $-5.19/2.45$ solutions of equation (3.38) cease to exist and the properties of the flow caused by such larger surface mounted obstacles are not known at present. Furthermore, the numerical computations indicate that the interaction equation admits two different solutions for values $-5.19 < \alpha < 1.77$, $-1.24 < \alpha < -0.93$ and $0.76 < \alpha < 2.45$ while four different solutions exist for $-1.77 < \alpha < -1.24$. So far numerical difficulties have prevented the calculation of separated flow solutions for $-0.93 < \alpha < 0.76$. The difficulties in continuing the solution to smaller values than 0.76 seem to be caused by the formation of a small loop similar to that for $-1.77 < \alpha < -1.24$ while the difficulty arising near $\alpha = -0.93$ is most probably associated with the abrupt changes of the field quantities near reattachment.

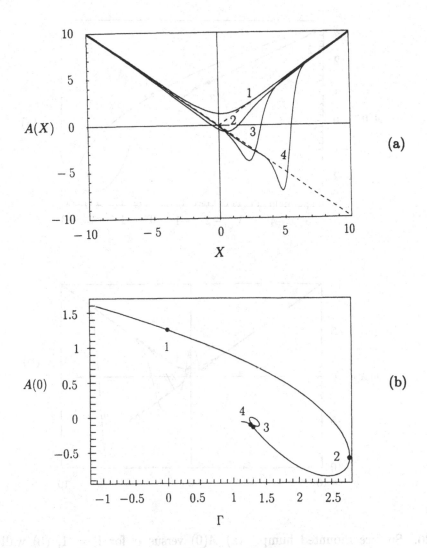

Figure 19: Numerical solutions of the interaction equation (3.38) with $h = 0$. (a) $A(X)$ for various values of Γ, (b) $A(0)$ versus Γ.

Wall shear-stress distributions for $\Gamma = 1$ and various values of $\alpha \leq 0$ are depicted in Fig. 20b. For $\alpha = 0$ the flow (described by the upper branch of Fig. 20a) is attached and incipient separation was found to occur at $\alpha \approx -2.87$. As α increases further a separation bubble forms at the leeward side of the protrusion and the limiting value $\alpha = -5.19$ is finally reached. Moving on along the lower branch of Fig. 20a leads to protrusions of decreasing height while the streamwise extent of the recirculation region still increases and, presumably, assumes its maximum length at $\alpha = 0$.

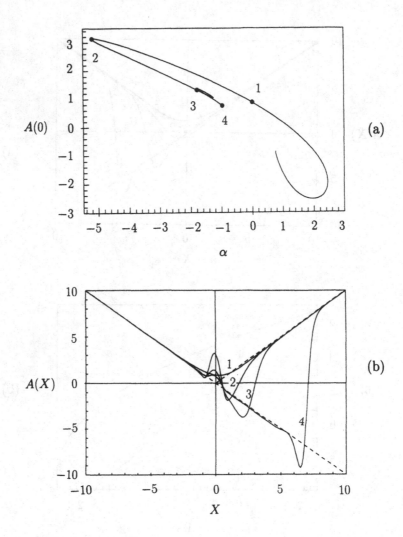

Figure 20: Surface mounted hump. (a) $A(0)$ versus α for $\Gamma = 1$, (b) wall shear distribution $A(X)$ for various values of α.

3.4 Three-dimensional effects

The theory of two-dimensional marginally separated flows outlined so far has been generalized in various ways to include three-dimensional effects.

Firstly, perturbations of a two-dimensional incoming boundary layer on the verge of separation caused by three-dimensional surface mounted obstacles have been considered by Hackmüller and Kluwick (1990), (1991), who showed that the appropriate generalization of equation (3.36) is given by

$$\overline{A}^2(\overline{X}, \overline{Z}) - \overline{X}^2 - 2a = -\frac{\lambda}{2} \int_{-\infty}^{\overline{X}} \frac{F(\overline{X}_1, \overline{Z})}{\sqrt{\overline{X} - \overline{X}_1}} d\overline{X}_1,$$

$$F(\overline{X}, \overline{Z}) = -\frac{1}{\pi} \int_{-\infty}^{\infty} \frac{f_\xi(\xi, \infty)}{\xi - \overline{X}} d\xi$$

$$+ \frac{1}{2\pi} \int_{-\infty}^{\infty} \int_{-\infty}^{\infty} \frac{d\xi d\zeta}{[(\overline{X} - \xi)^2 + (\overline{Z} - \zeta)^2]^{1/2}} \left[\frac{\partial^3}{\partial \xi^3} + \frac{\partial^3}{\partial \xi \partial \zeta^2} \right] [f(\xi, \infty) - f(\xi, \zeta)],$$

$$f = A - H$$

$$(3.40)$$

where $\overline{Z}$ denotes the stretched coordinate z_*: $\overline{Z} = \lambda^{-2/5} U_{00}^{-4/5} a_0^{-2/5} p_{00}^{1/5} z_*$ in the lateral direction. The related problem for axisymmetric boundary layers has recently been studied by Zametaev and Sychev (1995).

A different type of interaction equation arises if one considers three-dimensional disturbances of flows past swept wings. Introducing (suitably stretched) coordinates X, Z parallel and normal to the leading edge of the wing equation (3.36) is replaced by

$$A^2(X, Z) - X^2 + \Gamma + 2 \int_{-\infty}^{X} A_Z(\xi, Z) d\xi = \int_{X}^{\infty} \frac{A_{\xi\xi}(\xi, Z) - H_{\xi\xi}(\xi, Z)}{\sqrt{\xi - X}} d\xi. \quad (3.41)$$

In the limit of weak interaction $\Gamma \to -\infty$ the term on the right hand side can be neglected. It is convenient to rewrite the resulting equation in the equivalent form

$$A(X, Z) A_X(X, Z) + A_Z(X, Z) = X \quad (3.42)$$

which can easily be solved using the method of characteristics, Hackmüller (1991). The slope of characteristic curves $\chi(X, Z) = $ const is given by

$$\frac{\partial X}{\partial Z}\bigg|_{\chi = \text{const}} = A(X, Z) \quad (3.43)$$

and shows that lines $\chi = $ const coincide with wall streamlines. This result carries over unchanged to cases of interactive flows where lines $\chi = $ const play the role of subcharacteristics which, among others, determine domains of dependence and influence associated with the full equation (3.41), Fig. 21. This is seen most easily if this equation is recast in the form

$$\frac{\partial X}{\partial Z}\bigg|_{\chi=\text{const}} = \frac{1}{2}\int_X^\infty \frac{A_{\xi\xi\xi}(\chi, Z) - H_{\xi\xi\xi}(\chi, Z)}{\sqrt{\xi - X}} d\xi + X \tag{3.44}$$

using the definition (3.43).

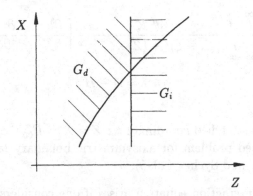

Figure 21: Regions of influence G_i and regions of dependence G_d of equation (3.41).

Equations (3.40) and (3.41) are considerably more complicated than their two-dimensional counterparts (3.36), (3.38). In the original study by Hackmüller and Kluwick (1990), therefore, disturbances which vary slowly in the lateral direction have been considered only. Fully three-dimensional flows associated with surface mounted obstacles have been obtained only recently by Kluwick, Reiterer and Hackmüller (1997). As an example of a quasi-two-dimensional flow Fig. 22 shows the wall streamline pattern generated by a protrusion whose height varies periodically in the spanwise direction of a swept wing. The dotted lines are contour lines indicating the height-distribution of the protrusion and the dot-dashed line corresponds to the critical height for which incipient separation would occur in the limiting case of strictly two-dimensional flows. As expected the onset of separation is delayed by the three-dimensional nature of the flow. Furthermore, the protrusion is seen to generate a periodic array of recirculation regions. Each such region represents a small open separation bubble with a mass flux into the bubble from the left. Similar to the triple deck solutions obtained for swept wing configurations by Gittler and Kluwick (1989) no lines which have the meaning of separation or reattachment lines can be identified in Fig. 22. There are regions of high density of wall streamlines as a consequence of the displacement of the fluid in the direction normal to the surface but no envelope exists.

Secondly, the structure of a three-dimensional boundary layer near a point of vanishing skin friction in a plane of symmetry has been investigated by Brown (1985), Duck

(1989), Zametaev (1989). The investigation carried out by Brown (1985) is based on the assumption that the induced pressure disturbances are independent of the lateral coordinate $z = \tilde{z}/\tilde{L}$. As a result, the cross flow pressure gradient is known in advance and obtained from inviscid theory. Furthermore, it is assumed that the velocity components have regular expansions for small distances from the plane of symmetry $z \to 0$:

$$u(x,y,z) = u_0(x,y_m) + O(z^2)\,,$$

$$Re^{1/2}v(x,y,z) = v_0(x,y_m) + O(z^2)\,,$$

$$w(x,y,z) = zw_0(x,y_m) + O(z^3)\,,$$

$$p(x,z) = p_0(x) + \frac{1}{2}z^2 s(x) + O(z^4)\,. \tag{3.45}$$

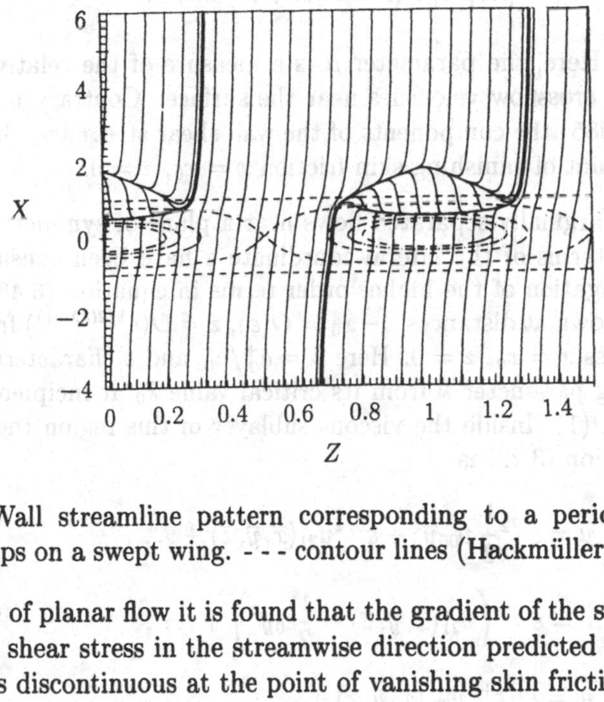

Figure 22: Wall streamline pattern corresponding to a periodic array of surface mounted humps on a swept wing. - - - contour lines (Hackmüller and Kluwick (1991)).

As in the case of planar flow it is found that the gradient of the streamwise component τ_w of the wall shear stress in the streamwise direction predicted by classical boundary layer theory is discontinuous at the point of vanishing skin friction $x = x_0$, $z = 0$:

$$\tau_w = \left.\frac{\partial u_0}{\partial y_m}\right|_{y_m=0} = \begin{cases} -a_0^-(x - x_0)\,, & (x - x_0) \to 0^-\,, \\ a_0^+(x - x_0)\,, & (x - x_0) \to 0^+\,. \end{cases} \tag{3.46}$$

In contrast to the case of planar flow, however, the values a_0^- and a_0^+ are not necessarily equal but satisfy the relationship

$$a_0^+ = a_0^- - c_1 \tag{3.47}$$

where c_1 determines the variation of the crossflow component σ_w of the wall shear stress with z at $x = x_0$:

$$\sigma_w = z \left. \frac{\partial w_0}{\partial y_m} \right|_{y_m=0} = c_1 z. \tag{3.48}$$

Marginally separated flows in which τ_w and σ_w vanish at the single point $x = x_0$, $z = 0$ only have been studied by Duck (1989). He showed that a self consistent description of such flows can be achieved if the following forms

$$\tau_w = \text{prop} \, [\mu^2(x - x_0)^2 + z^2]^{1/2} + \ldots,$$

$$\sigma_w = \text{prop} \, z[\mu^2(x - x_0)^2 + z^2]^{1/2} + \ldots \tag{3.49}$$

are assumed to hold. Here, the parameter μ is a measure of the relative magnitude of the streamwise and crossflow velocities near the surface: Contrary to the solution obtained by Brown (1985) the components of the wall shear stress are, therefore, non-differentiable at the point of vanishing skin friction $x = x_0$, $z = 0$.

The general form of marginally separated flows near a plane of symmetry which have regular expansions in terms of the lateral coordinate z have been considered by Za-metaev (1989). Investigation of the higher order terms in equation (3.45) shows that the expansions break down at distances $x - x_0 = O(\varepsilon)$, $z = O(\varepsilon^{1/2(3\lambda-1)})$ from the point of vanishing shear stress $x = x_0$, $z = 0$. Here $\lambda = a_0^+/a_0^-$ and ε characterizes the deviation of the controlling parameter k from its critical value k_0 at incipient separation: $k - k_0 = \Gamma \varepsilon^{1+\lambda}$, $\Gamma = O(1)$. Inside the viscous sublayer of this region the appropriate generalization of equation (3.22) is

$$u = \varepsilon^{1/2} \frac{1}{2} \, p_{00} \overline{y}^2 + \varepsilon^{5/4} u_{21}(\overline{x}, \overline{y}, \overline{z}) + \ldots,$$

$$v = \varepsilon^{1/2} \left(v_{21}(\overline{x}, \overline{y}, \overline{z}) - \frac{1}{2} c_0 \overline{y}^2 \right) + \ldots, \tag{3.50}$$

$$w = \varepsilon^{1/4+\beta} w_{20}(\overline{x}, \overline{y}, \overline{z}) + \ldots$$

with $\overline{x}$ and $\overline{y}$ as defined in equation(3.23) and

$$\overline{z} = \frac{z}{\varepsilon^\beta}, \quad \beta = \frac{1}{2}(3\lambda - 1). \tag{3.51}$$

The solutions for u_{21}, v_{21} and w_{20} take on the form

$$u_{21} = A_{21}(\overline{x}, \overline{z})\overline{y} - \frac{c_0}{40}p_{00}\overline{y}^5,$$

$$v_{21} = -\frac{1}{2}\frac{\partial A_{21}}{\partial \overline{x}}\overline{y}^2, \tag{3.52}$$

$$w_{20} = c_0\overline{y}\,\overline{z}.$$

Introducing transformed quantities S, Z and A

$$\overline{x} = \alpha S, \quad \alpha = \left(\frac{|\Gamma|}{a_0^-}\right)^{1/(1+\lambda)},$$

$$\overline{z} = \gamma Z, \quad \gamma = \left(\frac{a_0^- \alpha^{3\lambda-1}}{|b_{20}|}\right)^{1/2}, \tag{3.53}$$

$$A_{21} = a_0^- \alpha A$$

in place of $\overline{x}$, $\overline{z}$ and A_{21} the evolution equation for the streamwise component of the wall shear stress if given by

$$A\frac{\partial A}{\partial S} - \lambda S + (1-\lambda)A + (1-\lambda)Z\frac{\partial A}{\partial Z} = 0,$$

$$A(S \to -\infty, Z) = -S - \text{sign}(\Gamma)(-S)^{-\lambda} + \text{sign}(b_{20})Z^2(-S)^{2-3\lambda} \tag{3.54}$$

if $\lambda > 2/3$ which has been investigated so far only. The parameter b_{20} is positive/negative if the streamwise velocity component in the external flow region increases/decreases with increasing values of Z. Inspection of the first equation (3.54) shows that the slope of characteristic curves agrees with the slope of wall streamlines. As pointed out earlier a similar result has been obtained by Hackmüller and Kluwick (1990) in their study of swept wing configurations. This ties in nicely with the analysis of Wang (1971) who argued that the streamlines inside a classical boundary layer with imposed pressure gradient play the role of subcharacteristics. If the dependence of the field quantities in the direction normal to the wall is approximated by known functions as for example in integral methods these subcharacteristics become characteristics of the reduced set of equations. In the theory of marginally separated flows the variation of the field quantities with $\overline{y}$ following from equation (3.50) is known in advance also and is the same for all specific solutions - in a sense, therefore, this theory can be viewed as an asymptotically exact version of an integral method. Since the viscous sublayer in which equation (3.50) holds is so thin, the streamlines there almost collapse onto the wall streamline and, as a consequence, the evolution equation for the "shape function"

A contains a single characteristic only in contrast to usual integral methods used to determine the flow properties in the boundary layer as a whole.

Owing to the nonlinearity of equation (3.54) its formal solutions following from the integration of the equivalent set of slope - and compatibility conditions may develop regions of multivaluedness. These have to be eliminated by the insertion of jump discontinuities.

Typical results for the wall streamline patterns and the wall shear stress distributions obtained from equation (3.54) are depicted in Figs 23, 24 and 25, Kluwick and Reiterer (1998). In all three cases the parameter λ is taken to be smaller than one, which, according to equation (3.47), implies $c_1 > 0$ so that the lateral velocity component is positive for $z > 0$.

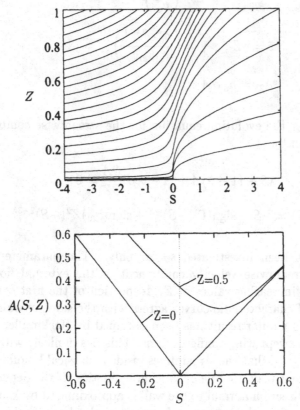

Figure 23: Wall streamline pattern. $\lambda = 3/4$, $\Gamma = 0$, $b_{20} = 1$.

Fig. 23 corresponds to a flow with $k = k_0$, e.g. $\Gamma = 0$ where the wall shear stress in the plane of symmetry exhibits the marginal separation singularity. Owing to the choice $b_{20} > 0$ the streamwise velocity component increases with increasing Z and the wall shear stress is strictly positive in planes $Z > 0$. Inspection of equation (3.52) then indicates that the distribution of the normal velocity component is discontinuous at $S = Z = 0$ but smooth elsewhere. Furthermore, it is seen that the local reduction of the wall shear stress leads to a significant displacement of the wall streamlines in the lateral direction.

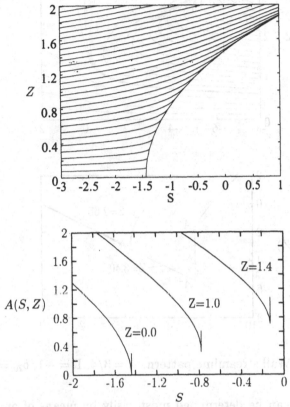

Figure 24: Wall streamline pattern. $\lambda = 3/4$, $\Gamma = 1$, $b_{20} = 1$.

Fig. 24 indicates how the flow pattern changes if the controlling parameter k increases beyond its critical value k_0, e.g. $\Gamma > 0$. In contrast to Fig. 23 the wall streamlines do not longer cover the whole S, Z-plane but form an envelope thus generating a region which is inaccessible from upstream. Along the envelope the wall shear stress distribution remains positive and ends with a vertical tangent. Consequently the normal velocity component v tends to $+\infty$ as in the case of a Goldstein singularity.

Finally, a case of subcritical flow $k < k_0$ is considered in Fig. 25. As a consequence of $\Gamma < 0$ the flow remains attached in the plane of symmetry. If $b_{20} < 0$, e.g. if the streamwise velocity component decreases with increasing distance Z as assumed here the flow may separate for sufficiently large Z. At the point of vanishing wall shear stress the wall streamlines turn back upstream to form an overlapping region which, as pointed out earlier, has to be obviated by the insertion of a shock front. Across this

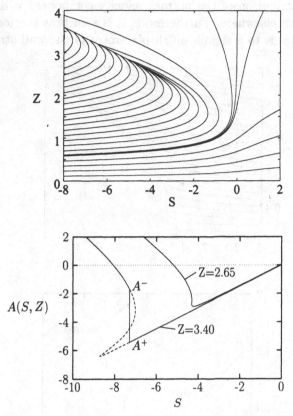

Figure 25: Wall streamline pattern. $\lambda = 3/4$, $\Gamma = -1$, $b_{20} = -1$.

front (whose position can be determined most easily by means of an equal area rule applied to wall shear stress distribution $A(S, Z)$, Kluwick and Reiterer (1998)) the wall shear stress changes discontinuously.

Similar to the case of one-dimensional flow the local solution of the boundary layer equations with imposed pressure gradient does not represent a uniformly valid approximation of the full Navier Stokes equations. If the deviations of the controlling parameter k from its critical value are sufficiently small

$$\varepsilon = O\left(Re^{-1/5}\right), \quad \lambda \leq 1, \qquad \varepsilon = O\left(Re^{-1/(3\lambda+2)}\right), \quad \lambda > 1 \qquad (3.55)$$

such an approximation can be obtained however from equation (3.52) if the evolution equation (3.54) is modified to account for the interaction between the boundary layer and the external inviscid flow. If $\lambda < 1$ inspection of equation (3.51) indicates that the lateral extent of the local interaction region is asymptotically large compared to its streamwise extent. As a consequence, the lateral variation of the induced pressure disturbances does not enter the leading order problem and the interaction equation derived by Brown (1985) is recovered on the symmetry line $Z = 0$. In contrast, the streamwise pressure gradient represents the dominating driving force of the evolution equation A if $\lambda > 1$. Finally, both components of the feedback pressure gradient are equally important if $\lambda = 1$ investigated by Duck (1989).

4 Axisymmetric laminar interacting boundary layers

4.1 Flows past flared cylinders

Axisymmetric interacting boundary layers have been investigated first by Horton (1971), Vatsa and Werle (1977), and using the triple-deck concept by Kluwick, Gittler and Bodonyi (1984), Duck (1984), Timoshin (1985), (1986). As an example of this type of problem the flow past a flared cylinder is sketched in Fig. 26. It will be assumed first that the radius $\tilde{a}$ of the cylinder is of the same order of magnitude as the length of the interaction region in the case of two-dimensional flow $\tilde{a}/\tilde{L} = O(\varepsilon^3)$. Owing to this assumption the thickness of the lower-deck is small compared to $\tilde{a}$ and as a consequence the flow inside the viscous layer adjacent to the wall is governed by equations (2.13). Similarly, axisymmetric effects do not affect the flow inside the main deck to leading order. They do, however, have to be taken into account in the upper deck, where they cause modifications of the pressure-displacement relationships. Introducing the scaled cylinder radius

$$a = C^{-3/8}\lambda^{4/5}|1 - M_\infty^2|^{1/8}\left(\frac{\tilde{T}_w}{\tilde{T}_\infty}\right)^{-3/2}\frac{\tilde{a}}{\varepsilon^3\tilde{L}}, \quad \lambda = 0.33206\ldots \quad (4.1)$$

and the Fourier transforms $\overline{P}$, $\overline{A}$ of $P(X)$, $A(X)$ the appropriate viscous inviscid coupling condition for subsonic flow is

$$\overline{P}(\omega) = -i\omega\overline{A}(\omega)H_0^{(1)}(i\omega a)H_1^{(1)}(i\omega a)\,, \quad \Re(\omega) > 0\,,$$

$$\overline{P}(\omega) = -i\omega\overline{A}(\omega)H_0^{(2)}(i\omega a)H_1^{(2)}(i\omega a)\,, \quad \Re(\omega) < 0\,. \quad (4.2)$$

If the external flow is supersonic the feedback pressure can be expressed in the form

$$P(X) = -A'(X) + \frac{1}{a} \int\limits_{-\infty}^{X} W\left(\frac{X-\xi}{a}\right) A'(\xi) \, d\xi \,,$$

(4.3)

$$W(z) = \int\limits_{0}^{z} \frac{e^{-\lambda z}}{K_1^2(\lambda) + \pi^2 I_1^2(\lambda)} \frac{d\lambda}{\lambda} \,.$$

Note that as $a \to \infty$ (4.2) and (4.3) reduce to the limiting forms (2.16), (2.17).

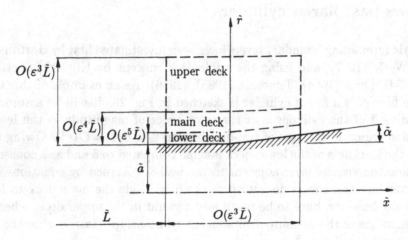

Figure 26: Triple-deck structure of the interaction region for axisymmetric flow.

As the scaled body radius tends to zero, $a \to 0$, both (4.2) and (4.3) yield

$$P(X) = A''(X) \, a \, \log a \,.$$

(4.4)

In agreement with the results following from slender body theory, the induced pressure disturbances are proportional to the curvature of the displacement body independent whether the external flow is subsonic or supersonic. In passing we note that the pressure/displacement - relationship (4.4) is similar to that encountered in jets, Smith and Duck (1977).

While axisymmetric effects are confined to the external flow region if $a = O(1)$ they eventually penetrate into the boundary layer as $a \to 0$. As pointed out by Duck (1984) two cases, $a = O(Re^{-1/2})$ and $a = o(Re^{-1/2})$ leading to interaction lengthscales of $O(Re^{-3/7}(\ln Re))$ and $O(Re^{-3/6}(\ln Re))$, respectively, have to be treated separately. The main difference between these regimes is the double structure of the unperturbed boundary layer which develops if the radius of the cylinder is considerably smaller than the classical boundary layer thickness as shown by Glauert and Lighthill (1955), Stewartson (1955), Bush (1976). In both of these cases, however, the effects of curvature are found to modify the flow properties inside the main deck only. To leading order, no matter how small a is, the governing equations for the viscous wall layer reduce to the two-dimensional boundary layer equations (2.13) supplemented by the pressure-displacement relationships (4.2), (4.3). As a consequence, equations (2.13) with (4.2) or (4.3) depending on whether the external flow is subsonic or supersonic provide a consistent description of viscous-inviscid interactions on axisymmetric bodies for arbitrary values of a.

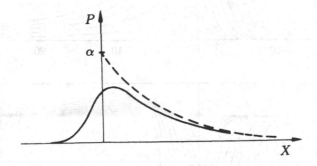

Figure 27: Pressure distribution an a flared cylinder. - - - inviscid flow, —— viscous flow.

Solutions to the interaction equations for supersonic external flow have been obtained by Kluwick et al (1984), (1985) and Gittler and Kluwick (1987a,b). In order to gain some insight into the effects caused by the additional integral term of equation (4.3) as compared to the case of planar flow it is useful to consider the pressure distribution on a flared cylinder according to inviscid theory, Fig. 27. In the vicinity of the corner the flow is approximately two-dimensional and the pressure jump at the corner can be calculated from the Ackeret relationship. Further downstream, however, the increase of the cross-section area of streamtubes due to the axisymmetry of the flow field becomes important and the pressure disturbances thus decrease (to leading order they even vanish in the limit $X = \infty$). This is the well known phenomenon of over-compression/overexpansion which occurs on axisymmetric bodies with concave/convex corners.

If viscous effects are taken into account, pressure disturbances make themselves felt well upstream of the corner and, consequently, the magnitude of the pressure peak decreases. One therefore expects that separation on axisymmetric compression ramps, $\alpha^* > 0$, will be delayed as compared to the case of two-dimensional flow and this is confirmed by the numerical results shown in Fig. 28.

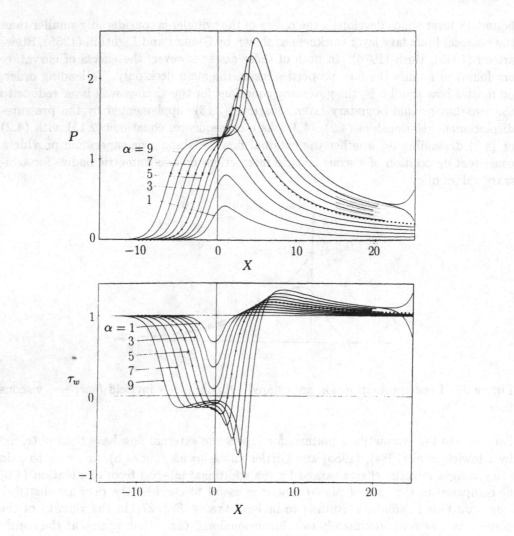

Figure 28: Pressure and wall-shear stress distributions for $a = 1$ and various positive cone angles α (Gittler and Kluwick (1987a)).

For sufficiently small values of the scaled ramp angle α

$$\alpha^* = \varepsilon^2 \lambda^{1/2} C^{1/4} (M_\infty^2 - 1)^{1/4} \alpha \qquad (4.5)$$

the boundary layer remains attached and due to the fact that the numerical computations were carried out for a slightly rounded rather than a perfectly sharp corner the wall shear stress assumes its minimum value upstream of $X = 0$. As α increases the minimum of the wall shear stress distribution decreases and incipient separation occurs at $\alpha_{is} \approx 3.39$ $(a = 1)$ which should be compared with the corresponding result holding in the case of planar flow $\alpha_{is} \approx 1.57$ $(a = \infty)$ obtained by Rizzetta, Burggraf and Jenson (1978). For lager values of α the boundary layer separates upstream of the corner and reattaches at the cone surface. As expected, the pressure increase near the reattachment point is smaller than in the case of two-dimensional flow. Moreover, the maximum of the pressure distribution differs only slightly from the value of the pressure disturbance at the reattachment point and the compression region upstream of this point is followed by a zone of rapid expansion caused by the increasing stream tube area. Finally, it should be noted that the formation of a plateau region of almost constant pressure for large values of α is clearly visible in Fig. 28.

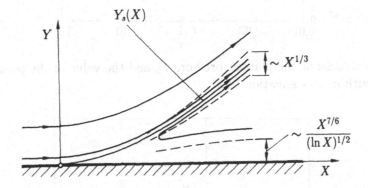

Figure 29: Asymptotic structure of the plateau region.

The structure of the flow inside the plateau region is sketched in Fig. 29. Similar to the case of planar flow the fluid is slowly moving upstream in the main part of this region. To leading order viscous effects are of importance only inside a thin boundary layer adjacent to the wall where the no slip condition has to be satisfied and inside a thin shear layer centred at the separation streamline. However, while in the case of two-dimensional flow the perturbation displacement thickness is a linear function of the distance X from the separation point, such a linear thickening of the displacement body is not sufficient to generate an axisymmetric separated flow region of constant pressure. Inversion of equation (4.3) with $P = P_0 = $ const yields, Kluwick et al (1985)

$$A(X) \sim \frac{P_0}{a}\left[\frac{X^2}{2\ln X} + \frac{X^2}{(\ln X)^2}\left(\frac{3}{4} + \frac{1}{2}\ln\frac{a}{2}\right)\right] \quad \text{as} \quad X \to \infty. \qquad (4.6)$$

It should be noted that the derivation of (4.6) requires that $X/a \gg 1$ and thus the above expression does not reduce to the two-dimensional result $A(X) \sim -P_0 X$ as $a \to \infty$. To recover the two-dimensional case it would be necessary to carry out the limit $a \to \infty$ with $X \gg 1$ fixed.

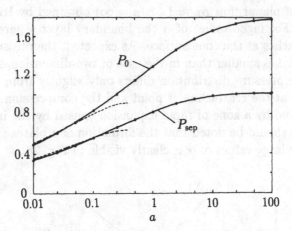

Figure 30: Variation of the plateau pressure P_0 and the value of the pressure at separation P_{sep} with a. - - - equation (4.9).

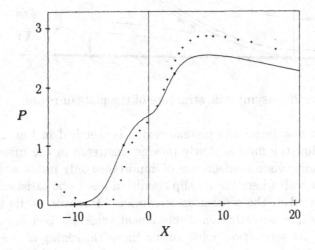

Figure 31: Comparison between experimental and theoretical results: +++ experimental data from Leblanc and Ginoux (1979, Fig. 4a); $M_\infty = 2.25$, $Re = 8.7 \times 10^4$, $\alpha^* = 7.5°$, $a = 34.6$; —— triple deck solution (Kluwick et al (1984)).

The variation of the plateau pressure, P_0, and the separation pressure, P_{sep}, with a is depicted in Fig. 30. It is seen that both P_0 and P_{sep} assume their largest values in the case of two-dimensional flow $a \to \infty$ while they tend to zero in the limit $a \to 0$. To study the properties of P_0 and P_{sep} for $a \to 0$ it is convenient to introduce the transformation

$$X = (-a \ln a)^{3/7} x^* , \quad Y = (-a \ln a)^{1/7} y^* ,$$

$$U = (-a \ln a)^{1/7} u^* , \quad V = (-a \ln a)^{-1/7} v^* , \tag{4.7}$$

$$A = (-a \ln a)^{1/7} A^* , \quad P = (-a \ln a)^{2/7} p^*$$

which leaves the boundary layer equations unchanged while the pressure-displacement relationship (4.4) reduces to

$$p^* = -d^2 A^* / dx^{*2} . \tag{4.8}$$

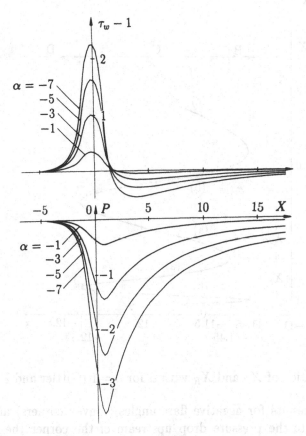

Figure 32: Pressure and wall shear stress distributions for $a = 1$ and various negative cone angles α (Gittler and Kluwick (1987)).

As mentioned earlier, interaction problems of exactly this form have been studied in a different context by Smith and Duck (1977), Smith and Merkin (1982). Combination of their numerical results $p_0^* = 1.222$, $p_{\text{sep}}^* = 0.855$ and (4.7) yields the dependence of the plateau pressure and the separation pressure on a in the limit $a \to 0$:

$$P_0 \sim 1.222(-a \ln a)^{2/7} , \quad P_{\text{sep}} \sim 0.855(-a \ln a)^{2/7} . \tag{4.9}$$

In Fig. 31 one set of experimental data obtained by Leblanc and Ginoux (1970) in the (16 in. × 16 in.) continuous supersonic wind tunnel of the von Kármán Institute is compared with a numerical solution of the triple-deck equations. The test conditions correspond to adiabatic flow over a flared cylinder with $\alpha^* = 7.5°$ at $M_\infty = 2.25$ with $Re = 8.7 \times 10^4$ and $a = 34.6$. As in the case of two-dimensional flow investigated by Rizzetta et al (1978) the initial pressure rise in the interaction region is overpredicted by the asymptotic theory. In addition, inspection of Fig. 31 shows that the pressure maximum is lower than the value determined experimentally by about 11%. This discrepancy seems to be caused mainly by the linearization of the governing equations in the upper deck.

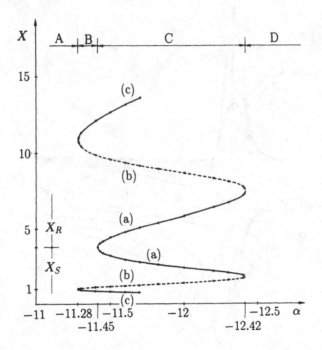

Figure 33: Variation of X_S and X_R with α for $a = 1$ (Gittler and Kluwick (1987a)).

Some numerical results for negative flare angles (convex corners) are summarized in Fig. 32. Owing to the pressure drop upstream of the corner the wall shear stress increases initially. The formation of a recompression zone downstream of the corner, however, causes the wall shear stress to decrease before it finally rises again to approach the unperturbed value $\tau_w = 1$ far downstream. Furthermore, the minimum of the wall shear stress distribution is seen to drop progressively as the flare angle $(-\alpha)$ increases indicating the possibility of boundary layer separation if $(-\alpha)$ is sufficiently large which was confirmed numerically. Since this effect is a direct consequence of the overexpansion and subsequent recompression of the fluid as it turns around the corner, it does not occur in the case of two-dimensional expansion ramps where the flow remains attached for all values of $-\alpha \geq 0$.

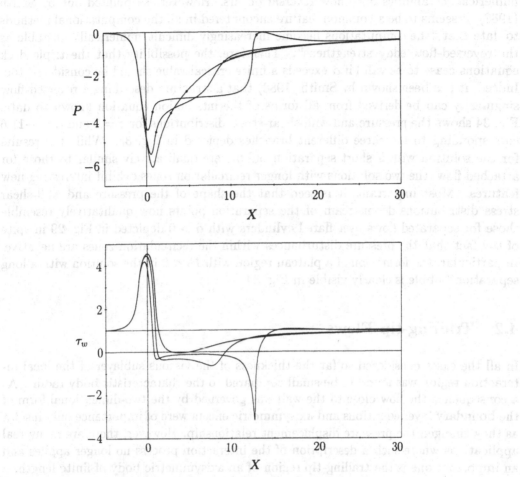

Figure 34: Pressure- and wall-shear stress distributions for $a = 1$ and $\alpha = -11.6$ (Gittler and Kluwick (1987a)).

In Fig. 33 the position of the separation and the reattachment points is plotted as function of the flare angle, α, for $a = 1$. If $\alpha \geq -11.28$ the boundary layer remains attached and the solutions of the interaction problem are unique. In contrast, two solutions exhibiting relatively long separated flow regions and a third one yielding attached flow are obtained if $-11.42 \leq \alpha \leq -11.28$. For values of the turning angle $\alpha < -11.28$ attached flow is no longer possible. The numerical results point to the existence of three different types of separated flow within the range $-11.42 < \alpha < -12.42$ and, furthermore, indicate that the triple-deck solutions are again unique if $\alpha < -12.42$. Unfortunately, the numerical scheme used by Gittler and Kluwick (1987a) did not yield converged solutions with long recirculation zones for $\alpha < -11.7$. This difficulty may be caused by the Flare approximation which was employed to prevent numerical instabilities once flow reversal occurs. However, as pointed out by Smith (1988), it seems to be a common feature encountered in all the computational methods to date that "the computations become increasingly difficult/numerically unstable as the reversed-flow eddy strengthens". Therefore, the possibility that the triple deck equations cease to be valid if α exceeds a finite critical value should be considered too. Indeed, it has been shown by Smith (1988) that a structure describing a reversed-flow singularity can be derived from all forms of the interaction equation known to date. Fig. 34 shows the pressure and wall-shear-stress distributions for $a = 1$ and $\alpha = -11.6$ corresponding to the three different branches depicted in Fig. 33. While the results for the solution with a short separation bubble are qualitatively similar to those for attached flow, the two solutions with longer recirculation zones exhibit interesting new features. Most important, it is seen that the shape of the pressure and wall-shear stress distributions downstream of the separation points now qualitatively resemble those for separated flows over flared cylinders with $\alpha > 0$ depicted in Fig. 29 in spite of the fact that the pressure disturbances within the recirculation zones are negative. In particular, the formation of a plateau region with $P_0 < 0$ in the solution with a long separation bubble is clearly visible in Fig. 34.

4.2 Trailing-tip Flows

In all the cases considered so far the thickness of the viscous sublayer of the local interaction region was found to be small compared to the characteristic body radius. As a consequence the flow close to the wall was governed by the two-dimensional form of the boundary layer equations and axisymmetric effects were of importance only insofar as they changed the pressure displacement relationship. However, there are many real applications where such a description of the interaction process no longer applies and an important one is the trailing-tip region of an axisymmetric body of finite length.

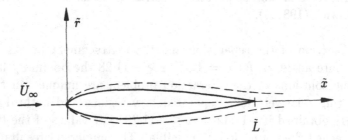

Figure 35: Boundary layer flow past a slender body of revolution.

In contrast to the case of two-dimensional flow near the trailing edge of a flat plate which has been studied in detail starting with the pioneering work of Stewartson (1969),

Messiter (1970), trailing-tip flows have received much less attention. A summary of studies dealing with such flows has been given by Bodonyi, Smith and Kluwick (1985) who investigated the local flow structure assuming that the body under consideration is so slender that the effect of the pressure gradient on the evolution of the boundary layer according to classical theory can be neglected. The boundary-layer equations can then be written in the form

$$\frac{\partial}{\partial x}(ru) + \frac{\partial}{\partial r}(rv) = 0, \qquad u\frac{\partial u}{\partial x} + v\frac{\partial u}{\partial r} = \frac{\partial^2 u}{\partial r^2} + \frac{1}{r}\frac{\partial u}{\partial r}. \tag{4.10}$$

Here (u, v) denote the non-dimensional velocity components of an incompressible viscous fluid in a cylindrical coordinate system (x, r), Fig. 35. Coordinates and velocity components are nondimensionalized with the length $\tilde{L}$ of the body and the freestream velocity $\tilde{U}_\infty$, respectively. In addition, r and v are scaled with $Re^{-1/2} = (\tilde{U}_\infty \tilde{L}/\tilde{\nu})^{-1/2}$ in the usual manner.

The boundary conditions of equations (4.10) include the no-slip condition at the body surface, the symmetry condition at the wake centerline and the requirement that the axial velocity component approaches its freestream value at large distances from the wall:

$$u = v = 0 \text{ on } r = r_b(x), \ 0 \le x \le 1; \quad v = \frac{\partial u}{\partial r} = 0 \text{ on } r = 0, \ x > 1; \quad u \to 1 \text{ as } r \to \infty.$$
$$\tag{4.11}$$

Solutions to equations (4.10), (4.11) have to be calculated numerically in general. To this end Bodonyi et al (1985) considered body shapes of the form

$$r_b(x) = a(1 - x)^n x^{1/2}, \quad 0 \le x \le 1 \tag{4.12}$$

where a is an arbitrary constant. It is perhaps tempting to expect the fluid motion in the trailing tip area of an axisymmetric body to be just a minor generalization of the two-dimensional case. In this latter case the boundary layer is essentially unperturbed upstream of the trailing edge interaction region where the velocity profile undergoes a rapid transition over a distance $O(Re^{-3/8})$ tending to zero in the limit $Re \to \infty$. In contrast, the boundary layer on an axisymmetric body is forced to adjust much more gradually, in fact well ahead of the trailing tip interaction region. In this pre-interaction region the evolution of the boundary layer is dominated by the large relative changes of the body radius which grow without bound as the trailing tip is approched. As a consequence one would expect viscous effects to be of importance only within small fraction of the oncoming classical boundary layer thus leading to the development of an interesting but rather complicated multilayered structure. As shown by Bodonyi et al (1985) three different ranges of the exponent n characterizing the trailing-tip shape have to be considered.

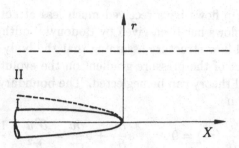

Figure 36: Multilayer structure of the trailing tip area for $1/4 < n < 1/2$ (Bodonyi et al (1985)).

A schematic of the flow stucture for $1/4 < n < 1/2$ is depicted in Fig. 36. In this case viscous effects are confined to a thin layer I comparable in thickness with the body radius $r_b(x)$. Here the stream function ψ can be expressed in terms of the similarity solution

$$\psi(x,r) = (1-x)f(\xi), \quad \xi = \frac{r}{a(1-x)^n} \tag{4.13}$$

while in region II outside the viscous sublayer the appropriate representation of ψ is given by

$$\psi = \psi_0(r) + O(1) , \quad r = O(1) \tag{4.14}$$

which describes a slightly perturbed uniform shear flow. The function $\psi_0(r)$ depends on the entire history of the boundary layer upstream of the trailing tip area and cannot be determined by local considerations. Substitution of the relationship (4.13) into equations (4.10), (4.11) yields

$$(2n - 1)\xi(f'/\xi)^2 + (f(f'/\xi))' = (\xi(f'/\xi)')' ,$$
$$f(1) = f'(1) = 0 \tag{4.15}$$

where $()'$ denotes differentiation with respect to ξ. A third boundary condition follows from the investigation of the asymptotic behaviour of $f(\xi)$ as $\xi \to \infty$. Matching with the solution (4.14) holding outside the viscous sublayer requires that the viscous terms of equation (4.15) vanish at large distances from the wall which leads to

$$f(\xi) \sim B\xi^{1/n} , \quad B > 0 \quad \text{as} \quad \xi \to \infty . \tag{4.16}$$

Equations (4.15), (4.16) have been solved numerically using both shooting and finite difference techniques. The variation of the effective wall shear stress $\gamma = f''(1)$ with n, depicted in Fig. 37, indicates that $\gamma \to 0$ as $n \to 1/2$ while $\gamma \to \infty$ as $n \to 1/4$

which can be confirmed analytically by means of asymptotic expansions. In addition, these analytical results provide useful information how the double structure holding for $1/4 < n < 1/2$ has to be modified if $n < 1/4$ or $n > 1/2$. In both cases the oncoming boundary layer develops a three tiered rather than a two tiered structure, viscous effects being confined to the innermost layer. However, while the thickness of the viscous sublayer is small compared to the local body radius if $n < 1/2$ this layer is much thicker than X^n if $n > 1/2$. Again it is found that the description of the flow in the outermost region of the boundary layer contains a certain amount of arbitrariness which reflects the influence of the entire flow between the leading and trailing tips. In contrast, the solutions in the thinner layers are determined uniquely to leading order.

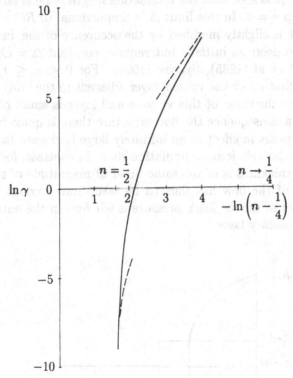

Figure 37: Variation of the effective wall shear stress $\gamma = f''(1)$ with n for $1/4 < n < 1/2$ (Bodonyi et al (1985)).

The results discussed so far are based on the assumption that the pressure gradient term of the boundary-layer equations is negligible small compared to the inertia and viscous terms. Owing to the boundary-layer displacement exerted on the external inviscid flow and the resulting pressure response, however, this assuption is violated at very small distances from the trailing tip $x = O(\Delta)$ where viscous inviscid interaction finally comes into play. Application of slender body theory (e.g. Cole (1968), pp 182) yields the order of magnitude estimate

$$p = O(Re^{-1} \ln Re) \frac{d^2\delta}{dx^2} \qquad (4.17)$$

for the feedback pressure where $\delta(x)$ is a representative displacement function such that $\psi \sim r^2/2 - \delta(x)$ in equations (4.1), (4.2) as $r \to \infty$ for all x. To determine the interaction length scale Δ the induced pressure gradient has to be compared with the smallest inertia term inside the boundary layer. The final results, plotted in Fig. 38, show that Δ depends on the parameter n which characterizes the body shape in the pre-interaction region. It is seen that the interaction length scale is largest in the case of a blunted trailing tip $n = 0$. In this limit Δ is proportional to $Re^{-3/8}$ but the result holding for planar flow is slightly modified by the occurence of the $\ln Re$ term. As n increases Δ is found to decrease initially but remains constant $\Delta = O(Re^{-1} \ln Re)^{1/2}$ for $n > 1/3$, Bodonyi et al (1985), Sychev (1990). For $0 < n \le 1/3$ the induced pressure disturbances first affect the viscous layer adjacent to the body surface. In the range $0 < n \le 1/4$ the thickness of this viscous wall layer is small compared to the local body radius. As a consequence the flow structure there is quasi-two-dimensional and the trailing tip appears in effect as an infinitely large backward facing step which is expected to force a relatively long recirculation zone. In contrast, for $1/4 < n \le 1/3$ the viscous wall layer thickness is of the same order of magnitude of the body radius and the axisymmetry of the flow has thus to be taken into account there. Finally, when $n > 1/3$ the effect of the feedback pressure is felt first in the outmost primarily inviscid part of the boundary layer.

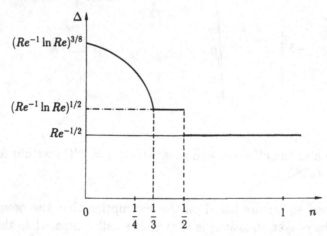

Figure 38: Variation of the interacting length scale Δ with n (Bodonyi et al (1985), Sychev (1990)).

The results summarized so far provide a consistent description of the classical boundary layer development and breakdown in the trailing tip area of a slender body of revolution.

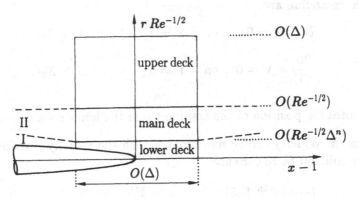

Figure 39: Asymptotic structure of the trailing tip local interaction region for $1/4 < n < 1/3$.

To complete the flow description the vicinity of the tip where the induced pressure disturbances represent a leading order rather than a higher order effect has to be considered next. As shown recently by Kluwick, Gittler and Bodonyi (1994), Gittler and Kluwick (1997) an interaction strategy can be applied successfully in the parameter range $1/4 < n \leq 1/3$. The interaction zone sketched in Fig. 39 then comprises a region of external inviscid flow and two wall layers which are the continuation of regions I and II of the pre-interaction zone. In contrast to the problems summarized in section 2 axisymmetric effects are of importance in all three layers of the interaction zone but, again, it is possible to express the solutions holding in the upper and main decks in closed form. The flow inside the lower deck is governed by the axisymmetric form of the boundary layer equations for an incompressible fluid:

$$\frac{\partial}{\partial X}(YU) + \frac{\partial}{\partial Y}(YV) = 0, \quad U\frac{\partial U}{\partial X} + V\frac{\partial U}{\partial Y} = -\frac{dP}{dX} + \frac{\partial^2 U}{\partial Y^2} \qquad (4.18)$$

where
$$(x - 1) = \Delta X, \quad r = \Delta^n Y,$$
$$u = \Delta^{1-2n}U, \quad v = \Delta^{-2n}V, \quad p = \Delta^{2-4n}P. \qquad (4.19)$$

The body surface is denoted by $Y = Y_b(X)$ and consistency with the results holding in the pre-interaction zone requires the power law behaviour of the body shape:

$$Y_b(X) \sim a(-X)^n \quad \text{as} \quad X \to -\infty. \qquad (4.20)$$

It should be emphasized, however, that the shape of the body inside the interaction zone is completely arbitrary otherwise. The boundary conditions at the body surface and the wake centreline are

$$U = V = 0 \quad \text{on} \quad Y = Y_b(X), \quad X \le X_T,$$

$$\frac{\partial U}{\partial Y} = V = 0 \quad \text{on} \quad Y = 0, \quad X > X_T. \tag{4.21}$$

Here X_T denotes the position of the trailing tip in the lower deck region.

Far upstream the velocity profile must match with the velocity distribution given by the similarity solution (4.13). Hence

$$U \sim \frac{(-X)^{1-2n}}{a^2} \frac{f'(\xi)}{\xi}, \quad \xi = \frac{Y}{a - (-X)^n} \quad \text{as} \quad X \to -\infty. \tag{4.22}$$

Finally, matching with the results holding in the main and upper deck yields

$$U \sim KY^{(1-2n)/n} \left[1 + \frac{1-2n}{n} \frac{A(X)}{Y^2} \right] \quad \text{as} \quad Y \to \infty,$$

$$K = \frac{B(n)}{na^{1/n}}, \quad P = -\frac{1}{2} \frac{d^2 A}{dX^2}. \tag{4.23}$$

The interaction problem defined by equations (4.18)-(4.23) differs from other triple deck problems in a number of points. Firstly, the fluid motion inside the lower deck is governed by the axisymmetric rather than the planar form of the boundary layer equations. Secondly, the velocity profile entering the upstream condition at the base of the main deck is not a linear Blasius type profile. In fact the velocity distribution has to be calculated numerically by solving the boundary value problem (4.15), (4.16) separately for each value of n and, as evidenced by the asymptotic relationship (4.23), the velocities are found to grow much more rapidly with increasing distance from the body surface. Thirdly, owing to this novel feature of the incoming classical boundary layer it is not possible to eliminate the parameter a which characterizes the "thickness" of the body in the trailing tip region from the interaction equations. In this connection it should be noted that a related interaction problem has been formulated recently by Sychev (1991), Korolev and Sychev (1993) who investigated the flow past bodies which are slightly thicker than the boundary layer scale $O(Re^{-1/2})$. As a consequence, the Mangler transformation can be applied upstream of the trailing tip local interaction zone and the velocity profile there is again of the linear Blasius type form.

As a typical example of a numerical solution for pure power law bodies the distributions of the pressure, the wall shear stress and the wake centreline velocity are displayed in Fig. 40. Also included in the figure are the distributions of these quantities following from the upstream and downstream similarity solutions determined by Bodonyi et al

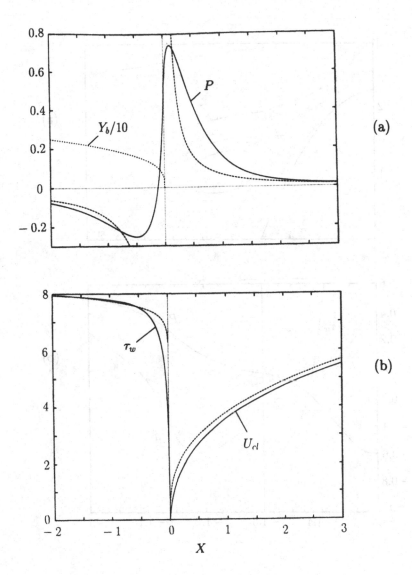

Figure 40: (a) Pressure P, (b) Wall shear τ_w and wake centreline velocity U_{cl} for a pure power-law body Y_b with $n = 0.32$ and $a = 1.5$, - - - asymptotes, $\cdots$ $Y_b/10$ (Gittler and Kluwick (1997)).

(1985) which represent the asymptotes of the numerical results for $|X| \to \infty$. The pressure distribution (4.17) predicted by second order boundary layer theory is characterized by the presence of a negative singularity at the tip $X = 0^-$. This singularity

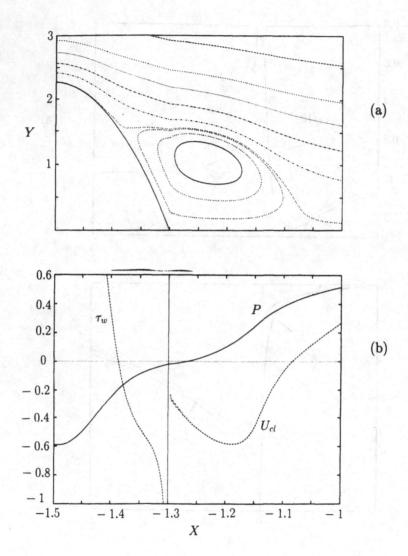

Figure 41: Power law body with blunted rear end with $n = 0.32$, $a = 2.0$, $X_A = -1.5$, $X_T = -1.3$: (a) streamline plot, (b) results for P, τ_w and U_{cl} (Gittler and Kluwick (1997)).

which results from a singularity of the displacement curvature is eliminated by taking into account the interaction between the boundary layer and the external inviscid flow.

In agreement with the results of classical theory the wall shear stress is found to decrease monotonically up to the tip. Although the inclusion of the induced pressure disturbances enhances the drop of the wall shear initially the boundary layer remains attached. In fact the numerical investigations carried out by Gittler and Kluwick (1997) indicate that separation occurs only if the rear end of a power law body is made extremely blunt. Specifically, they considered body contours which follow the power law (4.20) with a, n fixed up to $X = X_A$ while they are given by parabolic arcs further downstream.

Results for a power law body which terminates almost abruptly are depicted in Fig. 41a. Even so the boundary layer remains attached over more than one half of the deformed body contour before separation finally occurs. As can be seen from Fig. 41b this leads to the formation of a toroidal vortex extending beyond the tip.

4.3 Incompressible flow past cones at incidence

Experimental as well as numerical studies dealing with high Reynolds number flows past axisymmetric bodies at incidence have revealed the formation of strikingly complex streamline patterns if boundary layer separation occurs, Bippes and Turek (1984). It is not surprising, therefore, that such flows have received scant attention, theoretically, so far. Starting from the pioneering work of Legendre (1956), Oswatitsch (1958), Lighthill (1963) a number of interesting new results concerning the local behaviour and global topological structure of general three-dimensional separating flows have been obtained by Hornung and Perry (1984), Perry and Chong (1986), Dallmann (1982). However, attempts to calculate the complete flow field generated by an axisymmetric body at incidence are rare and appear to concentrate mainly on the case of conical tips.

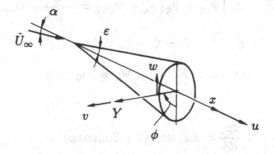

Figure 42: Flow past a slender cone at a small angle of attack.

Let (x, r, ϕ) denote cylindrical coordinates with corresponding velocity components (u, v, w) where x, r are nondimensional with the length $\tilde{L}$ of the cone and u, v, w are nondimensional with the free stream velocity $\tilde{U}_\infty$. Furthermore, let the cone axis coincide with the x-axis. Then the body surface is given by

$$r = \varepsilon x, \quad 0 \ll \varepsilon \ll 1 \tag{4.24}$$

where ε denotes the semi vertex angle of the cone which is assumed to be small and of the same order of magnitude as the angle of incidence $\alpha = \varepsilon \alpha_0$, $\alpha_0 = O(1)$, Fig. 42. Introducing the stretched distance from the cone surface $Y = Re^{1/2}(r - \varepsilon x)$, $Re = \tilde{U}_\infty \tilde{L}/\tilde{\nu}$, and asymptotic expansions of the form

$$u = u_1(x, Y, \phi) + \dots ,$$

$$v = \varepsilon u_1(x, Y, \phi) + Re^{-1/2} v_1(x, Y, \phi) + \dots ,$$

$$w = \varepsilon w_1(x, Y, \phi) + \dots , \tag{4.25}$$

$$p = -\varepsilon^2 \ln \varepsilon + \varepsilon^2 p_1(\phi) + \dots$$

for the velocity components and the pressure disturbances, solutions to the resulting boundary layer equations are sought in self-similar form

$$u_1 = \frac{\partial F}{\partial \eta}(\eta, \phi) , \quad w_1 = \frac{\partial \Psi}{\partial \eta}(\eta, \phi) , \quad \eta = \frac{Y}{\sqrt{x}} . \tag{4.26}$$

$F(\eta, \phi)$, $\Psi(\eta, \phi)$ satisfy the set of equations

$$-\left(\frac{3}{2}F + \Psi_\phi\right)F_{\eta\eta} + \Psi_\eta F_{\phi\eta} = F_{\eta\eta\eta} ,$$

$$-\left(\frac{3}{2}F + \Psi_\phi\right)\Psi_{\eta\eta} + \Psi_\eta\left(F_\eta + \Psi_{\phi\eta}\right) = -\frac{dp_1}{d\phi} + \Psi_{\eta\eta\eta} , \tag{4.27}$$

$$\eta = 0 \ : \ F = \Psi = F_\eta = \Psi_\eta = 0 ,$$

$$\eta = \infty \ : \ F_\eta = 1 , \quad \Psi_\eta = 2\alpha_0 \sin \phi$$

where

$$\frac{dp_1}{d\phi} = 2\alpha_0 \sin \phi(1 + 2\alpha_0 \cos \phi) . \tag{4.28}$$

Equations (4.27) simplify considerably in the windward plane of symmetry $\phi = 0$ and the leeward plane of symmetry $\phi = \pi$. For example, by adopting the assumptions

$w = 0$, $\partial w / \partial \phi \neq 0$ for $\phi = 0$ one obtains

$$F_{\eta\eta\eta} + \left(\frac{3}{2}F - G\right)F_{\eta\eta} = 0 \,,$$

$$G_{\eta\eta\eta} + \left(\frac{3}{2}F - G\right)G_{\eta\eta} + G_\eta{}^2 - F_\eta G_\eta - \frac{3}{2}k\left(\frac{3}{2}k + 1\right) = 0 \,,$$

$$\eta = 0 \ : \ F = G = F_\eta = G_\eta = 0 \,,$$

$$\eta = \infty \ : \ F_\eta = 1 \,, \ G_\eta = -\frac{3}{2}k$$

(4.29)

where $G = \Psi/\phi$ and $k = 4\alpha_0/3$ denotes the parameter introduced by Moore (1953). Exactly the same equation holds for $\phi = \pi$ if the definitions of G, k are modified according to $G = \Psi/(\pi - \phi)$, $k = -4\alpha_0/3$. Numerical investigations of equations (4.29) have been performed by Cheng (1961), Roux (1972), Murdock (1972), Cebeci, Stewartson and Brown (1983) and various other authors (the reader interested in this topic is referred to Rubin, Lin and Tarulli (1977) for a useful summary). It is found that (dual) solutions exist in the windward plane $\phi = 0$ for all values of k. In contrast, similarity solutions for the leeward plane $\phi = \pi$ could be obtained for $-0.292 < k < 0$ and $-1 < k < -0.666$ only. Furthermore, integration of the full set of equations (4.27), (4.28) by marching in the ϕ-direction starting with the (appropriate)windward similarity solution shows that the leeside similarity solution is recovered in the limit $\phi \to \pi$ if $0 > k > -0.292$. Moreover, these calculations indicate that the azimuthal velocity component w is no longer proportional to $(\pi - \phi)$ as $\phi \to \pi$ but rather remains finite if $-0.292 > k > -0.80$. Therefore, one of the assumptions leading to equations (4.29) is violated in this range of the parameter k measuring the incidence of the cone. As a consequence the boundary layers growing on either side of the conical tip do not blend smoothly at $\phi = \pi$ but give rise to a collision phenomenon similar to that encountered in studies dealing with high Dean number entry flows in curved ducts, Stewartson, Cebeci and Chang (1980), Stewartson and Simpson (1982) Kluwick and Wohlfart (1984, 1986). Finally, due to the formation of a region of adverse pressure gradient $dp/d\phi > 0$ near $\phi = \pi$ for $k < -2/3$, the boundary layer separates and the numerical integration terminates before the leeside symmetry plane is reached if $k < -0.800$. The leeside similarity solutions for $k < -0.666$ thus do not appear to be of relevance, e.g. do not seem to be embedded in a global boundary-layer solution.

A detailed investigation, both numerical and analytical, of the properties of the solutions to the boundary layer equations (4.27), (4.28) has been performed by Zametaev (1986). The distributions of the axial and azimuthal components of the wall shear stress $\tau_1 = F_{\eta\eta}(0, \phi)$, $\tau_2 = G_{\eta\eta}(0, \phi)$ at incipient separation $\alpha_0 = \alpha_{0c} = 0.6$ are depicted in Fig. 43. It is seen that τ_2 vanishes at $\phi = \phi_c \approx 3.01$ but the numerical integration of the boundary layer equations can be continued through this point. Furthermore the

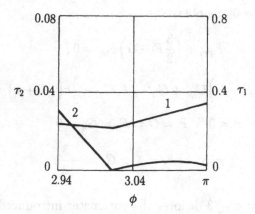

Figure 43: Graphs τ_1, τ_2 versus ϕ (Zametaev (1986)).

numerical results indicate that τ_2 varies linearly with ϕ as $\phi \to \phi_c$ and the rates of change being different for $\phi - \phi_c \to 0^-$ and $\phi - \phi_c \to 0^+$, however.

$$\tau_2(0, \phi) = \begin{cases} a_0(\phi_c - \phi), & \phi - \phi_c \to 0^-, \\ \lambda a_0(\phi - \phi_c), & \phi - \phi_c \to 0^+. \end{cases} \qquad (4.30)$$

Investigation of the local properties of the boundary layer equations for $|\phi - \phi_c| = O(\sigma)$, $|\alpha_0 - \alpha_{0c}| = O(\sigma^{1+\lambda})$, $\sigma \to 0$ yields

$$\tau_2(0, \phi_1) = \sigma a_0 A(\phi_1), \quad \phi_1 = \frac{\phi - \phi_c}{\sigma} \qquad (4.31)$$

where

$$(A + \phi_1)(A - \lambda\phi_1)^\lambda = G_1, \quad (A + \phi_1)(\lambda\phi_1 - A)^\lambda = G_2. \qquad (4.32)$$

It is interesting to note that exactly the same result has been derived by Brown (1985) who studied the marginal separation of a three-dimensional boundary layer on a line of symmetry, as discussed in section 3.4.

Using these relationships, which contain equation (4.30) as a special case $G_1 = 0$, it can be shown that $\tau_2(0, \phi)$ is a smooth function in the whole domain $-\infty < \phi_1 < \infty$ if $\alpha_0 < \alpha_{0c}$. In contrast, the wall shear stress distribution exhibits a singularity at a finite value $\phi_1 < 0$ and cannot be extended up to $\phi_1 = \infty$ if $\alpha_0 > \alpha_{0c}$. Since the strength of this singularity vanishes in the limit $\alpha_{0c} - \alpha_0 \to 0$ it can, however, be eliminated in a manner similar to the two-dimensional marginal separation case by taking into account the interaction between the boundary layer and the external inviscid flow. Comparison of the induced pressure gradient $\partial p / \partial(r\phi) = O(\sigma Re^{-1/2})$ and the viscous term $\partial^2 w / \partial r^2 = O(\varepsilon \sigma^{7/4} Re^{-1})$ entering the second order momentum equation following from the local analysis of the classical boundary-layer equations, indicates that the parameter range covered by the interaction concept is $\sigma = O(\varepsilon^{-2/5} Re^{-1/5})$. The appropriate asymptotic interaction theory has been formulated by Zametaev (1987). According to the similarity solution (4.26) the radial velocity component at the outer edge of the boundary layer is proportional to $x^{-1/2}$. As a consequence, the induced pressure disturbances inside the interaction region depend on both x and ϕ_1 which in turn forces x to occur explicitly in the expansions of the various field quantities. Viscous effects are found to be negligibly small in the main part of the boundary layer but they have to be taken into account in a thin sublayer adjacent to the wall. There the flow is governed by the three-dimensional form of the boundary-layer equations linearized with respect to the separation profile to leading order. Its solution involves an arbitrary function $B(x, \phi_1)$ characterizing the azimuthal component of the wall shear stress. In order to determine $B(x, \phi_1)$ it is necessary to derive the solvability condition for the second order equations holding in the viscous sublayer. Using suitably scaled variables $A, \bar{\phi}$ in place of B, ϕ this condition can be expressed in the form, Zametaev (1987)

$$\int_{-\infty}^{\bar{\phi}} \left[A \frac{\partial A}{\partial t} - \lambda t + (1-\lambda)A + \frac{3}{4}(1-\lambda)x \frac{\partial A}{\partial x} \right] dt = \frac{1}{2\sqrt{x}} \int_{\bar{\phi}}^{\infty} \frac{\partial^2 A}{\partial t^2} \frac{dt}{\sqrt{t - \bar{\phi}}} \qquad (4.33)$$

$$A \sim -\bar{\phi} - \Gamma(-\bar{\phi})^{-\lambda}, \quad \bar{\phi} \to -\infty \,; \quad A \sim \lambda\bar{\phi}, \quad \bar{\phi} \to \infty$$

where the parameter $\Gamma \propto (\alpha_0 - \alpha_{0c})/\sigma^{1+\lambda}$ measures the incidence of the cone. As shown by Zametaev (1987), solutions to equations (4.33) for $x \to 0$ can be expressed in terms of the similarity variable $y = x^{1/5}\bar{\phi}$: $T = x^{1/5}A$. Starting with the similarity solution equation (4.33) was then integrated numerically by marching in the x-direction for the two cases $\lambda = 0.5$, $\Gamma = -2$ and $\lambda = 0.5$, $\Gamma = 2$. Results for the latter case, corresponding to a value of α_0 which is larger than α_{0c}, are depicted in Fig. 44. It is seen that the minimum of the azimuthal component of the wall shear stress decreases and that the slope of the wall shear stress distribution at reattachment steepens as the distance x from the tip of the cone increases. Theoretical considerations indicate that a singularity is formed at finite values of $x = x_*$, $y = y_*$ and evaluation of the numerical data yields the estimate $x_* \approx 2.3$, $y_* \approx 0.7$ for $\Gamma = 2$, $\lambda = 0.5$. It does not seem unlikely that the formation of this singularity heralds the onset of global separation characterized by the appearance of a vortex sheet shed from the point $x = x_*$, $y = y_*$. Furthermore, it may be expected that the location of the singularity moves upstream as Γ increases thus eventually leading to the case of conical separation.

The interaction process associated with the separation of a vortex sheet from a smooth conical surface has been investigated by Riley (1979), Smith (1978), generalizing earlier work by J.H.B. Smith (1977) dealing with the properties of inviscid flow. Let x, y and z denote the distance along the separation line $\phi = \phi_s$, the distance in the body surface

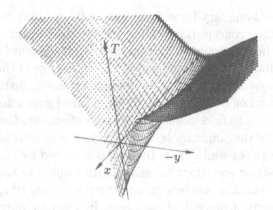

Figure 44: Azimuthal shear stress distribution for $\Gamma = 2$, $\lambda = 0.5$ (Zametaev (1987))

perpendicular to this line and the distance normal to the body surface, respectively. It then follows from Smith's (1979) work that, when y is sufficiently small, the shape of the vortex sheet leaving the surface at $y = z = 0$ can be expressed in the form $z = \mu(x)y^n$ with $n = (2M+1)/2$. Here $M = 0, 1, \ldots$ to ensure (i) that the departure of the sheet from the body surface is tangential and (ii) that the pressure p is continuous at the sheet.

If $M = 0$ the azimuthal component of the pressure gradient exhibits a singularity as the separation line is approached from upstream $\partial p/\partial y \sim c_1(-y)^{-1/2}$, $y \to 0^-$ while $\partial p/\partial y$ vanishes in the limit $y \to 0^+$. In contrast $\partial p/\partial y$ is a smooth function of y if $M > 0$ and $\partial p/\partial y \to 0$ for $y \to 0^\pm$. As a consequence, the attempt to incorporate viscous effects into the inviscid flow model faces a difficulty. Owing to the strong adverse pressure gradient present for $M = 0$ the boundary layer is expected to separate upstream of the inviscid separation line $y = 0$. On the other hand it is not clear why the boundary layer separates at all if the location of the separation line is chosen such that inviscid theory predicts smooth separation, e.g. if $M > 0$. A similar difficulty has been encountered in studies dealing with high Reynolds number flow past circular cylinders (Sychev (1972), Smith (1977, 1979)) and as in this problem the difficulty is resolved by taking μ and c_1, to be small rather than $O(1)$ quantities: $\mu = Re^{-1/16}\overline{\mu}(x)$, $c_1 = Re^{-1/16}\overline{c}_1$, $\overline{\mu} = O(1)$, $\overline{c}_1 = O(1)$. Physically, this means that the separation line is shifted a small distance $(\phi - \phi_s) = O(Re^{-1/16})$ from its inviscid position following from the requirement of smooth separation thus introducing a weak singularity in the pressure gradient which can be smoothed out by taking into account viscous-inviscid interaction. Since $y = O(Re^{-3/8})$, $z = O(Re^{-3/5})$ in the local interaction region, the variation of the field quantities in the x-direction is much smaller than in the circumferential direction as well as the direction normal to the surface, and derivations with respect to x thus drop out of the interaction equations to leading order. Similar to the case of viscous inviscid interactions on swept wing configurations (Gittler and

Kluwick (1989)), therefore, the continuity and y-momentum equations can be solved independently of the x-momentum equation which is linear in u. In terms of suitably scaled variables one then essentially recovers the two-dimensional interaction problem discussed in section 2.2.

$$W\frac{\partial W}{\partial Y} + V\frac{\partial W}{\partial Z} = -\frac{dP}{dY} + \frac{\partial^2 W}{\partial Z^2}, \qquad \frac{\partial W}{\partial Y} + \frac{\partial V}{\partial Z} = 0 \qquad (4.34)$$

$$V = W = 0 \quad \text{on} \quad Z = 0$$

$$W \sim Z + A(Y), \qquad Z \to \infty$$

$$W \sim Z, \qquad Y \to -\infty \qquad (4.35)$$

$$P = \frac{1}{\pi} \oint_{-\infty}^{\infty} \frac{A'(\xi)}{Y - \xi}\, d\xi$$

In addition to the boundary and matching conditions (4.35), the triple-deck solution must satisfy the relationship

$$P(Y) \sim -\alpha |Y|^{1/2} \quad \text{as} \quad Y \to -\infty \qquad (4.36)$$

following from the match with the pressure distribution upstream of the local interaction region. Here α denotes the scaled parameter $\bar{c}_1$ characterizing the strength of the pressure gradient singularity present there.

The success of the theoretical model outlined so far hinges on the existence of a solution to equations (4.34), (4.35) and (4.36), which arises also in studies dealing with the separation of a two-dimensional flow from a curved wall. As pointed out in section 2.2 a detailed numerical study of this nonlinear eigenvalue problem has been performed by Smith (1977). His results strongly support the conclusion that such a solution exists and is unique, the corresponding value of α being $\alpha \doteq 0.44$.

As pointed out earlier, the value of α determines the strength of the pressure gradient singularity occuring in the solution upstream of the triple deck region and which is smoothed out by the local interaction process. However, it also determines the local shape of the vortex sheet leaving the body surface as well as the position of the separation line, Fig. 45.

Equations (4.34), (4.35) and (4.36) provide a self consistent description of the local flow properties near the separation line of a conical vortex sheet. Unfortunately, however, attempts to embed this local structure into the global flow field are faced with severe difficulties. These difficulties are associated with the behaviour of the boundary

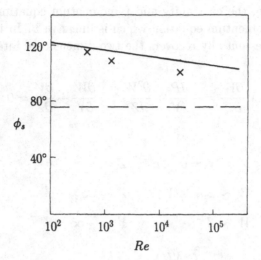

Figure 45: Variation of the separation position ϕ_s on a circular cone with Re (Fiddes (1980)), —— triple deck results, - - - smooth separation result, ××× experiments.

layer downstream of the triple-deck region which, due to the curvature of the external streamlines, develops a jet-like velocity profile. Smith (1978) argues that the boundary layer, therefore, must separate before it reaches the triple-deck region to shelter the decelerating fluid in the lower deck from the faster upstream moving fluid. This in turn leads to the formation of another separated sheet and it is not yet clear how the merging between this sheet and that emanating from the triple deck can be achieved.

5   Interacting turbulent boundary layers

5.1   Unperturbed turbulent boundary layer

The asymptotic structure of turbulent boundary layers in the limit of large Reynolds numbers formulated first by Yajnik (1970), Bush and Fendell (1970), Mellor (1972) and summarized in the sections: Turbulent Boundary Layers I, II of this volume are based on the assumption that the field quantities vary slowly in the streamwise direction. It thus breaks down if the flow is subjected to rapid changes of the boundary conditions at the wall or in the external inviscid flow region. Such disturbances may be caused, for example, by a sharp trailing edge, a short surface mounted obstacle or an impinging shock wave.

As a starting point we consider two problems associated with the boundary layer forming on a smooth flat plate immersed in an incompressible fluid, Fig. 46. In the case

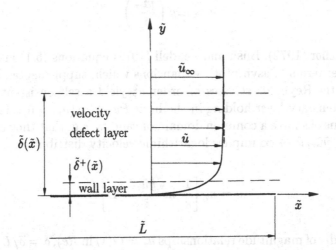

Figure 46: Asymptotic structure of unperturbed turbulent boundary layer on flat plate.

of laminar flow the velocity distribution inside the boundary layer is given by the Blasius similarity solution. If the Reynolds number $Re = \tilde{u}_\infty \tilde{L}/\tilde{\nu}$ is sufficiently large, however, transition occurs at some distance from the tip of the plate leading eventually to a fully turbulent boundary-layer flow whose effective starting location will be denoted by $\tilde{x}_1$. For $\tilde{x} > \tilde{x}_1$ no similarity solution can be found which describes the boundary layer as a whole. As shown already by von Kármán (1930), Prandtl (1933) it is necessary to distinguish between an outer (velocity defect) region which comprises most of the boundary layer and a thin wall layer which have to be treated separately. Inertia and Reynolds stress terms are the dominating terms of the momentum balance in the outer region while the properties of the wall region are characterized approximately by the equilibrium between Reynolds stresses and viscous stresses.

Theoretical considerations as well as experimental results indicate that in the outer part of the boundary layer the (time averaged) streamwise velocity component $\tilde{u}$ differs from the external flow value $\tilde{u}_\infty$ by a term proportional to the friction velocity $\tilde{u}_\tau = \sqrt{\tilde{\tau}_w/\tilde{\rho}}$. The similarity form of the velocity distribution is then given by the velocity defect law

$$u = 1 - u_\tau f\left(\frac{\tilde{y}}{\tilde{\delta}}\right), \qquad u = \frac{\tilde{u}}{\tilde{u}_\infty}, \qquad u_\tau = \frac{\tilde{u}_\tau}{\tilde{u}_\infty} \qquad (5.1)$$

where $\tilde{y}$ and $\tilde{\delta}$ denote the distance from the wall and the boundary layer thickness.

In the wall layer of thickness $\tilde{\delta}^+ \ll \tilde{\delta}$ the mean velocity is small compared to $\tilde{u}_\infty$ and the velocity distribution assumes the form of the law of the wall

$$u = u_\tau g \left(\frac{\tilde{y} \tilde{u}_\tau}{\tilde{\nu}} \right) \tag{5.2}$$

As shown by Mellor (1972), Bush and Fendell (1972) equations (5.1) and (5.2) provide the leading order terms of asymptotic expansions which, supplemented by appropriate expansions for the Reynolds stresses of order u_τ^2 yield a selfconsistent description of the turbulent boundary layer holding in the limit $Re \to \infty$ $(u_\tau \to 0)$. The requirement that these expansions have a common domain of validity, e.g. that they can be matched as $\tilde{y}/\tilde{\delta} \to 0$ and $\tilde{y} \tilde{u}_\tau/\tilde{\nu} \to \infty$ implies logarithmic velocity distribution

$$u = u_\tau \left[\frac{1}{\kappa} \ln \frac{\tilde{y} \tilde{u}_\tau}{\tilde{\nu}} + B \right] \tag{5.3}$$

as well as the order of magnitude relationships $u_\tau = O(1/\ln Re)$, $\delta = \tilde{\delta}/\tilde{L} = O(u_\tau)$, $\delta^+ = \tilde{\delta}^+/\tilde{L} = O(u_\tau \exp(-1/u_\tau))$. Here κ is the von Kármán constant. According to experimental observations $\kappa \approx 0.41$ and $B \approx 5.0$.

5.2   Flow near the trailing edge of an aligned flat plate

As in the outline of laminar viscous-inviscid-interactions it is instructive to consider first the problem of a boundary layer that evolves into a wake flow at the sharp trailing edge of a thin flat plate. The resulting flow structure obtained by Bogucz and Walker (1987) extending partial solutions by Robinson (1969), Alber (1980), Messiter (1980a) is shown in Fig. 47. Significant changes of the velocity distribution caused by the loss of the no slip condition occur downstream of the trailing edge only and are limited to a thin layer centred at the wake symmetry line. In this layer viscous effects are of importance over a short distance $O(\delta^+)$ leading to an $O(\delta^+)$ square region which describes the response of the wall region of the surface boundary layer to the sudden change of the boundary condition at $y = 0$. Further downstream the thickness Δ_n of this layer is much larger than δ^+ and, as a result, the acceleration of the fluid in the inner near wake is accomplished primarily by Reynolds stresses rather than viscous stresses. In the limit of large Reynolds number $Re \to \infty$ considered here the lateral and streamwise extent of the wall layer adjustment region is exponentially small and is, therefore, expected to be passive in the sense that it does not influence the leading order description of the trailing edge flow in the inner near wake and the outer portions of the surface boundary layer and the wake. As the outer part of the surface boundary layer passes into the wake region the streamwise velocity distribution remains almost unchanged. Owing to the (negative) displacement of the inner near wake the transverse velocity component in the outer near wake, however, differs in a significant way from that in the defect part of the surface boundary layer thus necessitating the introduction of an additional trailing-edge adjustment region having an $O(\delta)$ extent in the streamwise and the lateral direction to achieve a smooth variation of the transverse velocity component.

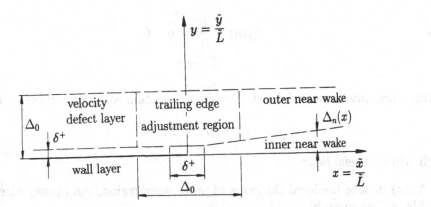

Figure 47: Near wake structure

We start a more detailed description of the trailing edge flow with the consideration of the inner wake region which acts as the driving force of the whole process. Here the proper form of the expansions of the various field quantities is motivated by the following assumptions:

(i) the streamwise velocity distribution in the outer near wake is essentially the continuation of the streamwise velocity distribution in the outer portion of the surface boundary layer,

(ii) the inner wake solution for the velocity distribution exhibits (as indicated in the studies by Robinson (1969), Alber (1980) although not fully expressed in an asymptotic framework) a similarity form,

(iii) the leading order term of the velocity distribution describes a nontrivial flow in the inner wake region.

Following Bogucz and Walker (1988) the velocity distribution in the outer velocity defect part of the surface boundary layer at the trailing edge of the flat plate $x = 0$ is written in the form

$$u = 1 + \varepsilon F_1'(\eta) + \dots, \quad v = \varepsilon^2(\eta F_1'' - F_1) + \dots, \quad -\overline{u'v'} = \varepsilon^2 \Sigma_1(\eta) + \dots \qquad (5.4)$$

where

$$\eta = \frac{y}{\Delta_0}, \quad \Delta_0 = \delta(x = 0) , \qquad (5.5)$$

$$\varepsilon = u_\tau(x = 0) .$$
(5.6)

In the limit $\eta \to 0$ the functions F_1 and Σ_1 have the properties

$$F_1'(\eta) \sim \frac{1}{\kappa} \ln \eta + C_0 ,$$
(5.7)

$$\Sigma_1 \sim 1 .$$
(5.8)

Furthermore, integration of the streamwise momentum equation yields the relationship

$$F_1 - \eta F_1' = \Sigma_1 - 1$$
(5.9)

which will be useful later.

Let $\Delta_n(x)$ denote the local thickness of inner wake region. An appropriate similarity variable is then given by

$$z = \frac{y}{\Delta_n(x)} .$$
(5.10)

Substitution into equation (5.7) shows that the limiting behaviour of the inner velocity distribution for $z \to \infty$ is given by

$$u \sim 1 + \varepsilon \left[\frac{1}{\kappa} \ln \left(\frac{\Delta_n}{\Delta_0} \right) + \frac{1}{\kappa} \ln z + C_0 \right] .$$
(5.11)

This indicates that the inner near wake velocity profile should be expanded as

$$u = u_{cl}(x) + \varepsilon f_w'(z) + \dots ,$$
(5.12)

$$u_{cl}(x) = 1 + \frac{\varepsilon}{\kappa} \ln \left(\frac{\Delta_n}{\Delta_0} \right) + \dots .$$
(5.13)

Here $u_{cl}(x)$ is the leading order approximation of the wake centreline velocity expressed in terms of the yet unknown quantity $\Delta_n(x)$.

Equations (5.12) and (5.13) have to be supplemented with expansion for the transverse velocity component, the pressure and the dominating Reynolds stresses. Evaluation of the continuity equation yields

$$y = -\Delta_n u_{cl}' z + \varepsilon \Delta_n'(z f_w' - f_w) + \dots .$$
(5.14)

Assuming that the leading order pressure disturbances result form the displacement effect of the surface boundary layer and that the Reynolds stress in the inner wake is of the same order of magnitude as in the boundary layer upstream of the trailing edge the expansion for p and $-\overline{u'v'}$ are written as

$$p = \varepsilon^2 p_1(x, z) + \dots , \tag{5.15}$$

$$-\overline{u'v'} = \varepsilon^2 \sigma_w(z) + \dots . \tag{5.16}$$

Substituting equations (5.12)-(5.16) into the streamwise momentum equation we obtain

$$\frac{\Delta_n u_{cl} u'_{cl}}{\varepsilon^2} - \frac{\Delta'_n u_{cl}}{\varepsilon} z f_w'' = \sigma'_w . \tag{5.17}$$

For a similarity solution to exist the first term and the coefficient of the second term on the left hand side have to be independent of x. The first requirement reproduces expression (5.13) for the wake centreline velocity while the second condition yields the yet unknown inner near wake thickness. Without loss of generality we take

$$\frac{\Delta'_n u_{cl}}{\varepsilon} = \kappa \tag{5.18}$$

which can easily be integrated resulting in

$$\Delta_n \left[1 + \varepsilon \left(\frac{1}{\kappa} \ln \frac{\Delta_n}{\Delta_0} - \frac{1}{\kappa} \right) \right] = \varepsilon \kappa (x - x_o) \tag{5.19}$$

where $x_0 = O(\Delta_i)$, $\Delta_i = \delta^+(x = 0)$ is a constant of integration. Inspection of equation (5.19) shows that the effect of x_0 on Δ_n is negligible small in the region $x \gg \Delta_i$ considered here.

Equations (5.13) and (5.19) which express the velocity at the wake centreline and the local thickness of the inner near wake in terms of the quantities Δ_0 and ε characterizing the properties of the unperturbed surface boundary layer at the trailing edge represent a fundamental result of the analysis of Bogucz and Walker (1988). It should be noted that these relationships are general, e.g. do not depend on any specific closure hypothesis.

Expansion of equation (5.19) for $\varepsilon \to O(Re \to \infty)$ keeping x fixed yields

$$\Delta_n = \varepsilon \kappa x + O(\varepsilon^2) \tag{5.20}$$

which shoes that the thickness of the inner near wake grows linearly with increasing distance from the trailing edge.

Using equation (5.18) the leading order streamwise momentum balance assumes the form

$$1 - \kappa z f_w'' = \sigma'_w . \tag{5.21}$$

The corresponding boundary conditions are

$$f_w = f_w'' = \sigma_w = 0 \quad \text{on} \quad z = 0, \tag{5.22}$$

$$f_w' \sim \frac{1}{\kappa} \ln z + C_0, \quad \sigma_w \sim 1 \quad \text{for} \quad z \to \infty. \tag{5.23}$$

Equations (5.21) and (5.22) contain the unknown $f_w(z)$ and $\sigma_w(z)$. In order to determine the velocity distribution in the inner near wake, therefore, a closure hypothesis expressing the Reynolds stresses in terms of the velocity distribution has to be adopted. A simple closure model compatible with the asymptotic trailing edge flow structure has been proposed in Bogucz and Walker (1988) and will later be outlined briefly. The results for u_{cl} and Δ_n obtained so far without relying on such a model, however, are sufficient to determine the large distance behaviour of the transverse velocity component which represents the leading order displacement effect exerted on the outer near wake. To this end equation (5.21) is integrated with respect to z using the boundary conditions (5.22) at $z = 0$

$$z - \kappa(z f_w' - f_w) = \sigma_w. \tag{5.24}$$

Using equation (5.23) we thus obtain

$$z f_w' - f_w \sim \frac{1}{\kappa}(z - 1) \quad \text{for} \quad z \to \infty. \tag{5.25}$$

Substitution into equation (5.14) then yields the desired result

$$v \sim -\frac{\varepsilon}{\kappa} \Delta_n' \quad \text{for} \quad z \to \infty \tag{5.26}$$

which describes the flux of mass directed forwards the wake centreline necessary to satisfy the continuity equation in the inner wake region where the fluid accelerates rapidly under the action of Reynolds stresses.

Next let us consider the outer near wake region in somewhat more detail. As pointed out earlier, the streamwise velocity distribution there is expected to be simply the continuation of the incoming surface boundary layer profile. Combination of the dominating momentum balance

$$\frac{\partial u}{\partial x} = \frac{\varepsilon^2}{\Delta_0} \Sigma_1'(\eta) + \dots \tag{5.27}$$

with equations (5.9) and (5.26) then suggests the following representation of the streamwise velocity profile

$$\begin{aligned} u &= 1 + \varepsilon F_1'(\eta) + \varepsilon^2 \frac{x}{\Delta_0} \Sigma_1'(\eta) + \dots \\ &= 1 + \varepsilon F_1'(\eta) - \varepsilon^2 \frac{x}{\Delta_0} \eta F_1''(\eta) + \dots. \end{aligned} \tag{5.28}$$

Evaluation of the continuity equation taking into account the matching condition $v \to -\varepsilon\Delta_n'(x)/\kappa$ as $\eta \to 0$ yields

$$v = -\frac{\varepsilon}{\kappa}\Delta_n'(x) - \varepsilon^2[F_1(\eta) - \eta F_1'(\eta)] + \dots \qquad (5.29)$$

which is seen to constitute the linear superposition of the transverse velocity profile in the velocity defect part of the surface boundary layer and the (negative) displacement term caused by the inner near wake. As a consequence, the transverse velocity profile holding, in the outer portions of the near wake and the surface boundary layer do not match as $|x| \to 0$. However, as pointed out by Bogucz and Walker (1988) this mismatch can be resolved through the introduction of a trailing edge adjustment region with streamwise extent $O(\Delta_0)$, Fig. 47. Herein pressure disturbances caused by the surface boundary layer and wake displacement which played no significant role so far have to be taken into account to achieve a smooth transition of the vertical velocity profiles from the surface boundary layer to the near wake form.

Equations (5.4), (5.26) and (5.29) for v indicate that the leading order velocity and pressure disturbances outside the boundary layer and wake regions are of $O(\varepsilon^2)$.

$$\begin{aligned} u &= 1 + \varepsilon^2 u_1(x,y) + \dots, \\ v &= \varepsilon^2 v_1(x,y) + \dots, \\ p &= \varepsilon^2 p_1(x,y) + \dots. \end{aligned} \qquad (5.30)$$

Substitution of equation (5.30) into the continuity and momentum equations taking into account equations (5.4) and (5.29) for $\eta \to \infty$ then results in the boundary value problem

$$\frac{\partial u_1}{\partial x} + \frac{\partial v_1}{\partial y} = 0, \qquad \frac{\partial u_1}{\partial y} - \frac{\partial v_1}{\partial x} = 0,$$

$$v_1(x,0) = \begin{cases} 0 & \text{for } x < x_1, \\ 1 & \text{for } x_1 < x < 0, \\ 0 & \text{for } x > 0. \end{cases} \qquad (5.31)$$

Accordingly, the leading order displacement effect on the outer inviscid flow caused by the boundary layer wake transition is equivalent to that exerted by a shallow corner with apex angle ε^2 (Messiter (1980a)). The solution of the problem (5.31) obtained by standard techniques is given by

$$u_1 = -p_1 = -\frac{1}{2\pi} \ln (x^2 + y^2) + \frac{1}{2\pi} \ln \left((x - x_1)^2 + y^2\right),$$

$$v_1 = \frac{1}{\pi} \arg\left(\frac{y}{x}\right) - \frac{1}{\pi} \arg\left(\frac{y}{x - x_1}\right).$$

(5.32)

The appropriately stretched streamwise coordinate in the trailing edge adjustment region is

$$\hat{x} = \frac{x}{\Delta_0}.$$

(5.33)

Comparison of equation (5.32) expanded for $y \to 0$, $\hat{x} = O(1)$ and equation (5.28) indicates the following form of the expansions for the velocity components and the pressure inside the boundary layer and wake

$$u = 1 + \varepsilon F_1'(\eta) - \varepsilon^2 \left[\frac{1}{\pi} \ln \Delta_0 + \frac{1}{2\pi} \ln (\hat{x}^2 + \eta^2) - u_2(\hat{x}, \eta)\right] + \ldots,$$

$$v = \varepsilon^2 \left[\frac{1}{\pi} \arg\left(\frac{\eta}{\hat{x}}\right) + v_2(\hat{x}, \eta)\right] + \ldots,$$

(5.34)

$$p = \varepsilon^2 \left[\frac{1}{\pi} \ln \Delta_0 + \frac{1}{2\pi} \ln (\hat{x}^2 + \eta^2) + p_2(\hat{x}, \eta)\right] + \ldots.$$

Equations (5.34) have to be supplemented with expansions for the Reynolds stresses. Inspection of the Reynolds stress transport equations shows that the variations of the Reynolds stresses over distances $O(\Delta_0)$ are negligible small. As a consequence the leading order terms of these stresses are independent of $\hat{x}$ and of the same form as in the surface boundary layer upstream of the trailing edge:

$$-\overline{u'^2} = \varepsilon^2 Q_1(\eta) + \ldots,$$

$$-\overline{v'^2} = \varepsilon^2 R_1(\eta) + \ldots,$$

(5.35)

$$-\overline{u'v'} = \varepsilon^2 \Sigma_1(\eta) + \ldots.$$

One then obtains the solutions

$$u_2 = \hat{x}\Sigma'(\eta) = \eta F_1''(\eta),$$

$$v_2 = \eta F_1'(\eta) - F_1(\eta) - 1,$$

(5.36)

$$p_2 = R_1(\eta)$$

which are seen to match with the results holding in the surface boundary layer and the near wake region.

The results obtained so far are expected to be general in the sense that they do not depend on special features of specific turbulence models. In order to evaluate these results, e.g. to calculate the velocity fields inside the surface boundary layer and the wake such a model has to be adopted however. As shown in Bogucz and Walker (1988) the simplest possible choice compatible with the general asymptotic structure of the flow field is an algebraic eddy viscosity model of the form

$$-\overline{u'v'} = \varepsilon_m^*(x, y)\frac{\partial u}{\partial y} . \tag{5.37}$$

The Reynolds stress function $\Sigma_1(\eta)$ in the outer portion of the surface boundary layer then satisfies the relationship

$$\Sigma_1(\eta) = \varepsilon_m(\eta)F_1''(\eta) , \quad \varepsilon_m = \frac{\varepsilon_m^*}{\varepsilon\Delta_0} . \tag{5.38}$$

Furthermore, it is assumed that the eddy viscosity distribution can be approximated sufficiently well by the ramp function

$$\varepsilon_m = \begin{cases} K & \text{for } \eta \geq K/\kappa , \\ \kappa\eta & \text{for } \eta < K/\kappa . \end{cases} \tag{5.39}$$

where $K \approx 0.016$ (Mellor and Gibson (1966)).

Equations (5.38) and (5.39) are taken unchanged to model the Reynolds stress distribution in the outer near wake where, as shown earlier, the solution is simply the continuation of the surface outer layer to the order considered here.

Inspection of equations (5.12), (5.16) and (5.37) shows that the Reynolds stress distribution in the inner near wake has to be represented in the form

$$\sigma_w(z) = \varepsilon_{mw}(z)f_w''(z) , \quad \varepsilon_m^* = \varepsilon\Delta_n(x)\varepsilon_{mw}(z) . \tag{5.40}$$

Matching with the results holding in the outer near wake requires

$$\varepsilon_{mw} \sim \kappa z \quad \text{as } z \to \infty . \tag{5.41}$$

To investigate the limiting behaviour of the eddy viscosity function near the wake axis equations (5.24) and (5.40) are expanded about $z = 0$ to yield

$$z \sim \varepsilon_{mw}(0)zf_w'''(0) \quad \text{as } z \to 0 \tag{5.42}$$

where the boundary conditions $f_w(0) = f_w''(0)$ have been used. It thus follows that $\varepsilon_{mw}(z)$ remains finite at the wake centreline. The simplest function which is compatible with equations (5.41) and (5.42) is again a ramp function

$$\varepsilon_{mw} = \begin{cases} \kappa z & \text{for } z \geq \alpha, \\ \kappa\alpha & \text{for } z < \alpha \end{cases} \tag{5.43}$$

where α is a constant to be determined by comparison with experimental data.

Combination of equations (5.39) and (5.43) results in the following composite representation of the eddy viscosity distribution in the near wake region which satisfies all the constraints imposed by the asymptotic analysis

$$\varepsilon_m = \begin{cases} K, & \eta \geq \eta_1 = K/\kappa, \\ \kappa\eta, & \eta_w < \eta < \eta_1, \\ \kappa\alpha\Delta_n/\Delta_0, & \eta \leq \eta_w \end{cases} \tag{5.44}$$

where

$$\eta_w(x) = \alpha\frac{\Delta_n(x)}{\Delta_0}. \tag{5.45}$$

The ramp distributions for the eddy viscosity have the advantage that the velocity distributions in the inner and outer near wake regions can be obtained in closed form.

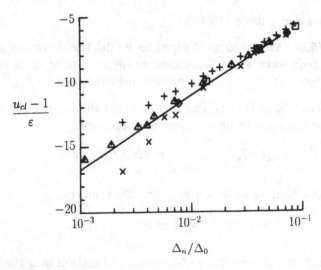

Figure 48: Defect centreline velocity. Comparison between theoretical prediction (——) and experimental data (Bogucz and Walker (1988)).

Careful comparison of the results with existing experimental data by Chevray and
Kovasznay (1969), Pot (1979), Andreopoulos and Bradshaw (1980), Ramapiran, Patel
and Sastry (1981) indicate that $\alpha = 0.6$ is a representative value of the inner-wake eddy
viscosity constant. Typical results displayed in Figs 48 and 49 strongly support the
predictions of the asymptotic theory. For a more detailed comparison between theory
and experiment the reader is recommended to the original publications.

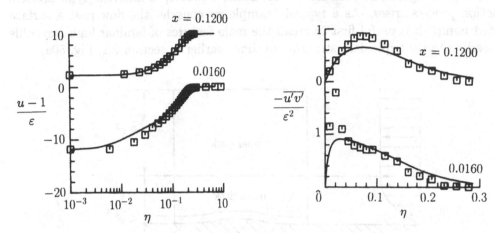

Figure 49: Velocity-defect and Reynolds stress profiles in the near wake. Comparison
between theoretical prediction and experimental data (—— Bogucz and Walker (1988),
□ Pot (1979)).

The analysis outlined so far reveals that the turbulent and laminar trailing edge problems for an aligned thin plate are significantly different. Although the near wake
regions exhibit a two layer structure in both cases the displacement effects exerted by
the turbulent inner wake are too small to cause a substantial departure of the outer
near wake velocity distribution from the surface boundary layer form. In fact the velocity profile in the outer near wake is a continuation of the surface outer boundary
layer. As a consequence, the pressure disturbances induced in the outer inviscid flow
region are small in the whole trailing edge region and do not initiate a local interaction
process near the trailing edge as in the laminar case which requires the simultaneous
treatment of outer inviscid and inner viscous regions. Rather it is sufficient to account
for pressure-displacement effects in a classical hierarchical manner, e.g. to calculate
first the pressure disturbances generated in the outer inviscid region and second the
boundary layer response which resolves the nonuniformity in the transverse velocity
distribution causing a mismatch of the surface boundary layer and near wake regions.

5.3 Flow past a surface mounted obstacle and related problems

The trailing edge problem discussed so far represents a very special case in so far as the flow remains completely undisturbed if viscous effects are neglected. If the inviscid solution carries significant pressure disturbances, however, a different type of local interaction process arises. As a typical example we consider the flow past a surface mounted hump. It is useful first to recall the main features of laminar high Reynolds flows leading to the triple deck structure outlined earlier in section 2.1, Fig. 50a.

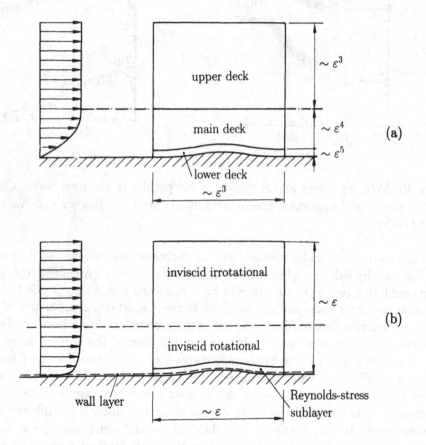

Figure 50: Structure of interacting (a) laminar ($\varepsilon = Re^{-1/8} \ll 1$), (b) turbulent ($\varepsilon = u_\tau \ll 1$) boundary layers.

As the boundary layer approaches the hump viscous and pressure forces cause the layer of low speed fluid close to the wall - the lower deck - to decelerate. This in turn leads to changes of the stream tube area much larger than in the outer layer of high speed fluid - the main deck. The fluid in this region, therefore, is shifted almost passively in the lateral direction thus generating pressure disturbances inside the layer of external inviscid flow - the upper deck - which are experienced also by the fluid in the lower deck. The pressure gradient driving the flow near the wall thus is induced by the interaction between the boundary layer and the external inviscid flow rather than imposed as in classical boundary layer theory. Asymptotic analysis shows the appropriate perturbation parameter is $Re^{-1/8}$ and that the length of the interaction region $O(Re^{-3/8})$ is much larger than the boundary layer thickness $O(Re^{-1/2})$. As a consequence, the induced pressure disturbances do not vary across the boundary layer to leading order.

Next let us turn to the case of a turbulent boundary layer encountering a surface mounted obstacle, Fig. 50b. As pointed out before the region of low speed fluid close to the wall then is extremely thin. In fact, it is transcendentally small $O(u_\tau \exp(-1/u_\tau))$ and, therefore, an asymptotic interaction model similar to that for laminar flows is immediately ruled out as long as the boundary layer remains attached (only such flows have be treated successfully so far). Owing to the lack of significant displacement effects exerted by the viscous wall layer, the velocity and pressure disturbances in the outer region of the boundary layer have to be of the same order of magnitude as those in the external inviscid region in marked contrast to the case of laminar flow. As a consequence, the lateral extent of the local interaction region is found to be of the same order of magnitude as the boundary layer thickness. If the outer flow is purely subsonic or supersonic the same result holds for the streamwise extent of the local interaction region which indicates (i) that u_τ is the appropriate perturbation parameter of an asymptotic interaction theory and shows (ii) that the variation of the pressure disturbances normal to the wall can no longer be neglected as in the related problem of laminar flow.

Summarizing the results obtained so far we are lead to the following model of the turbulent interaction process. Outside the boundary layer the disturbances generated by the hump satisfy the equations of inviscid and irrotational flow. Over most of the boundary layer the pressure gradient is large compared to the Reynolds stress gradients. The perturbations of the field quantities there are of the same order of magnitude as in the outer region but they are superimposed on a shear flow. Therefore, they are governed in general by equations for inviscid rotational flow. Since viscous effects are confined to a transcendentally thin wall layer the boundary conditions which have to be satisfied by the solutions to these equations can be formulated at the surface of the hump and, moreover, simply require that the normal velocity component vanishes there. Again, since the viscous wall layer is so thin, pressure variations across this layer are negligible small. As a result specific closure models do not enter explicitly the solutions for the leading order pressure and velocity disturbances in- and outside the boundary layer. Reynolds stress models affect these solutions only insofar as they determine the velocity profile of the unperturbed boundary layer.

The interaction model outlined so far is still incomplete. As pointed out already, viscous and Reynolds stresses do not directly influence the disturbances caused by the hump in the outer portion of the boundary layer. In contrast, only viscous and Reynolds stress terms are of importance inside the wall layer. The governing equations for these regions, therefore, do not contain a common mechanism for the exchange of momentum in the lateral direction and the solutions cannot be matched. This suggests that an additional layer has to be considered in which the flow is governed by a boundary layer type equation such that the inertia term is balanced by the streamwise pressure gradient and the variation of the Reynolds stresses normal to the wall. It is found that the normal velocity component in this so called blending layer (Melnik and Grossman (1974)) or Reynolds stress sublayer (Adamson and Feo (1975)) is too small to influence the leading order results holding in the outer inviscid region.

We are thus lead to the following asymptotic interaction model of an attached turbulent boundary layer. Over most of the boundary layer the perturbations caused by the local interaction process are essentially inviscid since the pressure gradient is much larger than the gradient of the Reynolds stresses. As we move closer to the wall the flow eventually is influenced also by Reynolds stresses which finally, inside the wall layer, determine the values of the viscous stresses and therefore also the wall shear stress distribution.

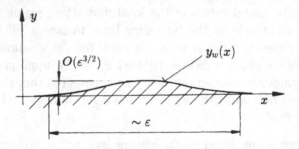

Figure 51: Surface mounted hump.

Following this brief outline of the general physical properties of the interaction process let us consider next the structure of the asymptotic expansions for the various layers in somewhat more detail. As in the study by Sykes (1980b) it will be assumed that the flow is incompressible and that the length and height of the hump are $O(\varepsilon)$ and $O(\varepsilon^{3/2})$, respectively, where as before ε denotes the friction velocity in the unperturbed boundary layer nondimensional with the freestream velocity at the hump location. Nondimensional Cartesian coordinates x and y are measured along and normal to freestream direction with $y = 0$ at the undistorted wall, Fig. 51. The reference length is taken as the distance between the leading edge of the plate and the origin of the coordinate system. Finally, u, v, t_{ij} and p denote the velocity components, the Reynolds stresses and the pressure disturbances nondimensional with the freestream velocity and the dynamic pressure in the external flow. The shape of the hump then is represented in the form

$$\eta_w = \varepsilon^{1/2} f(X),$$

$$X = x/\varepsilon, \quad \eta = y/\varepsilon.$$

(5.46)

The magnitude of the pressure and velocity disturbances in the outer portion of the boundary layer and the external flow region affected by the interaction process is of the same order as the slope of the hump surface $O(\varepsilon^{1/2})$. This suggests the following form of the asymptotic equations for u, v, p and t_{ij}:

$$u = 1 + \varepsilon^{1/2} u_1(X, \eta) + \varepsilon \left[U_{BL} \left(\eta - \varepsilon^{1/2} f(X) \right) + u_2(X, \eta) \right] + \cdots,$$

$$v = \varepsilon^{1/2} v_1(X, \eta) + \varepsilon v_2(X, \eta) + \cdots,$$

$$p = \varepsilon^{1/2} p_1(X, \eta) + \varepsilon p_2(X, \eta) + \cdots,$$

$$t_{ij} = \varepsilon^2 t_{ij}^{BL}(\eta) + \varepsilon^{5/2} t_{ij}^{(1)}(X, \eta) + \cdots$$

(5.47)

where $U_{BL}(\eta)$ and $t_{ij}^{BL}(\eta)$ characterize the velocity profile and the Reynolds stress distributions in the unperturbed boundary layer at $x = 0$.

Insertion of equation (5.47) into the Reynolds equation yields that $p_1(X, \eta)$ satisfies Laplaces equation

$$\frac{\partial^2 p_1}{\partial X^2} + \frac{\partial^2 p_1}{\partial \eta^2} = 0$$

(5.48)

subject to the boundary condition

$$\frac{\partial p_1}{\partial \eta} = -f''(X) \quad \text{on} \quad \eta = 0.$$

(5.49)

As a consequence, the $p_1(X, \eta)$ is independent of any specific turbulence model. As shown by Sykes (1980b) this is true also for $p_2(X, \eta)$ while higher order terms involve the rotational boundary layer profile $u_{BL}(\eta - \varepsilon^{1/2} f(X))$.

Next let us consider the flow properties inside the wall layer where

$$y^+ = \left[\eta - \varepsilon^{1/2} f(X) \right] \varepsilon^2 Re$$

(5.50)

represents an appropriately stretched normal coordinate, which measures the distance from the hump surface.

Owing to the fact that the thickness of the wall layer is transcendentally small the total stress gradient, e.g. the gradient of Reynolds plus viscous stresses, is zero to all orders of ε across the wall layer. This implies that the velocity component u_t tangential to the surface satisfies the logarithmic law

$$u_t \sim \frac{\varepsilon}{\kappa}\sqrt{\tau_w^+(X)}\,\ln y^+ \quad \text{as} \quad y^+ \to \infty \tag{5.51}$$

where $\tau_w^+(X)$ denotes the local wall shear stress. Motivated by the results holding in the outer layer the first two terms in the expansion of τ_w^+ are assumed to be of order $O(\varepsilon^{1/2})$ and $O(\varepsilon)$, respectively:

$$\tau_w^+(X) = 1 + \varepsilon^{1/2}\tau_1^+(X) + \varepsilon\tau_2^+(X) + \dots . \tag{5.52}$$

Indeed, by rewriting the expression for u_t in terms of the outer variable η one finds that the two leading contributions of the resulting asymptotic expansion agree with the corresponding result holding in the velocity defect region. However, higher order contributions in the expansions of the velocity components do not match and to resolve this mismatch it is necessary to introduce an additional layer, the Reynolds-stress sublayer, in which inertia, pressure and Reynolds stress terms are of importance (as they are in the outer portion of the undisturbed boundary layer):

$$
\begin{aligned}
u &= 1 + \varepsilon^{1/2}\hat{u}_1(X,\hat{\eta}) + \frac{\varepsilon\ln\varepsilon}{k} + \varepsilon[U_{BL}(\hat{\eta}) + \hat{u}_2(X,\hat{\eta})] \\
&\quad + \varepsilon^{3/2}\ln\varepsilon\,\hat{u}_{31}(X,\hat{\eta}) + \varepsilon^{3/2}\hat{u}_3(X,\hat{\eta}) + \dots , \\
p &= \varepsilon^{1/2}\hat{p}_1(X,\hat{\eta}) + \varepsilon\hat{p}_2(X,\hat{\eta}) + \varepsilon^{3/2}\ln\varepsilon\,\hat{p}_{31}(X,\hat{\eta}) + \varepsilon^{3/2}\hat{p}_3(X,\hat{\eta}) + \dots , \\
t_{ij} &= \varepsilon^2 t_{1j}^{BL}(\hat{\eta}) + \varepsilon^{5/2}t_{ij}^{(1)}(X,\hat{\eta}) + \dots , \\
\hat{\eta} &= (\eta - \varepsilon^{1/2}f(X))/\varepsilon .
\end{aligned}
\tag{5.53}
$$

Substitution of these expansions into the momentum equations yields that $\hat{u}_1$, $\hat{u}_2$, $\hat{p}_1$, $\hat{p}_2$ are functions of X only and matching with the defect region requires that they agree with the first two terms of the outer expansion evaluated at the wall $\eta = \varepsilon^{1/2}f(X)$:

$$
\begin{aligned}
\hat{u}_1 &= u_1(X,0), \quad \hat{u}_2 = u_2(X,0) + f(X)\frac{\partial u_1}{\partial\eta}\bigg|_{\eta=0} , \\
\hat{p}_1 &= p_1(X,0), \quad \hat{p}_2 = p_2(X,0) + f(X)\frac{\partial p_1}{\partial\eta}\bigg|_{\eta=0} .
\end{aligned}
\tag{5.54}
$$

As a consequence, not only the first two terms of the pressure distribution inside the boundary layer but also the first two terms of the wall shear stress distribution are independent of a closure hypothesis

$$\tau_1^+ = \hat{u}_1 , \quad \tau_2^+ = 2\hat{u}_2 + \hat{u}_1^2 + f'^2 . \tag{5.55}$$

The first term in the expansion for u influenced by such a hypothesis is $\hat{u}_3$ and it is only at this stage that a special closure scheme has to be adopted. Inspection of the time averaged equations for the components of the Reynolds-stress tensor indicates that turbulent diffusion is negligible in the outer region while it has to be taken into account inside the Reynolds-stress sublayer "This confirms the role of the Reynolds-stress sublayer as a blending layer between the wall region where the stresses are in local equilibrium and the outer flow where the stresses are rapidly distorted. It also prohibits the use of any simple turbulence closure scheme, since the solutions must encompass both the mixing-length result near the surface and the rapid distortion result at the outer edge", Sykes (1980b).

Comparison of equations (5.47) and (5.52) shows that the perturbations of the wall shear stress caused by the hump are of the same order of magnitude as the associated pressure disturbances. The same conclusion has been reached by all other local interaction studies based on the asymptotic structure of an attached turbulent boundary layer. This indicates that pressure disturbances of order one will be necessary to separate such a boundary layer.

The results summarized so far indicate that a four tiered structure emerges in studies dealing with the flow of a turbulent boundary layer past surface mounted obstacles. It is found, however, that this multilayer structure is not restricted to these special type of flow but occurs in similar form in a variety of turbulent boundary layer interactions. These include other cases of incompressible flows such as flows past cusped trailing edges at an angle of attack or wedge shaped trailing edges, Melnik, Chow and Mead (1977), Melnik (1978), (1980), Melnik, Chow, Mead and Jameson (1985) as well as compressible flows. In the latter case a difficulty not present in the incompressible limit is encountered insofar as the density variations across the unperturbed boundary layer prohibit a direct match of the wall layer and outer layer expansions. As shown by Adamson and Feo (1975), Melnik and Grossman (1974), (1976) this difficulty can be overcome by using a generalized representation of the velocity profile which is not a limit function expansion and where the density terms are retained to all orders. Calculations using the Prandtl mixing length approximation (with the added simplifications of uniform total enthalpy) then lead to van Driests (1951) solution. This solution is valid in the fully turbulent part of the wall layer only but as pointed out in Melnik and Grossman (1974), Messiter (1980a) its domain of validity can be extended to include $\eta = O(1)$ in a manner originally suggested by Maise and McDonald (1968). A different approach to obtain an asymptotically correct large Reynolds number description of compressible turbulent boundary layers which is based on the Howarth-Dorodnitsyn transformation has recently be proposed by Kazakia and Walker (1995). For a brief outline of the method the reader is referred to the section: Turbulent Boundary Layers II of this volume.

A second new feature of compressible turbulent interactions occurs in supercritical transonic flow and is related to the position of the sonic line inside the unperturbed

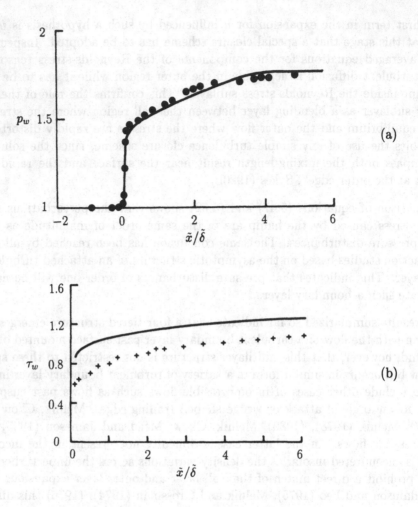

Figure 52: Flow past a compression ramp: $M_\infty = 2.85$, $u_\tau = 0.022$, $\alpha^* = 8°$.
(a) wall pressure (Melnik et al (1985)), (b) wall shear (Agrawal and Messiter (1984)).

boundary layer. If the interaction parameter $\chi_t \equiv (M_\infty^2 - 1)/\varepsilon \ll 1$ the sonic line is very close to the outer edge of the boundary layer while it is located in the main part of the defect region of $\chi_t = O(1)$. Finally, if $\chi_t \gg 1$ the distance of the sonic line from the wall is (transcendentally) small compared to the boundary layer thickness.

Shock wave-turbulent boundary layer interactions have been investigated first by Adamson and Feo (1975) who considered the reflection of oblique shocks in the range $\chi_t \ll 1$ and Melnik and Grossman (1974) who studied normal shocks assuming $\chi_t = O(1)$. In the latter case the flow disturbances in the velocity defect region are governed by a nonlinear transonic small perturbation equation generalized to account for the rotational part of the undisturbed boundary layer profile which has to be solved numerically. Partial analytical results can be derived for stronger normal and oblique shocks having $\chi_t \gg 1$ as shown by Adamson and Messiter (1977), Messiter (1980b), Liou and Adamson (1980), Kluwick and Stross (1984). The flow near the foot of the shock, however, is again governed by a generalized transonic small perturbation equation there necessitating a numerical treatment.

In addition to the shock induced phenomena mentioned so far local interaction processes caused by compression ramps in supercritical external flows have been investigated by Agrawal and Messiter (1984), Melnik, Siclari and Cusic (1985). Theoretical and experimental results are compared in Fig. 52 where p_w and τ_w are nondimensionalized with the unperturbed values of the pressure and the wall shear stress immediately upstream of the corner, respectively. The agreement between the calculated and measured wall pressure distribution is seen to be very good. In contrast, the theoretical prediction of the wall shear stress is less accurate. Although the theory produces the trend of the experimental data quite well, the values are about 20% too high.

Acknowledgements

This research was supported in part by the Austrian FWF under grant number P5557 and jointly by FWF and the National Science Foundation under grant number P5825. The author also want to thank Mrs. Hubert, Dr. Braun und Dr. Reiterer for their assistance during the preparation of the manuscript.

6   References

Ackeret, J., Feldmann, F. and Rott, N. (1946) Untersuchung an Verdichtungsstößen in schnell bewegten Gasen. Mitteilungen aus dem Institut für Aerodynamik der ETH Zürich, Nr. 10.

Adamson, T.C.Jr. and Feo, A. (1975) Interaction between a shock wave and a turbulent boundary layer in transonic flow. SIAM J. Appl. Math. **29**, 121-145.

Adamson, T.C.Jr. and Messiter, A.F. (1977) Normal Shock Wave-Turbulent Boundary Layer Interactions in Transonic Flow near Separation. in: Transonic Flow Problems in Turbomachinery, (eds.: T.C. Adamson, Jr. and M.F. Platzer), Hemisphere Publ. Co., 392-414.

Adamson, T.C.Jr. and Messiter, A.F. (1980) Analysis of two-dimensional interactions between shock waves and boundary layers. Ann. Rev. Fluid Mech. **12**, pp. 103-138.

Agrawal, S. and Messiter, A.F. (1984) Turbulent boundary layer interaction with a shock wave at a compression corner. J. Fluid Mech. **143**, 23-46.

Alber, I.E. (1980) Turbulent Wake of a Thin, Flat Plate. AIAA J. **18**, 1044-1051.

Andreopoulos, J. and Bradshaw, P. (1980) Measurements of interacting turbulent shear layers in the near wake of a flat plate. J. Fluid Mech. **100**, 639-668.

Bippes, H. and Turek, M. (1984) Oil flow patterns of separated flow on a hemisphere-cylinder at incidence. DFVLR FB 84-20.

Bodonyi, R.J. and Kluwick A. (1977) Freely interacting transonic boundary layers. Phys. Fluids **20**, 1432-1437.

Bodonyi, R.J. and Kluwick A. (1982) Supercritical transonic trailing-edge flow. Q. J. Mech. Appl. Math. **35**, 265-277.

Bodonyi, R.J. (1979) Transonic laminar boundary-layer flow near convex corners. Q. J. Mech. Appl. Math. **22**, 63-71.

Bodonyi, R.J., Smith, F.T. and Kluwick, A. (1985) Axisymmetric flow past a slender body of finite length. Proc. R. Soc. Lond. A **400**, 37-54.

Bodonyi, R.J. and Smith, F.T. (1986) Shock-wave laminar boundary-layer interaction in supercritical transonic flow. Computers and Fluids **14**, 97-108.

Bodonyi, R.J. and Kluwick, A. (1997) Transonic Trailing-Edge Flow. Q. J. Mech. Appl. Math. (in press)

Bogucz, E.A. and Walker, J.D.A. (1988) The turbulent near wake at a sharp trailing edge. J. Fluid Mech. **196**, 555-584.

Brilliant, H.M. and Adamson, T.C.Jr. (1973) Shock Wave-Boundary Layer Interactions in Laminar Transonic Flow. AIAA Paper 73-239.

Brown, S.N. and Stewartson, K. (1970) Trailing edge stall. J. Fluid Mech. **42**, 561-584.

Brown, S.N. and Stewartson, K. (1983) On an integral equation of marginal separation. SIAM J. Appl. Math. **43**, 5, 1119-1126.

Brown, S.N. (1985) Marginal separation of a three-dimensional boundary layer on a line of symmetry. J. Fluid Mech. **158**, 95-111.

Bush, W.B. and Fendell, F.E. (1972) Asymptotic analysis of turbulent channel and boundary-layer flow. J. Fluid Mech. **56**, 657-681.

Bush, W.B. (1976) Axial incompressible flow past a body of revolution. Rocky Mt. J. Math. **6**, 527.

Cassel, K.W., Ruban, A.I. and Walker, J.D.A. (1995) An instability in supersonic boundary-layer flow over a compression ramp. J. Fluid Mech. **300**, 265-285.

Cebeci, T.C., Stewartson, K. and Brown, S.N. (1983) Nonsimilar boundary layers on the leeside of cones at incidence. Computers and Fluids **11**, 175-186.

Chen, H.C. and Patel, V.C. (1987) Laminar Flow at the Trailing Edge of a Flat Plate. AIAA J. **25**, 920-928.

Cheng, H.K. (1961) The Shock Layer Concept and Three-Dimensional Hypersonic Boundary Layers. Rep. AF-1285-A-3, Cornell Aeronautical Laboratory.

Chernyshenko, S.I. (1988) The asymptotic form of the stationary separated circumfluence of a body at high Reynolds number. Prikl. Matem. Mekh. **52**, 958-966.

Chernyshenko, S.I. and Castro, Ian P. (1993) High-Reynolds-number asymptotics of the steady flow through a row of bluff bodies. J. Fluid Mech. **257**, 421-449.

Chevray, R. and Kovasznay, L.S.G. (1969) Tubulence measurements in the wake of a thin flat plate. AIAA J. **7**, 1641-1643.

Chow, R. and Melnik, R.E. (1976) Numerical Solutions of Triple-Deck Equations for Laminar Trailing-Edge Stall. Grumman Research Dept. Report RE-526J.

Cole, J.D. (1968) Perturbation methods in applied mathematics, Blaisdell Publishing Company.

Cramer, M.S., Park, S.H. and Watson, L. (1997) Numerical Verification of of Scaling Laws for Shock-Boundary Layer Interactions in Arbitrary Gases. J. Fluids Engg. **119**, 67-73.

Dallmann, U. (1982) Topological structures of three dimensional separations. DFVLR FB 221-82-A07.

Daniels, P.G. (1974) Numerical and asymptotic solutions for the supersonic flow near the trailing edge of a flat plate. Q. J. Mech. Appl. Math. **27**, 175-191.

Dennis, S.C. (1973) private communication to Melnik, R.E. and Chow, R.

Duck, P.W. (1984) The effect of a surface discontinuity on an axisymmetric boundary layer. Q. J. Mech. Appl. Math. **37**, 57-74.

Duck, P.W. and Burggraf, O. (1986) Spectral solutions for three-dimensional triple-deck flow over surface topography. J. Fluid. Mech. **162**, 1-22.

Duck, P.W. (1989) Three-dimensional marginal separation. J. Fluid Mech. **202**, 559-575.

Fiddes, S.P. (1980) A theory of separated flow past a slender elliptic cone at incidence. AGARD C-P paper 30, Sept/Oct., Colorado Springs, USA.

Gersten, K. (1989) Die Bedeutung der Prandtlschen Grenzschichttheorie nach 85 Jahren. Z. Flugwiss. Weltraumforsch. **13**, 209-218.

Gittler, Ph. and Kluwick, A. (1987a) Triple-deck solutions for supersonic flows past flared cylinders. J. Fluid Mech. **179**, 469-487.

Gittler, Ph. and Kluwick, A. (1987b) Nonuniqueness of triple-deck solutions for axisymmetric supersonic flow with separation. in: Boundary Layer Separation; IUTAM Symposium London 1986, Springer-Verlag, Berlin-Heidelberg-New York.

Gittler, Ph. and Kluwick, A. (1989) Interacting laminar boundary layers in quasi-two-dimensional flow. Fluid Dynamics Research **5**, 29-47.

Gittler, Ph. and Kluwick, A. (1997) Viscous-inviscid interaction on a slender axisymmetric trailing tip flow. Proc. R. Soc. Lond. A **453**, 963-982.

Glauert, M.B. and Lighthill, M.J. (1955) The axisymmetric boundary layer on a long thin cylinder. Proc. R. Soc. Lond. A **230**, 188-203.

Goldstein, S. (1930) Concerning some solutions of the boundary-layer equations in hydrodynamics. Proc. Camb. Phil. Soc. **26**, 1-30.

Goldstein, S. (1948) On laminar boundary layer flow near a point of separation. Q. J. Mech. Appl. Math. **1**, 43-69.

Gurevich, M.I. (1966) The Theory of jets in an ideal fluid. Pergamon Press.

Hackmüller, G. and Kluwick A. (1989) The effect of a surface mounted obstacle on marginal separation. Z. Flugwiss. Weltraumforsch. **13**, 365-370.

Hackmüller, G. and Kluwick A. (1990) Effects of 3-D surface mounted obstacles on marginal separation, in: Separated Flows and Jets, IUTAM Symposium Novorsibirsk/USSR, eds. Kozlov, V.V., Dovgal, A.V., Springer-Verlag, Berlin-Heidelberg, 55-65.

Hackmüller, G. (1991) Zwei- und dreidimensionale Ablösung von Strömungsgrenzschichten. Dissertation, TU Wien.

Hackmüller, G. and Kluwick A. (1991) Marginal Separation in Quasi-two-dimensional Flow, in: Trends in Applications of Mathematics to Mechanics, Part III, eds. Schneider, W., Troger, H., Ziegler, F., Longman Scientific & Technical, England, 143-149.

Hakkinen, R.J., Greber I., Trilling, L. and Ababanel, S.S. (1959) The Interaction of an Oblique Shock with a Laminar Boundary Layer. Nat. Aeronaut. Space Admin. Memo 2-18-59W.

Herwig, H. (1982) Die Anwendung der asymptotischen Theorie auf laminare Strömungen mit endlichen Ablösegebieten. Z. Flugwiss. Weltraumforsch. 6, 266-279.

Hornung, H.G. and Perry, A.E. (1984) Some aspects of three-dimensional separation, Part I: Streamsurface bifurcation. Z. Flugwiss. Weltraumforsch. 8, 77-78.

Horton, H.P. (1971) Adiabatic laminar boundary-layer/shock-wave interactions on flared axisymmetric bodies. AIAA J. 9, 2141-2148.

Janour, Z. (1951) Resistance of a plate in parallel flow at low Reynolds-numbers. NACA TM 1316.

Jobe, C.E. and Burggraf, O.R. (1974) The numerical solution of the asymptotic equations of trailing edge flow. Proc. R. Soc. A. 340, 91-111.

Karman, Th. von (1930) Mechanische Ähnlichkeit und Turbulenz. Nachr. Ges. Wiss. Göttingen, Math. Phys. Klasse 58, 58-68.

Katzer, E. (1989) On the length scales of laminar shock/boundary layer interaction. J. Fluid Mech. 206, 477-496.

Kazakia, He.J. and Walker, J.D. (1995) An asymptotic two-layer model for supersonic turbulent boundary layers. J. Fluid Mech. 295, 159-198.

Kirchhoff, G. (1869) Zur Theorie freier Flüssigkeitsstrahlen. J. reine angew. Math. 70, 289.

Kluwick, A. (1979) Stationäre, laminare wechselwirkende Reibungsschichten. Z. Flugwiss. Weltraumforsch. 3, 157-174.

Kluwick, A. (1985) On the nonlinear disortion of waves generated by interacting supersonic boundary layers. Acta Mechanica 55, 177-189.

Kluwick, A. (1989) Marginal separation of laminar axisymmetric boundary layers. Z. Flugwiss. Weltraumforsch. **13**, 254-259.

Kluwick, A. (1994) Interacting laminar boundary layers in dense gases. Acta Mechanica [Suppl.] 4, 335-349.

Kluwick, A. (ed.) (1991) Nonlinear Waves in Real Fluids. Springer-Verlag, Wien-Heidelberg-New York.

Kluwick, A. and Reiterer, M. (1998) On Three-dimensional Marginal Separation, To appear in ZAMM.

Kluwick, A. and Stross, N. (1984) Interaction between a weak oblique shock wave and a turbulent boundary layer in purely supersonic flow. Acta Mechanica **53**, 37-56.

Kluwick, A. and Wohlfart, H. (1984) Entry flow in weakly curved ducts. Ingenieur-Archiv **54**, 107-120.

Kluwick, A. and Wohlfahrt, H. (1986) Hot-wire-anemometer study of the entry flow in a curved duct. J. Fluid Mech. **165**, 335-353.

Kluwick, A., Gittler, Ph. and Bodonyi, R.J. (1984) Viscous-inviscid interactions on axisymmetric bodies of revolution in supersonic flow. J. Fluid Mech. **140**, 281-301.

Kluwick, A., Gittler, Ph. and Bodonyi, R.J. (1985) Freely interacting axisymmetric boundary layers on bodies of revolution. Q. J. Mech. Appl. Math. **38**, 575-588.

Kluwick, A., Gittler, Ph. and Bodonyi, R.J. (1994) Viscous-inviscid laminar interaction near the trailing tip of an axisymmetric body, in: Advances in Analytical Methods in Modeling of Aerodynamic Flows, Walker, J.D.A.; Burnett, M.; Smith, F.T. eds, American Institute of Aeronatuics and Astronautics.

Kluwick, A., Reiterer, M. and Hackmüller, G. (1997) Marginal separation caused by three-dimensional surface mounted obstacles. Proc. of the 2nd International Conference on Asymptotics in Mechanics, St. Petersburg 1996, pp. 113-120.

Korolev, G.L. (1980) Numerical solution to the asymptotic problem of separation of a laminar boundary layer from a smooth surface. Uch. Zap. TsAGI, **11**, 27.

Korolev, G.L. (1983) Flow in the neighbourhood of the trailing edge of a plate in a transonic flow of viscous gas. Fluid Dynamics **18**, 355-360.

Korolev, G.L. (1989) Contribution to the theory of thin profile trailing edge separation. Izv. Akad. Nauk. SSSR, Mekh. Zhidk. Gaza **4**, 55-59.

Korolev, G.L. and Sychev, Vik.V. (1993) Asymptotic theory of the flow near the rear end of a slender axisymmetric body. Izv. Akad. Nauk SSSR, Mekh. Zhidk. Gaza **5**, 68-77.

Leblanc, R. and Ginoux, J. (1970) Influence of cross flow on two dimensional separation. Von Karman Institute for Fluid Dynamics, TN 62.

Legendre, R. (1956) Séparation de l'ecoulement laminaire tridimensionnel. La Recherche Aéronautique No 54, 3-8.

Lighthill, M.J. (1953) On boundary layers and upstream influence II. Supersonic flows without separation. Proc. R. Soc. A **217**, 478-507

Lighthill, M.J. (1963) Attachement and separation in three-dimensional flow, in: Laminar boundary layers (ed.: L. Rosenhead), 72-82, Oxford Univ. Press, Oxford 1963.

Liou, M.S. and Adamson, F.C.Jr. (1980) Interaction between a normal shock wave and a turbulent boundary layer at high transonic speeds. Part II: Wall shear stress, ZAMP **31**, 227-246.

Maise, G. and McDonald, H. (1968) Mixing Length and Kinematic Eddy Viscosity in a Compressible Boundary Layer. AIAA J. **6**, 73-80.

McLachlan, R.I. (1991) The boundary layer in a finite plat plate. Phys. Fluids A **3**, 341-348.

Mellor, G.L. and Gibson, D.M. (1966) Equilibrium turbulent boundary layers. J. Fluid Mech. **10**, 225-253.

Mellor, G.L. (1972) The large Reynolds number asymptotic theory of turbulent boundary layers. Int. J. Engr. Sci. **10**, 851-873.

Melnik, R.E. (1978) Wake curvature and trailing edge interaction effects in viscous flow over airfoils. Advances Technology Airfoil Research Conference, NASA CP-2045, Part I.

Melnik, R.E. (1980) Turbulent interactions on airfoils at transonic speeds - recent developments. AGARD CP-291: Computations of viscous-inviscid interactions, paper No. 10.

Melnik, R.E. and Grossman, B. (1974) Analysis of the Interaction of a Weak Normal Shock Wave with a Turbulent Boundary Layer. AIAA Paper No. 74-598.

Melnik, R.E. and Chow, R. (1975) Asymptotic Theory of Two Dimensional Trailing Edge Flows. NASA SP-347.

Melnik, R.E. and Grossman, B. (1976) Further developments in an analysis of the interaction of a weak normal shock wave with a turbulent boundary layer. Symposium Transsonicum II, (eds.: K. Oswatitsch and D. Rues), Springer-Verlag, 262-272.

Melnik, R.E., Chow, R. and Mead, H.R. (1977) theory of Viscous Transonic Flow Over Airfoils at High Reynolds Number. AIAA Paper No. 77-680.

Melnik, R.E., Chow, R., Mead, H.R. and Jameson, A. (1985) An Improved Viscid/Inviscid Interaction Procedure for Transonic Flow Over Airfoils. NASA Contractor Report 3805.

Melnik, R.E., Siclari, M.J. and Cusic, R.L. (1985) An Asymptotic Theory of Supersonic Turbulent Interactions in a Compression Corner. Grumman Research Department Report RE-711.

Messiter, A.F. (1970) Boundary layer flow near the trailing edge of a flat plate. SIAM J. Appl. Math. 18, 241-257.

Messiter, A.F., Feo, A. and Melnik, R.E. (1971) Shock-wave strength for separation of a laminar boundary layer at transonic speeds. AIAA J. 9, 1197-1198.

Messiter, A.F. (1980a) Asymptotic Theory of Turbulent Boundary-Layer Interactions. 4th Canadian Symposium on Fluid Dynamics, Calgary, June 9-12.

Messiter, A.F. (1980b) Interaction between a normal shock wave and a turbulent boundary layer at high transonic speeds. Part I: Pressure distribution. ZAMP 31, 204-226.

Messiter, A.F. (1983) Boundary-layer interaction theory. ASME J. Appl. Mech. 50, 1104-1113.

Moore, F.K. (1953) Laminar boundary layer on cone in supersonic flow at large angle of attack. NACA Rept. No. 1132.

Murdock, J.W. (1972) The solution of sharp-cone boundary-layer equations in the plane of symmetry. J. Fluid Mech. 54, 665-678.

Neiland, V.Ya (1969) Towards a theory of separation of the laminar boundary layer in a supersonic stream. Izv. Akad. Nauk SSSR, Mekh. Zhidk. Gaza 4.

Neiland, V.Ya (1981) Asymptotic theory of the separation and the boundary layer supersonic gas flow interaction. Advances in Mechanics 4, 1-62.

Oswatitsch, K. and Wieghardt, K. (1941) Theoretische Untersuchungen über stationäre Potentialströmungen und Grenzschichten. Bericht der Lilienthal- Gesellschaft für Luftfahrtforschung.

Oswatitsch, K. (1958) Die Ablösebedingungen von Grenzschichten. In Grenzschichtforschung (ed. H. Görtler). IUTAM Symposium, Freiburg 1957, 357-367. Springer-Verlag, Berlin-Heidelberg-New York.

Park, S.H. (1994) Viscous-Inviscid Interactions in Dense Gases. Ph. D. Dissertation, Virginia Polytechnic Institute and State University, Blacksburg, Virginia.

Perry, A.E. and Chong, M.S. (1986) A Series-Expansion Study of the Navier-Stokes Equations with Applications to Three-Dimensional Separation Patterns. J. Fluid Mech. **173**, 207-223.

Pot, P.J. (1979) Measurements in a 2D wake and in a 2D wake merging into a boundary layer. Rep. NLR TR 79063 U. National Aerospace Laboratory NLR, Netherlands.

Prandtl, L. (1904) Über Flüssigkeitsbewegung bei sehr kleiner Reibung. Verhandl. d. III. Intern. Mathem. Kongresses, Heidelberg, 484-491.

Prandtl, L. (1933) Neuere Ergebnisse der Turbulenzforschung. Z. VDI **77**, 105-114.

Ramaprian, B.R., Patel, V.C. and Sastry, M.S. (1981) Turbulent wake development behind streamlined bodies. Iowa Institute of Hydraulic Research Rep. 231. University of Iowa, Iowa City.

Riley, N. (1979) Separation from a smooth surface in a slender conical flow. J. Eng. Math., **13**, 75-91.

Rizzetta, D.P., Burggraf, O.R. and Jenson, R. (1978) Triple-deck solutions for viscous supersonic and hypersonic flow past corners. J. Fluid Mech. **89**, 535-552.

Robinson, J.L. (1969) Similarity solutions in several turbulent shear flows. Rep. 1242. National Physical Laboratory, Teddington, UK.

Roux, B. (1972) Supersonic laminar boundary layer near the plane of symmetry of a cone at incidence. J. Fluid Mech **51**, 1-14.

Rubin, S.G., Lin, T.C. and Tarulli, F. (1977) Symmetry Plane Viscous Layer on a Sharp Cone. AIAA J. **15**, 204-211.

Ruban, A.I. (1978) Numerical solution of the local asymptotic problem of the unsteady separation of a laminar boundary layer in a supersonic flow. USSR Comput. Maths Math. Phys. **18**, 175-187.

Ruban, A.I. (1981a) Singular solution of boundary layer equations which can be extended continously throug the point of zero surface friction. Izv. Akad. Nauk. SSSR, Mekh. Zhidk. Gaza **6**, 42-52.

Ruban, A.I., (1981b) Asymptotic theory of short separation regions on the leading edge of a slender airfoil. Izv. Akad. Nauk. SSSR, Mekh. Zhidk. Gaza **1**, 42-51.

Smith, F.T. and Duck, P. (1977) Separation of jets on thermal boundary layers from a wall. Q. J. Mech. Appl. Math. **30**, 143-156.

Smith, F.T., Sykes, R.I. and Brighton, P.W.M. (1977) A two-dimensional boundary layer encountering a three-dimensional hump. J. Fluid Mech. **83**, 163-176.

Smith, F.T. (1977) The laminar separation of an imcompressible fluid streaming past a smooth surface. Proc. R. Soc. Lond. A **356**, 433-463.

Smith, F.T. (1978) Three-dimensional viscous and inviscid separation of a vortex sheet from a smooth non-slender body. RAE TR 78095, 33.

Smith, F.T. (1979) Laminar flow of an incompressible fluid past a bluff body: the separation, reattachement, eddy properties and drag. J. Fluid Mech. **92**, 171-205.

Smith, F.T. (1982) On the high Reynolds number theory of laminar flows. IMA J. of Appl. Maths. **28**, 207-281.

Smith, F.T. and Merkin, J.H. (1982) Triple deck solutions for subsonic flow past humps, steps, concave or convex corners, and wedged trailing edges. Computers and Fluids **10**, 7-25.

Smith, F.T. (1983) Interacting flow and trailing edge separation - no stall. J. Fluid Mech. **131**, 219-249.

Smith, F.T. (1985) A structure of laminar flow past a bluff body at high Reynolds number. J. Fluid Mech. **155**, 175-191.

Smith, F.T. (1987) Theory of High-Reynolds-Number Flow past a Blunt Body, in: Studies of Vortex Dominated Flows, (eds.: M.Y. Hussaini and M.D. Salas), Springer-Verlag. 87-107.

Smith, F.T. (1988) A reversed-flow singularity in interacting boundary layers. Proc. R. Soc. Lond. A **420**, 21-52.

Smith, F.T. and Khorrami A. Farid (1991) The interactive breakdown in supersonic ramp flow. J. Fluid Mech. **224**, 197-215.

Smith, J.H.B. (1977) Behaviour of a vortex sheet separating from a smooth surface, RAE TR 77058, 1-62.

Stewartson, K. (1955) The asymptotic boundary layer on a circular cylinder in axial incompressible flow. Q. appl. Math. **13**, 113-122.

Stewartson, K. (1969) On the flow near the trailing edge of a flat plate II. Mathematika **16**, 106-121.

Stewartson, K. and Williams, P.G. (1969) Self-induced separation. Proc. Roy. Soc. A **312**, 181-206.

Stewartson, K. (1970) On laminar boundary layers near corners. Q. J. Mech. Appl. Math **23**, 137-152.

Stewartson, K. (1970) Is the singularity at separation removeable. J. Fluid Mech. **44**, 347-367.

Stewartson, K. (1971) Corrections and an addition. Q. J. Mech. Appl. Math **24**, 387-389.

Stewartson, K. (1974) Multistructured boundary layers on flat plates and related bodies. Advances in Appl. Mech. **14**, 145-139.

Stewartson, K., Cebeci, T.C. and Chang, K.C. (1980) A boundary-layer collision in a curved duct. Q. J. Mech. Appl. Math **33**, 59-75.

Stewartson, K., Smith, F.T. and Kaups, K. (1982) Marginal separation. Stud. Appl. Math. **67**, 45-61.

Stewartson, K. and Simpson, C.J. (1982) On a singularity initiating a boundary-layer collision. Q. J. Mech. Appl. Math **35**, 1-16.

Sychev, V.V. (1972) Concerning laminar separation. Izv. Akad. Nauk. SSSR, Mekh. Zhidk Gaza **3**, 47-59.

Sychev, V.V, Ruban, A.I., Sychev, Vik.V. and Korolev, G.I. (1987) Asymptotic theory of separating flows. Nauka.

Sychev, Vik.V. (1990) Flow near the rear end of a slender body. Izv. Akad. Nauk SSSR, Mekh. Zhidk. Gaza **5**, 10-18.

Sychev, Vik.V. (1991) Asymptotic theory of a flow near the trailing tip of a slender axisymmetric body. In: Separated Flows and Jets. IUTAM Symposium Novosibirsk/USSR 1990, eds. Kozlov, V.V., Dovgal A.V., Springer-Verlag, Berlin-Heidelberg-New-York.

Sykes, R.I. (1980a) On the three-dimensional boundary layer flow over surface irregularities. Proc. R. Soc. Lond. A **373**, 311-329.

Sykes, R.I. (1980b) An asymptotic theory of incompressible turbulent boundary-layer flow over a small hump. J. Fluid Mech. **101**, 647-670.

Taganov, G.I. (1968) Contribution to the theory of stationary separation zones. Fluid Dyn. **3**, 1-11.

Taganov, G.I. (1970) On limiting flows of a viscous fluid with stationary separation for $Re \to \infty$. Uch. Zap. TsAGI **1**, 1-14.

Timoshin, S.N. (1985) Laminar flow in the vicinity of a surface slope discontinuity on an elongated body of revolution. Utschenic Zapiski Tsagi **16**, 10-21.

Timoshin, S.N. (1986) Interaction between the boundary layer on an elongated body of revolution and the external flow. Utschenic Zapiski Tsagi **17**, 33-41.

Van Driest, E.R. (1951) Turbulent Boundary Layer in Compressible Fluids. J. Aero. Sci. **18**, 145-160.

Vatsa, V. and Werle, M. (1977) Quasi-three-dimensional laminar boundary layer separations in supersonic flow. Trans. ASME J. Fluid Eng. **99**, 634-639.

Veldman, A.E.P. and Van de Vooren A.I. (1975) Drag of a Finite Flat Plate. Lecture notes in Physics **35**, 423-430, Springer-Verlag, Berlin-Heidelberg- New York.

Wang, K.C. (1971) On the determination of zones of influence and dependence for three-dimensional boundary-layer equations. J. Fluid Mech. **48**, 397-404.

Whitlock, S.T. (1992) Compressible Flows of Dense Gases. M.S. Thesis, Virginia Polytechnic Institute and State University, Blacksburg, Virginia.

Yajnik, K. (1970) Asymptotic theory of turbulent shear flows. J. Fluid Mech. **42**, 411-427.

Zametaev, V.B. (1986) Singular solution of equations of a boundary layer on a slender cone. Izv. Akad. Nauk SSSR, Mekh. Zhid Gaza **2**, 65-72.

Zametaev, V.B. (1987) Local separation on a slender cone preceding the appearance of a vortex sheet. Izv. Akad. Nauk SSSR, Mekh. Zhid Gaza **6**, 21-28.

Zametaev, V.B. (1989) Formation of singularities in a three-dimensional boundary layer. Izv. Akad. Nauk. SSR. Mekh. Zhidk. Gaza 58-64.

Zametaev, V.B. and Sychev, Vik.V (1995) Three-dimensional separation in the neighbourhood of roughness on the surface of an axisymmetric body. Fluid Dynamics **30**, 387-398.

Zieher, F. (1993) Laminare Grenzschichten in schweren Gasen. M.S. Thesis Technical University of Vienna.

Printed in the United States
by Bookmasters

Printed in the United States
By Bookmasters